Securing What Feeds Us

Securing What Feeds Us

Cybersecurity in Food and Agriculture

KRISTIN KING

***FOREWORD BY* DR. DARIN DETWILER**

WILEY

Published by John Wiley & Sons, Inc., Hoboken, New Jersey.

Library of Congress Control Number: 2026938485

ISBN: 9781394391004 (Paperback)
ISBN: 9781394391028 (ePDF)
ISBN: 9781394391011 (ePub)

Cover Design: Wiley
Cover Images: Emma/stock.adobe.com, Anja Kaiser/stock.adobe.com
SKY10164912_080626

For everyone who keeps the world fed and the food supply safe.

A special dedication to my grandfathers: you taught me so much about the natural world.
I will always think of you during my morning coffee or my afternoon fika.

Contents

Foreword

Over the years, I've come to recognize a simple question that reveals how organizations truly think about risk: why do we wait until something breaks before we decide it was worth protecting?

For many of us, our earliest ideas about technology, risk, and cost were shaped by the stories we grew up with. For me, one of those stories was the 1970s television show *The Six Million Dollar Man*. Each episode began the same way:

> *"Steve Austin, astronaut. A man barely alive. . . Gentlemen, we can rebuild him. We have the technology. We have the capability to make the world's first bionic man. Steve Austin will be that man. Better than he was before. Better. . . stronger. . . faster."*

In the show, Steve Austin barely survives a catastrophic test flight crash. The government rebuilds him with then-advanced technology: bionic limbs, enhanced vision, and superhuman strength. The cost: *$6 million*. At the time, that number sounded almost unimaginable. The technology was futuristic. The threats were global. And the message was clear: when something is truly a priority, we invest whatever it takes to achieve it.

But here's the deeper lesson the show unintentionally taught us. In the premise of that television series (much like what we often see following today's foodborne illness outbreaks or cyber-attacks), the investment typically comes only after disaster has already struck.

Today, the world that feeds us faces a similar dilemma.

The food and agriculture system is one of the most complex, interconnected infrastructures ever created. It spans continents, technologies, economies, and ecosystems. Increasingly, it relies on digital systems (such as industrial control systems, operational technology, sensors, data platforms, and interconnected networks) to function.

In other words, the modern food system runs not only on soil, water, and human expertise but also on code.

That reality creates enormous opportunity. Technology allows us to produce more food, move it more efficiently, and monitor safety and quality in ways unimaginable just decades ago. But it also creates new vulnerabilities. Every digital connection is a potential point of entry.

Cybersecurity is no longer just an IT issue.

It is a food safety issue.

It is a food defense issue.

It is a public health issue.

And, increasingly, it is a national security issue.

Perhaps most importantly, cybersecurity is a trust issue.

When consumers pick up a product at the grocery store, they trust that it is safe and authentic. They trust that the systems behind it—from farm to fork—are resilient enough to protect them. That trust is fragile. And once broken, it is extraordinarily difficult to rebuild.

This is where the lesson of *The Six Million Dollar Man* becomes especially relevant. Perhaps the dollar amount in the show's title is meant to remind us not of the cost of prevention but of the cost of doing nothing.

In the 1970s, $6 million represented an extraordinary investment. Today, in the context of global food systems, it seems almost modest. Yet the principle still holds: rebuilding after catastrophe is always more expensive than preventing it in the first place.

Across industries, we see the same pattern again and again. Companies hesitate to invest in preventive measures because the costs feel immediate: cybersecurity upgrades, system protections, operational safeguards, training, monitoring, and resilience planning.

These investments can feel expensive . . . until the moment they aren't.

When vulnerabilities are ignored, the cost of doing nothing multiplies rapidly. History has shown us time after time how a few million dollars saved by postponing proactive security measures can quickly become hundreds of millions spent on crisis response, recalls, operational shutdowns, regulatory consequences, legal liability, and lost consumer trust.

But the issue is not just financial.

The stakes within food and agriculture extend far beyond corporate balance sheets. Food quality, food safety, food security, and food authenticity are all vulnerable when defenses fail. Economically motivated adulteration, deliberate sabotage, supply chain manipulation, and even the possibility of food bioterrorism represent real risks.

These are not hypothetical scenarios: they are threats within a system that touches every person, every day.

The heroes confronting these risks today may not have superhuman strength, but their work is no less critical. They are engineers, cybersecurity professionals, food safety leaders, risk managers, and operational experts working behind the scenes to protect the integrity of the food supply.

Kristin King is one of those leaders.

What makes her perspective particularly valuable is that she understands the intersection where technology, operations, and risk converge. She understands the world of operational technology and industrial control systems that quietly power many processes in food production, agriculture, zoos, aquariums, and related industries. These systems often operate outside the spotlight, yet they are essential to maintaining continuity, safety, and trust.

More importantly, Kristin understands that cybersecurity is not merely a technical challenge. It is a leadership challenge.

Effective cybersecurity requires culture, awareness, and organizations willing to treat resilience not as an afterthought but as a responsibility.

It provides insight into the risks facing modern food and agriculture systems while offering a path forward: helping organizations understand vulnerabilities, strengthen defenses, and build systems that are not only efficient but resilient.

Because resilience is not about fear. It is about preparation.

If the lesson from *The Six Million Dollar Man* was that we have the technology to rebuild after catastrophe, the lesson from today's food system should be even more powerful: we have the knowledge to prevent catastrophe in the first place.

But only if we choose to use it.

The threats facing food and agriculture are real. They are evolving. And they are increasingly digital. Ignoring them will not make them disappear.

What Kristin King offers in the pages ahead is more than expertise. She offers a roadmap for protecting one of the most essential systems on Earth.

And that is why this book matters.

Because the goal is not simply to rebuild after something breaks.

The goal is to ensure that the systems feeding the world remain strong, resilient, and worthy of the trust we place in them every single day.

The real question is whether we will choose to protect those systems before they break.

The pages that follow offer a critical guide for anyone committed to protecting that future.

That work begins with the knowledge ahead.

—Dr. Darin Detwiler, Professor of Food Policy and Leadership, Distinguished Global Fellow in Food System Integrity, and author of Food Safety: Past, Present, and Predictions

Introduction: Unsecured and Feeding Us Anyway

It began, like many tragedies do, in a place that felt safe.

In the region of Zug, Switzerland, between Zurich and Lucerne, lies the village of Hagendorn. It sits among rolling hills, not far from lakes and forests, with views that stretch out toward the distant Swiss Alps. Farmland is woven into everyday life there, quiet but essential. And that's where you'll find Vital Bircher's dairy farm.[1]

Vital Bircher, a dairy farmer in his 70s, has been farming longer than some milking robots on the market have been around. Labor is harder to find, margins are tighter, and keeping a herd healthy these days involves more than a good pair of boots and instincts passed down through generations. It also involves operational equipment that runs software connected to the Internet.

And then one day, that software was hacked.

The software running his robotic milker was locked out and encrypted. A digital screen demanded $10,000 in ransom. No details. No reason. Just a digital shakedown in a world that didn't use to need computers.[2]

Bircher didn't pay. And maybe that sounds heroic. Maybe you want to cheer for him, standing tall, refusing to bend to criminals hiding behind a screen. But even though the hackers walked away with nothing, Bircher still lost.

The robot was able to continue milking the cows offline, but the data about the cows themselves, which is critical to his operation, was no longer available. The real-time tracking of each cow's health, behavior, and stress. The ability to see what he couldn't, like a pregnant cow showing subtle signs of distress. Without that insight, he didn't catch it in time. The calf died inside the mother. The mother sadly had to be euthanized.

This wasn't a spreadsheet, an email mail-merge issue, or a broken printer.

This was two lives. Gone.

No one got rich from this attack, but Bircher paid a price anyway: about 6,000 Swiss francs in immediate losses in vet bills, new equipment,[3] and up to $5,000 more for the value of a mature dairy cow.[4] But money's the easy part to measure. What about the grief? The guilt? The shame? The hollow echo in a barn where life used to be?

And all of it because someone, somewhere halfway across the world, decided his farm was worth the hack. But honestly? That's not even the worst part. The worst part is how preventable it was.

No one ever found the attackers. It could've been a cybercriminal halfway across the globe, a nation-state poking around, or a bored kid with a script. It doesn't matter. The damage was done. And it wasn't just a digital inconvenience. It was a breakdown of trust in systems, in safety, in the promise that technology will always help more than it harms.

I wish I could tell you this was a one-off or a weird glitch in the matrix. But it's not. This happened in November 2023. And since then, I've spoken with farmers across sectors and continents. Their response is the same every time: "That would be devastating." Not just the money. Not just the data. The animals. The people. The life they've built with callused hands and dawn-colored mornings.

We've spent too much time treating cybersecurity like it's just an information technology (IT) thing. This is about the gut-punch of losing something you love. About what happens when the systems you trust betray you. When machines meant to lighten your load become silent witnesses to tragedy.

This is about what really matters: the food on our tables, the people who grow it or raise it, and those who could bring it all down without a trace.

"Kristin, are you seriously saying cybersecurity is life or death?"

Yes. I am.

Because I'm tired of acting like it isn't.

I'm tired of hearing that food systems are "just agriculture" or "not a likely target." I'm tired of being told that it's "too dramatic" to connect cyberattacks to real-world harm. Tell that to the farmer who buried a cow that didn't need to die. Tell that to the processor whose production line was halted mid-shift over a food safety issue caused by an insider threat. Tell that to the families who won't get to eat because the supply chain cracked under pressure.[5]

This book is not about fear.

It's about clarity and getting you ready.

It's about understanding what's truly at stake when we ignore the digital threads now woven through every inch of our food system. It's about stepping into the storm, not to panic but to prepare.

Because the villains aren't cloaked in shadows, they're in the code. And the heroes? They're the ones in coveralls and steel-toed boots, on tractors and forklifts, in labs and kitchens, doing the work, day in and day out, to keep the rest of us fed.

And they deserve to be protected.

To be clear, I'm not a farmer, nor has any of my family been part of the practice since my grandparents raised chickens in their backyard. I know how to garden, I'm a decent cook, but I'm not about that measurement life, so I don't bake. I grew up in one of the last generations of feral children before the internet really took off in the middle of high school. I was always outside, in nature; sticks and leaves were my canvas and doorway of imagination. I'm grateful I have a balance between digital and analog lifestyles. I believe I can see systems better because I spent my afternoons trying to understand the social dynamics of ladybugs rather than leveling up a character in a video game.

As a waitress and bartender for many years, I witnessed firsthand how the food service industry operates, which taught me not only about customer service but also about food safety and the inner workings of the food supply chain. I also worked in the commissary at the National Zoo while pursuing my degree. It was there that a deeper understanding of food safety grew, both in food handling and safety. We prepared all the cut fruits, vegetables, and proteins for the various animals in the zoo. My favorite to make was the "gorilla salad," basically a ginormous salad for the gorillas to graze on every day. Later, that knowledge served me, and I was able to support the gorilla house of the Louisville Zoo when I was a research associate studying the then-bachelor troop of gorillas. After dozens of zoo kitchen tours, I understood how interconnected they were with the rest of the zoo and the importance of securing them on both the physical and digital fronts. Eventually, I graduated with my degree in Environmental Management, which was another piece in the puzzle of how the food and agricultural sector showed up in my life.

After I finished my degree, I was lucky enough to work in a bakery company for several years, in the IT, operational technology (OT), and cybersecurity divisions. It taught me many things, but I am still scratching the surface of that knowledge even after more than a decade in the industry.

One key difference between food production and agriculture is the role of corporations and compliance. I'm sure some of my readers come from food production or manufacturing (hi, friends!), and we often spend our time writing policy, ensuring that we meet the standards set by organizations such as the National Institute of Standards and Technology (NIST), the NIS2 Directive, the International Organization for Standardization (ISO), or even the General Data Protection Regulation (GDPR). Some of these organizations even have their own SOC or Red teams, which is fantastic. But not all of them are large corporations; some of them are small to medium-sized food companies that may or may not have an in-house IT team, let alone a designated cybersecurity professional.

Farms and ranches look incredibly different from food production organizations. I've driven thousands of miles and collected many calluses and blisters while learning about how these farming operations work. Suffice it to say, my education was top-notch and hands-on. I toured various dairy farms (robotic, rotary, traditional), a vertical farm, pig farms, cattle feedlots, chicken farms a wheat farm, a seed breeding facility, vineyards, an apple orchard, and even an edible insect farm. The one thing they all have in common is their mission: to feed people. It sounds simple and beautiful because it is. The other thing they have in common is their dependency on technology to accomplish this goal; but that is the duality of things, and I'll unpack that throughout the book.

In all seriousness, do you know where your food comes from? Do you ever stop to think about that when you're in the grocery store or grabbing an ice cream with your friends or family, or pouring your nonnegotiable morning coffee? Do you know all the steps your food took to reach its current state and all the people who helped it along the way? Even more importantly, do you know all the connected technology that controls the steps to your plate?

Food starts (unless it's lab-grown) in a field, pasture or body of water. Even the ingredients for your favorite snack or food fixation started on a farm. Someone, aka the farmer, cared for those plants or animals, essentially raising them for you. Wild, right? Someone you don't even know is part of the most important thing you do every day.

We all eat; it's the second thing we do after we scream into existence. So why don't we put as much care into securing our food as we do the rest of the systems in our lives? We talk about food, we cook, we shop, we dine, and we consume social media content about the latest recipe trends, but we still don't talk about securing it. I started wondering why that was, so I started asking questions. The first responses were "There's no money in the food industry," "There isn't enough technology or no technology," "They aren't a target," "No one would hack a tractor," and, my personal favorite, "There's nothing secret in the food industry."

All of these are incorrect, and I grew frustrated that no one was paying attention to the most basic and common need in our lives: food.[6]

After working closely with the agricultural community, I began to find these responses infuriating because working in the agricultural sector is thankless at times. It is almost a silent battleground, and it's the most important battle we should care about as humans. Sure, the rest of the critical infrastructure sectors are important; we can't have one without the other, but as my good friend Andrew Rose always points out, "How long would it take you to commit a crime to get food for your family if there wasn't any?" How many days, even hours, would it take you to decide you were going to break the law to feed your family if there was a massive food shortage? Two days? One day? Eight hours? If you have an infant or a toddler, maybe one hour? It really gets you to think about the basics, what is needed, and then what happens if it is simply not there. Worst-case scenarios for cyberattacks on the food and agricultural industries are no food or poisoned food. We have many buzzwords in cybersecurity; many industries do, but in the food and agricultural industry, it's pretty simple: life and death. No fancy dashboards, no color-coded vulnerability management, just a simple "Is it alive or is it dead?" I often wonder if this is why cybersecurity hasn't focused on the food and agricultural industry as much as other critical infrastructure: because it requires a lot of empathy not only for the people in the industry but also for the products. When something breaks in the food industry, death is a real possibility; add the digital or cyber-physical component, and we are talking about not only human life but also animal and environmental life. These are huge, complex systems built by humans that need to work in harmony with the world's natural cycles.

"Wait, Kristin, whoa, now you're telling me I need to understand cybersecurity, agriculture, manufacturing, environmental science, and people?" Yup, you've got to wear many hats, just like a farmer. They are farmers but also mechanics, chemists, drivers, meteorologists, pedologists, accountants, electricians, plumbers, fabricators, biologists, entomologists, grant writers, logistics coordinators, labor managers, negotiators, midwives, ethologists, and the list goes on and on. And now we need to add cybersecurity and risk officer, as well as IT technician. I don't think that's overly fair, do you?

I was interviewing a large-operation wheat farmer the other day for my *Bites and Bytes Podcast*, and I asked him, "Do you ever think about cybersecurity?" His response surprised me: "I do, and I think other farmers also have it in the back of their minds that we need to do something because we know we are at risk. But we don't have that expertise; we don't know where to start." In that moment, I felt like we had failed the agricultural community; we do so much security awareness generally that we never thought about the people who make our food and how specialized that training needs to be. On behalf of the entire cybersecurity community, I would like to extend a formal apology to the agricultural community. We should be there for you more. We need to be there for you and thank you for keeping us fed. This is why I'm writing this book: to remind the cybersecurity community that we need to do better in this critical infrastructure, and also to provide the food and agricultural community with a resource they can reference to help them manage their cyber-physical risks. My hope is that this will help open up conversations, create common ground, and start securing what feeds us.

We'll look at the farmer trying to manage sensors and security. The fishery that's vulnerable to data tampering. The bakery where a server crash leads to empty shelves.

This isn't sci-fi.

This is Tuesday.

Strategies that aren't just cookie-cutter (we will continue to use all food puns, you've been warned) but rooted in practical expectations of what needs to be secured and what matters.

I'm going to tell real stories from the folks who are part of the massive global food supply chain and how they view cybersecurity and technology; their real, valid concerns about their roles; and what is at stake. I want you to understand that they can't wait for policies, regulations, or someone to swoop in and save them; they need real action, real support, and not just more fear tactics or more technology. And I know my teams in food production will also agree that you don't need anything else that isn't about production or employee safety—you don't have time for it.

We will go over some strategies that are as simple as what to do in a fire. If your mind just said "stop, drop, and roll," congrats: go get a cookie of your choice. We need to make the proactive steps as easy as 1, 2, 3 and also confront the inevitable cyber incident that requires cool heads to keep everyone safe.

To make sense of all this, I've broken the book into six sections. We'll start with the big picture, zoom in on real-world sectors, explore the cybersecurity obstacles that trip us up, look at the big societal themes no one's talking about, untangle how all the sectors connect, and, finally, start mapping the way forward.

Part I: Foundations & Context

Before we talk defense, we need to understand what's at stake, how food systems connect people, culture, and machines, and why cybersecurity has become part of keeping us fed and functioning.

Part II: Sector Spotlights—Where the Risks Live

Every link in the food system, whether it's a tractor, a tuna boat, or a vending machine, has its own unique vulnerabilities, and we can't protect what we don't understand.

Part III: Core Cybersecurity Challenges in the Food Sector

From legacy equipment, insider threats, policy, and regulations to crisis response and insurance gaps, this section unpacks the cyber hurdles that food systems face every day.

Part IV: Strategic & Societal Themes

Food isn't just a supply chain. It's politics, religion, trust, equity, and culture, and the threats to it extend far beyond the equipment.

Part V: The Way Forward

Securing what feeds us means building cyber-aware teams, rethinking risk beyond compliance, and creating a future where resilience is the recipe for survival.

Cyberattacks aren't coming; they're already here.

It's time we all start securing what feeds us.

Notes

1. Luzerner Zeitung. (2024). "Cyberattacke in Cham: Melkroboter gehackt – Kuh gestorben." **https://www.luzernerzeitung.ch/zentralschweiz/zug/cham-cyberkriminelle-fordern-von-schweizer-bauer-10000-dollar-danach-stirbt-eine-kuh-ld.2651065**
2. Bayer, M. (2024). "Ransomware attack paralyzes milking robots—cow dead." *CSO*. **https://www.csoonline.com/article/3484349/ransomware-attack-paralyzes-milking-robots-cow-dead.html**
3. Ibid.
4. Schmitz, A. (2024). "Replacement cow prices reach all-time high." *AgProud*. **https://www.agproud.com/articles/60141-replacement-cow-prices-reach-all-time-high**
5. FBI. (2021). *Cyber Criminal Actors Targeting the Food and Agriculture Sector with Ransomware Attacks*. **https://www.cisa.gov/sites/default/files/publications/PIN_20210901.pdf**
6. Food and Agriculture-Information Sharing and Analysis Center. (2024). *2024 Ransomware Cyber Threat Report*.

PART I

Foundations & Context

CHAPTER 1

Why Cybersecurity in Food Matters Now

A Menu Hack That Could Have Killed

Imagine being on vacation at Disney World. Maybe it's with kids in tow, maybe it's an adult-only escape, maybe it's your first solo trip splurging on a little magic. You sit down at a restaurant, tired, hungry, and ready to trust that the menu in front of you is accurate. If you live with a severe food allergy, those little symbols marking allergens aren't just conveniences; they're survival tools. You check, double-check, maybe speak to the server, and order with confidence.

In 2024, that confidence was put at risk. A former Disney employee, bitter about being fired, hacked into the company's menu system.[1] He tampered with wine lists, added profanity, slipped in political jabs, and, worst of all, altered allergen information. Suddenly, items that contained peanuts were flagged as safe. For a person with a peanut allergy, one mislabeled entrée is enough to trigger anaphylaxis—a life-threatening emergency that can kill within minutes.

Fortunately, Disney identified and removed the altered menus before they were shipped to restaurants.[2] But the damage to the company's systems was significant. Disney had to take its Menu Creator application offline while it reverted to backups, and manual processes had to be used to fix the menus. For a couple of weeks, staff in one of the world's most controlled environments scrambled to ensure menu accuracy. The company estimated direct damages of at least $150,000, and the employee was ultimately sentenced to three years in federal prison and ordered to pay $687,776.50 in restitution.[3]

Think about that for a moment. A company that has engineered an empire around precision and perfection was reduced to running restaurants with manual processes because one fired employee still had a working login. This wasn't some shadowy, nation-state hacker team. It was a single human oversight, a basic process failure. And that single gap could have ended in tragedy.

The Disney sabotage is a warning about how fragile our trust in food systems really is. Cybersecurity in food is about the trust you place in what you eat, the assumption that the safety information is accurate, and the belief that the system protecting you is sound. When that trust falters, the consequences aren't theoretical. They're immediate, physical, and potentially deadly.

The Hidden Vulnerabilities

If a cyber-sabotage can happen in the "Happiest Place on Earth," it can happen anywhere. And in recent years, it has happened across farms, factories, and food outlets. Food systems have quietly become high-tech marvels, relying on sensors, software, and networks to feed millions around the world. Unfortunately, that makes them attractive targets for cyber mischief and crime. Hackers and malicious insiders have found that food businesses from agriculture to processing to retail often have gaping vulnerabilities. In fact, as other sectors (like finance and healthcare) tightened their digital defenses, attackers turned to the comparatively easier targets of food and agriculture. One industry analysis noted that cyberattacks against the food/ag sector jumped by 607% in 2020.[4] Why? Often, because food companies have lagged in cybersecurity investment, whether due to an "it won't happen to us" attitude or razor-thin profit margins that leave tech as an afterthought. In other words, what we eat has become the new playground for cyber villains precisely because it's under-protected and unregulated.

This blind spot has roots in how food has historically been viewed. For years, when government agencies classified as "critical infrastructure," food was listed alongside energy, water, and healthcare, but was rarely taken as seriously.[5] Banking systems were upgraded, hospitals received new regulations, and pipelines were stress-tested. Food, meanwhile, was assumed to be steady and self-reliant. After all, farmers have fed us through droughts, depressions, and wars. The assumption that food is somehow more permanent or resilient has bled into how we treat its digital and cyber-physical risks. The reality is different; food systems are now just as digitally dependent as finance or healthcare, but without equivalent investment in security protections.

Economics don't help, either. Food companies, especially processors and producers, operate on margins so thin that any cost not directly tied to production is seen as expendable. Updating software or running penetration tests rarely outranks production quotas or safety audits. But ignoring those invisible vulnerabilities often ends up being more expensive than prevention. One successful cyberattack can shut down operations, spoil products, and cost millions in downtime and recovery. And unlike a stolen credit card or a leaked email, the fallout lands on dinner plates and in grocery aisles.

Beyond Technology: People and Processes

Technology tends to get the spotlight in cybersecurity, but people and processes are where most food system risks originate. The Disney hack wasn't a sophisticated cyber-exploit; it was a process failure. A fired employee still had credentials that should have been revoked at termination. If his credentials were still active, he was a "trusted" login, and no firewall would have been able to prevent his access.

These failures repeat across the sector. Passwords get shared because "it's easier that way." A plant manager leaves the company, but their login still works a month later. A rushed employee plugs in a USB drive from home, not realizing it carries malware. Warehouse staff click on phishing emails because training is inconsistent or nonexistent. And even when security policies exist on paper, enforcement is often spotty in practice. In my own time working in the food industry, I saw this up close: the same password used for production spreadsheets, equipment logins, and even customer pickups at the "will call" window. Long after I left that company, the IT staff would still call me to ask if I remembered "the legacy passwords" so they could get into old systems. The disturbing part? A year after I walked out the door, those passwords still worked. And they weren't even clever ones, just generic stuff. They've changed since, so don't get any ideas, but I'll never forget them: Baking123 and Bake#123.

Culture also plays a big role here. Many food companies still view cybersecurity as an IT issue rather than an operational one. The line between food safety and cybersecurity remains blurred even though they overlap constantly. If employees don't see themselves as part of the protection system, they won't recognize red flags. That cultural divide can be as dangerous as an open network port.

Processes add another layer of risk. Companies delay updates or patches because shutting down equipment means halting production. Stronger security protocols get skipped because they're inconvenient. Vendor access isn't properly vetted because relationships are long-standing and trusted. Each of these shortcuts might make sense in the moment—after all, time is money in food production—but together they form cracks that attackers can pry open.

Cybersecurity in food runs on the everyday decisions, behaviors, and habits of the people who run the systems. The most high-tech protection won't hold if the people and processes behind it aren't aligned.

Why Cybersecurity in Food Matters Now

Food is a critical infrastructure.[6] You can skip streaming services or social media for a week; you can't skip eating. That makes food's vulnerabilities different from those of almost every other industry. The risks reach past finances and reputation. They strike at safety, trust, and continuity.

When cybersecurity in food fails, the fallout lands in the real world. A mislabeled allergen can become a medical emergency. A production line paralyzed by ransomware can leave grocery shelves empty. A disabled refrigeration system damages products, wastes food, drives up costs, and contributes to food insecurity. These are lived consequences, not just abstract risks.

And yet the systems behind our food often remain invisible until they break. Most people assume that expiration dates are correct, storage systems are reliable, and supply chains will simply work. The Disney incident ripped away that illusion. It revealed how fragile that trust can be and how quickly it can collapse when cyber touches food.

Trust is the currency of food systems, and once it's shaken, it's slow to rebuild. Consider how quickly panic spreads during food recalls, even when the issue is contained. A single viral post claiming tampering or even just raising suspicion can ignite outrage.

Panic buying, empty shelves, and social media storms follow. Add to that the power of misinformation and disinformation spreading online—false recalls, manipulated photos of "contaminated" products, or activists pushing misleading claims—and the damage multiplies. Even if the food itself is safe, perception becomes reality, and consumer confidence evaporates. Hackers and bad actors know this. Sometimes the easiest way to cause disruption isn't by touching the food at all, but by manipulating the story people believe about it.

This is why cybersecurity in food has to account for relationships—with consumers, with regulators, and with communities—not only the IT infrastructure. Every vulnerability in a system is a potential crack in that trust. It's also why resilience must extend beyond technical defenses. Firewalls and security software won't catch every insider threat or process gap. Building resilience means cultivating a culture where employees understand their role in security, where policies are enforced consistently, and where leadership recognizes cybersecurity as central to safety and continuity.

A Systems Lens

One of the reasons food cybersecurity has been left behind is that it doesn't fit neatly into silos. IT teams handle networks, quality assurance handles safety, and operations handle production. On an org chart, those divisions look tidy. But hackers don't care about org charts. To an attacker, it's all one system. A vulnerable refrigeration unit is as good as a weak password: it's just a way in.

And once a hacker is inside, the effects ripple. A single breach in a vendor's software can halt deliveries. A corrupted database can throw off food safety testing. A disabled logistics platform can strand products in warehouses. Now imagine that happening during a summer heatwave, when cold storage has no backup, or during peak holiday season, when grocery shelves are already strained. I'm sure we all remember the pandemic and the strain on the supply chain, right? That's the reality of food as a system: every link is connected, and when one breaks, others follow.

This is why systems thinking is essential. Cybersecurity requires understanding the web of connections—technology, people, processes, and culture—and anticipating how failures will cascade. Ask: if one piece fails, what else goes with it? Because in food, the answer is rarely "nothing."

Looking Ahead

Cybersecurity in food matters now because the risks are already here. The vulnerabilities are baked into systems that have outgrown their old safety nets. And the consequences don't stay behind screens; they land on plates, in grocery stores, and in public trust.

Food is more than calories. It's memory, identity, and community. Protecting it means protecting all of that. As we move forward, the conversation shifts. In the next chapter, we'll examine food as a cultural and emotional anchor. Because before we can talk about protecting technology, we need to understand what food itself means to us and why losing trust in it cuts so deep.

Notes

1. Alfonsi, S. (2024). "Former Disney employee accused of hacking menu system and changing peanut allergy information." *CBS News*. **https://www.cbsnews.com/news/disney-employee-michael-scheuer-hacking-menus-peanut-allergy**
2. CNN Business. (2024). "Fired Disney employee allegedly hacked into company system to change allergy info on menus." **https://www.cnn.com/2024/10/30/business/fired-disney-employee-allegedly-hacked-into-company-system-to-change-allergy-info-on-menus/index.html**
3. U.S. Department of Justice. (2025). "Former Disney employee sentenced for hacking menus, altering allergens." Press release, April 25. **https://www.wesh.com/article/former-disney-employee-sentenced-for-hacking-menus/64587578**
4. Malwarebytes. (2021). "FBI warns of ransomware threat to food and agriculture." **https://www.malwarebytes.com/blog/news/2021/09/fbi-warns-of-ransomware-threat-to-food-and-agriculture**
5. CISA. (2021). "Food and agriculture sector." **https://www.cisa.gov/topics/critical-infrastructure-security-and-resilience/critical-infrastructure-sectors/food-and-agriculture-sector**
6. CISA. (2023). "Food and agriculture sector." **https://www.cisa.gov/topics/critical-infrastructure-security-and-resilience/critical-infrastructure-sectors/food-and-agriculture-sector**

CHAPTER 2

The Human Element: Food, Culture, and Identity

From Hunger to Hearth

The story of food is fascinating: how we went from gathering to cultivation, which led to cooking and then opened up modifications for production, new flavors, and ultimately our modern global food supply chain.

If you want to understand how complicated the global food system is today, you have to go way back to when food wasn't something you bought or scheduled, but something you hoped to find before something else found you.

For most of human history, whether you would eat tomorrow was genuinely unclear. Hunting and gathering worked, sort of, but it didn't leave much margin. When people started planting crops and keeping animals close, the whole calculus shifted. Farming and herding didn't just change dinner: they changed everything. Once people had some control over their food supply, they had to start organizing themselves differently. You can't store grain for the winter if no one has built a place to put it. You can't raise animals at scale without figuring out who's watching the herd and who's patching the fence. That's when food production started demanding cooperation, planning, and a whole new set of roles, many of which still exist today (in much more complex forms).

The same instincts that drove people to dig pits for grain, build smokehouses, and post guards at granaries push us to protect food today—only now, the threats look different. Food has always needed protecting, because food has always been more than calories. It's culture, trust, and continuity. And when we follow the story of how humans secured food over time, we're really following the story of how we built the foundations of identity, community, and, yes, eventually technology.

Food as Identity and Belief

Every society builds its own food grammar. What people eat or refuse to eat says something about who they are, where they belong, and what they believe. Rice in Japan is a cultural anchor. Maize in Mesoamerica signifies creation itself. The Mediterranean's bread, wine, and olive oil are the region's worldview in three ingredients. Food has always been shorthand for identity, and cultures have written entire belief systems around what ends up on the table.

Take Japan, where rice cultivation stretches back thousands of years. By the 19th century, rice had become a sacred symbol of being Japanese. To be Japanese was to be a rice-eater. The government went to extraordinary lengths to protect domestic rice, banning foreign imports for decades. When a disastrous harvest in 1993 forced Japan to import foreign rice, consumers rejected it outright, calling it impure and tasteless.[1] That wasn't about flavor but about protecting identity. Even today, Japanese families share rice from the same pot to show fellowship, and they place cooked rice at family altars to honor ancestors. Rice is nourishment, yes, but it is also memory, belonging, and spiritual continuity.

Cross the ocean to Mesoamerica, where maize is so central that it is woven into origin myths. Mayans believed the gods formed humans out of cornmeal dough.[2] To eat maize was not simply to survive; it was to reaffirm existence. In Mexico today, tortillas remain a birthright, the thread connecting ancient belief to modern identity. A taco is never just lunch; it is the daily retelling of an origin story.

The Mediterranean world crafted its own identity around a trinity of wheat, wine, and olive oil. Christianity adopted bread and wine as a sacred ritual. Olive oil carried symbolic weight as purity and blessing. These foods represented civilization itself, distilled into ingredients.

And then there are food taboos, powerful cultural markers that separate communities as effectively as borders. Jewish and Muslim prohibitions on pork have for centuries reinforced identity through shared restraint. Hindu vegetarianism often signals devotion, spirituality, and caste alignment. To eat or not to eat is rarely just about preference. Instead, it's a declaration of who you are and where you stand.

Indigenous traditions underscore this truth even more deeply. Food was a sacred relationship. Native American tribes held ceremonies of thanks for animals and plants, treating them as gifts from the spirit world. Hunts weren't complete until the animal was honored and its sacrifice acknowledged. Colonization disrupted this balance brutally: tribes were forced from their lands, buffalo herds decimated, and fishing and hunting rights stripped. Traditional diets were replaced with rations of flour, lard, and sugar, and along with those substitutions came devastating health crises and cultural disconnection. Yet traditions endured underground: seeds saved, recipes whispered, ceremonies carried out in secret.

Today, Indigenous food sovereignty movements are reviving these practices. "Three Sisters" gardens of corn, beans, and squash reappear. Salmon and wild rice harvest rituals return. Elders describe traditional food as "spirit food," healing both body and culture. Passing down recipes in native languages is as much about preserving vocabulary as it is about flavor. These revivals are not nostalgic throwbacks. They are survival strategies for identity itself.

Migration and Transformation

Migration has always carried food across borders, and with it, identity. When people move, recipes are packed as carefully as heirlooms. Kitchens in immigrant households become sacred ground, where a grandmother teaching her grandchild how to fold dumplings or spice curry is passing on belonging, not just making dinner.

Even when language fades, food remains a strong tether. A second-generation child may not speak their parents' mother tongue, but they know how to bake bread or fry plantains the family way. These acts keep identity alive even as distance and time stretch the connection.

Adaptation shapes these stories, too. Chinese immigrants in North America created dishes like General Tso's chicken not because they were traditional, but because they were possible. They worked with what they had and what their customers wanted, turning survival into innovation. These adapted recipes tell their own cultural stories. One granddaughter described her grandmother's Egg Foo Young as unrecognizable compared to restaurant versions because it carried the grandmother's lived journey across multiple regions of China. Food, in that sense, is autobiography.

Colonization added a harsher layer to this story. The Columbian Exchange redistributed crops on a global scale: potatoes to Ireland, tomatoes to Italy, chilies to India, and cacao to Switzerland. Entire cuisines that we take for granted today are built on foods indigenous to the Americas and other regions. At the same time, colonization deliberately erased traditional diets, dismissing them as primitive and replacing them with wheat, sugar, and livestock. Boarding schools severed children from their food knowledge, weaponizing meals as tools of assimilation.

And yet resilience persisted. Caribbean and Brazilian cuisines blended African, Indigenous, and European influences into creole masterpieces. Colonizers themselves adopted local foods: chicken tikka masala is now called Britain's national dish.[3] Globalization accelerated this exchange, and now cities like London and Vancouver serve sushi one night, tacos the next, and biryani the day after. That diversity enriches lives but also threatens local distinctiveness, which is why cultures double down on protection. France defends Champagne with legal muscle. Naples won UNESCO recognition for its pizza-making. Japan still fiercely guards its rice. These aren't quirks. They're survival strategies in a homogenized world.

Food in the Digital Mirror

If migration carried recipes across oceans, social media carried them across time zones in seconds. TikTok and Instagram are the new spice routes. A grandmother in Oaxaca, Mexico, can go viral teaching tamale-making. A teenager in Seoul, South Korea, can make kimchi pancakes a global craving. During the pandemic, sourdough starters became international companions, nurtured like pets in kitchens from Melbourne to Milwaukee. Food has always been identity, but now identity comes with hashtags.

This digital exchange democratized food knowledge. Recipes that once stayed within families now circle the globe. Foodies build communities around veganism, keto, halal, or gluten-free living. Posting a plate online is a declaration: this is who I am; find me if you are, too.

But the digital mirror has cracks. Misinformation spreads faster than facts. Viral diet fads promise miracle cures. Dubious "hacks" encourage unsafe canning or food storage that could land someone in the hospital. Fake recalls tank businesses. Even well-intentioned influencers can cause harm when nuance is sacrificed for clicks. Trust in food systems can be shaken in seconds by a trending hashtag.

Food culture is now louder, more global, and more visible than ever, but also more vulnerable. Social media made every kitchen a stage and turned trust into a fragile currency. We've always protected food from pests, rot, and fraud. Now we have to protect it from misinformation too, and from the way entertainment sometimes overshadows food safety. Do your favorite food creators practice proper food handling, or are they more concerned with making a video go viral? Are they washing their hands, avoiding cross-contamination, and cooking meat to safe temperatures, or are they prioritizing bright lighting and a catchy soundtrack? Best practices that keep people healthy often get sidelined in the digital space, raising real questions about how much we should trust what we see. Those methods that are shared, good or bad, have consequences once they're copied in home kitchens around the world.

Food identity in the modern age also shows up in what people refuse to eat. Veganism and vegetarianism, once niche lifestyles, are now mainstream enough that people claim them as core identity markers. Saying "I'm vegan" today functions the same way someone might say "I'm Catholic" or "I'm Irish." What you eat has become a way of telling people who you are before you say a word. Like religious food taboos before them, these choices create boundaries and declarations. For some, it's about animal welfare or sustainability. For others, it's personal health or ethics. Whatever the motivation, it's proof that food has never stopped being shorthand for belief.

The challenge is that any strong identity can be co-opted or twisted. Online spaces where people celebrate plant-based eating can also turn into places where misinformation spreads quickly, or where extreme rhetoric takes hold. That doesn't make veganism itself dangerous, but it makes it powerful, and power can always be manipulated. Later in this book, we'll talk about how radical activism sometimes uses food identity as a weapon. For now, it's enough to say this: whether it's rice in Japan, maize in Mesoamerica, or vegan hashtags on Instagram, what we eat or refuse to eat has always been one of the clearest ways humans tell the world, "This is who I am."

Technology at the Table

Technology has always sat at the table with us. Clay pots and grinding stones were early tech. Fermentation was a miraculous hack, transforming milk into cheese, grain into beer, and cabbage into something that could last the winter. Salt, smoke, and sugar preserved meat and fruit long before refrigeration. Water and windmills multiplied output, freeing human arms for other work.

Refrigeration rewrote perishability, first with icehouses, then with mechanical cooling, and eventually with the global cold chain that moves strawberries across oceans and keeps vaccines viable. Canning fed navies and made peaches available in January. Barcodes transformed every tomato into a traceable data point, making inventory, safety recalls, and supply chain management possible at scale.

And alongside these tools, food safety science grew. Pasteurization made milk safe. Inspection systems formalized trust. NASA's need to feed astronauts safely pushed forward hazard analysis techniques that became the foundation of the Hazard Analysis Critical Control Point (HACCP), a food safety framework now used worldwide.[4] These were systems built so families could eat without fear. At every stage, the point of food technology has been the same: protection. Keeping food safe enough, long enough, to share. Technology doesn't erase culture. It scales it. It takes a grandmother's soup and makes sure the ingredients get to her kitchen intact. It keeps Friday fish on the table, ensures that rice reaches the altar, and guards the birthday cake waiting in a freezer. We don't call that cybersecurity, but the instinct is the same. To protect what feeds us.

Looking Ahead

The reason to tell all these stories—the hearth, the rice, the maize, the diaspora kitchens, the UNESCO lists, the hashtags, the vegans—is simple: food has always been culture, and culture has always needed protection. Technology has been our tool for that protection for thousands of years. Today, the tools are digital, but the stakes haven't changed. When food systems fail, we don't just lose nutrition; we lose memory, identity, and trust.

When a label is wrong, a child with an allergy pays for it. When a cold store is locked by ransomware, it's somebody's holiday dinner canceled. When misinformation about a food product trends online, it erodes the trust of entire communities—and that trust, once gone, rarely comes back quietly. Food is a physical expression of who we are: protecting it means protecting us.

If food is identity and technology is how we deliver it, then the next question is obvious: how exactly did technology reshape the food world we inherited into the modern system we depend on? Chapter 3 will walk through that journey, from mills to refrigeration to the connected systems we now take for granted, so we can see how innovation gave us abundance but also left us with new vulnerabilities.

Notes

1. Etonomics. (2025). "Japan and its rice crises." **https://etonomics.com/2025/12/02/japan-and-its-rice-crises**
2. Britannica. (2024). "Popol Vuh." **https://www.britannica.com/topic/Popol-Vuh**
3. Cook, R. (2001). "Robin Cook's chicken tikka masala speech." *The Guardian.* **https://www.theguardian.com/world/2001/apr/19/race.britishidentity**
4. NASA. (2024). "Food safety program for space has taken over on Earth." **https://www.nasa.gov/directorates/stmd/tech-transfer/spinoffs/food-safety-program-for-space-has-taken-over-on-earth**

CHAPTER 3

A (Human) History of Technology in the Food Industry

Food Tech Has Always Been Human Tech

Say "food technology," and most people picture robots planting lettuce or an AI greenhouse in the Netherlands. Shiny. Futuristic. But here's the thing: food tech is older than written language. The first time a human dropped a hunk of meat onto a fire, dinner became an engineering problem. Survival moved beyond what nature handed us to what we could engineer to make dinner happen. And that level of food production demanded cooperation, planning, and a new set of roles.

Surpluses really changed the game. When there was more food than a family could eat in a day, people settled, specialized, and invented. More babies lived past childhood. Communities grew. With that came trade, storage systems, early economies, and all the moving parts that now form the backbone of every modern food system we depend on.

Even our bodies evolved in response. Cooking unlocked more calories with less effort, which fueled bigger brains. Our digestive systems shifted. In a very real way, we reprogrammed ourselves around what and how we ate. And once food became culture: meals to share, rituals to mark time, and the earliest version of what we now call "hospitality."

This chapter is a journey through that story. We'll move from smoky campfires to the hiss of steam engines, from clanking conveyor belts to the quiet click of programmable controllers, and into today's world, where algorithms decide when to water tomatoes or feed salmon. At each step, technology rewired how we lived. And with every leap, we got faster, more ambitious, and a lot more dependent on the machines we built. So, pull up a chair and get a snack; let's see how we got here.

From Fire to Farming: Muscle, Tools, and Community

Picture it: tens of thousands of years ago, a small group huddles around a fire. Smoke curls upward, roots roast in the embers, and for the first time, dinner is more than raw chewing. Cooking made food safer, more digestible, and a lot tastier. That fire wasn't just a campfire; it was humanity's first food-tech upgrade. Once we realized we could bend food to our will, there was no going back.

Fire and sharpened stones carried us far. Coastal communities strung nets and carved boats to chase fish. By 3000 BCE, Egyptians were sailing the Nile with wooden ships stacked with catches. Inland, we built traps, corralled goats, and began experimenting with planting wild grains. Eventually, farming took root. Farming was humanity's first real biotech project. Over generations, we picked seeds that grew stronger, bred animals that were calmer, and shaped the dinner table without even realizing we were doing it.

Tools multiplied: digging sticks, clay pots, grinding stones. None of them glamorous, but revolutionary all the same. Surplus food meant fewer people went hungry, and more of them could do something other than farming. Civilization sprouted from surplus grain.

Then came animal power. Oxen and water buffalo were the horsepower before horsepower, dragging crude wooden plows through the soil. The Sumerians, around 4000 BCE, left us clay tablets with instructions for plowing and even a prayer to Ninkilim, goddess of field mice, because even with new tech, mice are mice. Farming becomes knowledge shared across generations, canals dug by whole villages, and granaries stocked for everyone.

And food tech wasn't limited to dry land. Aquaculture goes back at least 3,000 years in China, where carp were raised in ponds. By 500 BCE, Fan Li, a Chinese scholar, wrote manuals on how to manage those ponds. The Romans farmed oysters, too, corralling them behind coastal lagoons. Food was now something that could be cultivated, engineered, and scaled.

The Age of Mechanization: Steam, Steel, and Scale

For millennia, food production was powered by backs, hooves, and the occasional gust of wind. Then came the Age of Mechanization, the 18th and 19th centuries, and with it, the hiss of steam and the clang of steel. Suddenly, farming wasn't a matter of endurance; it was a matter of horsepower, the iron kind.

On land, the game-changer was the mechanical reaper. Before it, harvesting meant long days swinging scythes until your arms felt like lead. In 1831, Cyrus McCormick demonstrated his horse-drawn reaper, a machine with rotating blades and rakes that cut and gathered in one sweep.[1] A single farmer could now harvest what once took a crew. As one observer put it, the reaper "broke the harvest bottleneck." Translation: farms exploded in size, and cities swelled with people leaving the fields for factory jobs.

Steel plows carved through tough prairie sod. Seed drills planted in neat rows. Threshing machines separated grain faster than a team of flailing farmhands ever could. Farming shifted from a pace dictated by muscle to one set by machines.

Processing followed the same rhythm. Nicolas Appert's canning method in 1809, born of Napoleon's plea to feed his armies, meant food could be sealed in tins and stored far beyond its season. Chicago's Union Stock Yards turned butchering into an industrial spectacle, with carcasses moved by conveyor belts and hooks in what was essentially a disassembly line. Roller mills powered by steam spat out fine flour by the barrel. Gail Borden's condensed milk became a U.S. Civil War staple, proof that food could travel and endure.

And the sea wasn't left behind. In the 1870s, Britain launched steam trawlers like the *Pioneer*. Unlike sailboats, they didn't wait for wind. They steamed straight into deep waters, dragging heavy nets and hauling them back with winches. Ports like Grimsby boomed, unloading mountains of fish onto trains headed for London. The catches were bigger, faster, and fresher; and yes, this also meant local fish stocks crashed. Progress always comes with a shadow. Mechanization rewrote the entire pace of food and nature.

The Rise of the Tractor

If the reaper was the spark, the tractor was the engine that roared. Early models in the late 1800s looked like runaway locomotives, steam puffing from their stacks. By the early 20th century, gasoline and diesel had slimmed them into machines farmers could actually handle. Horses and oxen began their long retirement as tractors became the universal muscle of the farm. Hook up a plow, and dig deeper furrows than any ox can. Add a seed drill, sprayer, or baler, and the same machine runs the whole show. Over the years came hydraulics, more powerful diesel engines, and cabs with heat and air conditioning, a small mercy during a blazing July harvest or a January freeze. By mid-century, the tractor was the farm's beating heart, often with a radio playing inside the cab.

And then came the 1990s. Tractors stopped being just iron giants and became computers on wheels. GPS receivers, yield monitors, and variable-rate controls transformed them into platforms for precision agriculture farming guided by data, not just diesel. But before tractors turned digital, another revolution was humming in the background: the glow of electricity and the power of chemistry.

Electricity, Chemistry, and the Rise of Modern Processing

At the turn of the 20th century, food tech got two new fuels: electricity and chemistry. Together, they built the backbone of modern food. Electricity brought pumps for irrigation, milking machines that replaced hand labor, and refrigeration that kept milk from spoiling overnight. Rural electrification in the United States during the 1930s and 1940s lit up barns and turned them into efficient workshops. Factories swapped clunky steam belts for electric motors. Suddenly, bakeries could run day and night, and meatpacking plants wielded electric saws that sliced faster and more safely.

Chemistry cracked food open at the molecular level. The Haber–Bosch process, developed in the early 1900s, pulled nitrogen from the air and turned it into fertilizer. That single breakthrough feeds billions today; a 2008 study in *Nature Geoscience* estimates that without it, about half the world's population wouldn't have enough food.[2] Pesticides, preservatives, and vitamins all changed what we grow and how long it lasts. Food became as much a laboratory project as a harvest.

And then came advancement in cold storage. Refrigeration changed everything. In 1882, the *Dunedin* sailed from New Zealand to London with a full load of frozen lamb. Ninety-eight days later, the meat arrived in perfect condition, and suddenly, oceans weren't barriers; they were trade routes.[3] Refrigerated trains and ships globalized dinner. By the 1920s and 1930s, home refrigerators replaced iceboxes and daily milk panics.

Clarence Birdseye pushed it even further in the 1920s. Inspired by Inuit quick-freezing techniques he observed in Labrador, he developed industrial flash-freezing.[4] His first frozen products hit stores in Springfield, Massachusetts, in March 1930, looking like witchcraft: bright green peas in January. Shoppers hesitated but then caved. By the end of World War II, frozen food was everywhere. One fur trader's curiosity had turned into a global industry.

Electricity, chemistry, and cold storage stretched food's reach beyond seasons, regions, and shelf life. Food was manufactured, preserved, and shipped at scales our ancestors couldn't have imagined.

The Digital Turn: PLCs and Data

By the late 20th century, farms and factories were mechanized and electrified, but they were still mostly mechanical. The next leap was quieter: digital. The real action shifted to control rooms and circuit boards.

Enter the PLC, or programmable logic controller. First built in 1969 for General Motors by Bedford Associates, a PLC is basically a rugged computer built to run machines.[5] Unlike your office laptop, these thrive in heat, dust, and noise. They listen to sensors (is the bottle in place? is the oven hot enough?), follow programs written by engineers, and send commands (open the valve, stop the mixer). They're the conductors of industrial orchestras, making sure machines hit their notes at the right time.

Food companies grabbed hold. Breweries used PLCs to keep fermentation tanks at the perfect temperature. Bakeries timed kneading and baking cycles with them. Bottling plants lined up thousands of glass bottles and filled each to the milliliter. By the 1980s, PLCs were everywhere, and some managers joked about "silicon chefs" running their kitchens.

Meanwhile, farms started their own digital makeover. The 1990s brought precision agriculture: a farming approach that uses data, sensors, and GPS to manage fields with pinpoint accuracy instead of treating them as one giant plot. In other words, farming by the square meter, not the acre. Yield monitors on combines—the massive machines that cut, thresh, and clean grain in one pass (direct descendants of the old reapers)—mapped which parts of a field produced more. GPS-guided tractors drove straighter rows. Variable-rate spreaders adjusted fertilizer on the fly. Farmers were no longer relying on gut feel but farming with data and satellite images.

Livestock and aquaculture joined in. Dairy cows wore electronic tags tracking health and feeding. Fish farms modeled populations with computer data. Grocery stores ditched paper ledgers for barcodes, giving supply chains real-time visibility. Machines were thinking, in their own limited way.

Food Goes Smart: Automation, Analytics, and Cyber-Physical Systems

Now we've hit the present, and farms, barns, and factories run on systems that think and act together. Welcome to automation, AI, and cyber-physical systems.

A cyber-physical system is a connected system in which digital controls trigger real-world consequences. Its machines and computers are so tightly linked that they act like one organism. Picture a greenhouse where sensors feed data into software that tweaks lights, heaters, fans, and irrigation on the fly. Or a salmon farm where underwater cameras and oxygen sensors tell an algorithm when to feed and when to stop, saving pellets and keeping water clean.

On modern farms, tractors steer themselves, drones scan fields for pests, and soil sensors cue irrigation systems to drip water in the exact spots needed. Barns look like science fiction: cows wearing smart collars wander into robotic milking stalls when they feel like it, snack on custom feed, and leave behind a stream of data. Farmers don't milk by hand anymore; they check their dashboards.

Factories and warehouses are equally wired. High-speed cameras with AI catch bruised fruit or stray debris faster than human eyes. Cobots (collaborative robots) pack delicate foods alongside human workers. Distribution centers predict demand before it happens. Even vending machines order their own refills and flag a human when maintenance is needed. It's dazzling. Food, once bound by muscle and season, now flows through a web of machines that monitor, adjust, and optimize at every step. But every layer of tech adds another layer of dependency.

We Built the System, and Now We Rely on It

From fire to frozen peas, from oxen to algorithms, food technology has been one long, messy, astonishing experiment in human ingenuity. Each leap has solved problems and created new ones, feeding more people and binding us more tightly to our own systems.

Today, when you bite into an apple shipped across an ocean or pour milk tracked by a dozen sensors, you're tasting thousands of years of invention. The timeline matters because it shows how deeply food and technology are entwined and how much we stand to lose if those systems falter.

Our ancestors knew how to hunt, fish, and coax crops from the earth. Most of us now only know how to swipe a debit card at the grocery store. That's not criticism, it's the privilege of modern life. But it means the reliability of these systems isn't optional. It's survival. And that sets the stage for the next question: if food tech has always been human tech, what does it mean to protect it? That's where we're headed next.

Notes

1. Wisconsin Historical Society 2012. (n.d.). "Brief biography of Cyrus Hall McCormick." **https://www.wisconsinhistory.org/Records/Article/CS3399**
2. Erisman, J. W., Sutton, M. A., Galloway, J., Klimont, Z., & Winiwarter, W. (2008). "How a century of ammonia synthesis changed the world." *Nature Geoscience*, 1, 636–639. **https://doi.org/10.1038/ngeo325**
3. New Zealand History. (2021). "First frozen meat shipment leaves New Zealand." **https://nzhistory.govt.nz/first-shipment-of-frozen-meat-leaves-nz**
4. History.com. (2016). "How the modern frozen food industry took inspiration from Inuits." **https://www.history.com/articles/history-of-frozen-food-birdseye**
5. Control Engineering. (2008). "How programmable logic controllers emerged from industry needs." **https://www.controleng.com/how-programmable-logic-controllers-emerged-from-industry-needs**

CHAPTER 4

The Evolving Role of Cybersecurity in Food Systems

When most people hear the word "cybersecurity," they picture stolen bank accounts, fraudulent credit cards, hospitals paralyzed by ransomware, or even the stereotypical "hacker in a hoodie." Rarely does anyone think about a farm, a grain co-op, or the cookie factory where their favorite snacks are made. Yet food has always needed protection, and not just from the obvious enemies like pests or bacteria. What has changed is the addition of a new, invisible threat: attackers who exploit digital pathways with real-world consequences. What was once treated as just an IT issue or another sector's problem is now front and center in the food and agriculture sector. Cybersecurity in this sector is not just about protecting data. It is about safeguarding life, trust, and continuity.

In the last chapter, we traced the story of technology in food, from the muscle power of humans and animals to tractors, robotics, and data-driven decision-making. Each leap made us faster, more productive, and more interconnected. But with each advancement came a new set of risks. The tools that gave us efficiency also opened cracks in the foundation. Now we face the reality that the same systems that help feed the world can also be turned against us. Before jumping into the specifics of operational technology (OT) and industrial control systems (ICSs) in Chapter 5, it is essential to understand how cybersecurity has become a crucial component of food system resilience.

From Plows to Passwords: The Double-Edged Sword of Progress

The last four decades have significantly reshaped our global food systems in ways that many people may not recognize. As mentioned in the previous chapter, tractors now follow GPS coordinates, robots milk cows and package dairy products, crops are monitored by drones, and irrigation is made smarter with precision agriculture tools. Processors use predictive analytics to forecast demand and sensors to track storage conditions down to the last detail. These are no longer experimental technologies but standard tools of the trade.

Although this connectivity has brought enormous benefits, it has also created new exposure. Systems that used to be self-contained are now online, and vendors connect remotely to troubleshoot machines from a distance. Data is uploaded to the cloud, and entire operations are run on networks that can be accessed through smart devices, such as phones or tablets. For years, people in the industry thought their distance or their niche protected them. A farm in rural Iowa or a brewery in Australia did not feel like a target, but hackers do not need to be nearby; they only need an open window into the network.

By 2020, the food and agriculture sector had seen a clear surge in cyberattacks. Incidents that once felt rare were now appearing regularly in industry threat reports and FBI alerts. Why the sudden shift? Because industries like finance and retail had invested heavily in security protection, whereas food and agriculture had prioritized differently. In food and agriculture, thanks to tight margins, the priority was moving product without pausing for security software updates, and, sadly, the backburner problem came to the forefront. Hackers did their research, took notice, and adapted. They identified a sector of global importance and vulnerabilities that could be exploited. To paraphrase one expert, food companies often had security on par with basic shipping operations, which is not a standard that inspires confidence. Cybersecurity was still thought of as an IT responsibility, not an operational concern. That illusion did not last long once attackers figured out how much leverage food provided and how lucrative it could be for them.

Why Hackers Care About Food

The obvious question is: why food? Why would a hacker in another country care about milk, meat, or grain? The answer is money, leverage, and disruption.

Criminal groups quickly realized that food companies are ideal targets. Unlike some industries, downtime here does not mean an idle office. It means spoiled product, ruined batches, halted deliveries, food safety concerns, and panicked retailers. Every lost hour translates into lost revenue. Hackers know companies will pay quickly if it means avoiding waste and shortages. The fact that everyone depends on food only makes the pressure more intense. Hackers are well-researched and know their targets intimately; they often understand the global food supply chain better than we do. This makes them uniquely dangerous to the food and agricultural sectors because we are still playing catch-up in security controls and processes.

Their most common tactic is ransomware. I'm sure you've heard about ransomware, but here's a refresher: attackers encrypt systems and demand payment or a ransom to unlock them. Modern ransomware often includes double extortion, where attackers also steal sensitive files and threaten to release them. The choice becomes grim: pay the ransom or face both operational paralysis and a data leak. And even if the ransom is paid, there is no guarantee that the data will not be sold later. It's particularly nasty and time-consuming, and the operational costs can take years to untangle.

Financial gain is not the only motive for bad-threat actors. For some attackers, food is a weapon. Nation-states and extremist groups understand that disrupting food supply chains destabilizes populations faster than almost anything else. The logic is brutal: hungry people are desperate people. Communities that cannot trust their food supply lose stability, and nations lose strength. Ukraine has become the clearest example of this reality. In 2015, a cyberattack on its power grid left approximately 225,000 customers without electricity.[1] It was a demonstration of how digital tools could disrupt civilian lifelines, and it read like something out of a science fiction book. By 2023, the strategy had escalated to food; Russian drone strikes on Ukrainian grain ports along the Danube destroyed tens of thousands of tons of grain in a single night, with one attack at Izmail damaging nearly 40,000 tons destined for Africa, China, and Israel.[2] Those attacks were deliberate attempts to weaponize hunger, not only against Ukraine but also against countries in Africa and the Middle East that rely on Ukrainian wheat. Global markets felt the shock, with prices spiking and food insecurity worsening. In modern conflicts, just as in ancient times, food is not collateral damage; it is a target. And now it has become digitalized.

This kind of weaponization is not limited to war zones. Disinformation campaigns have been used to shake consumer confidence. Fake recall notices, false contamination reports, and manipulated social media campaigns spread doubt faster than any pathogen. Hacktivist groups have also attacked agriculture to make political points rather than to collect ransom. The outcome is the same: disruption, fear, and erosion of trust.

Food systems are particularly vulnerable because of their biology. Milk cannot wait; produce spoils quickly. Processors cannot idle lines for long without loss, and retailers cannot keep shelves empty without backlash. Attackers understand this urgency. Every assumption that food will "always be there" becomes a pressure point waiting to be exploited.

The Wake-Up Calls the Industry Could Not Ignore

The years 2020 and 2021 marked a turning point. A string of attacks made it impossible to dismiss cybersecurity as irrelevant to food. In June 2020, the Australian company Lion, a major producer of milk and beer, was hit with ransomware.[3] Production halted, and orders were delayed. The company reverted to manual processes just as pubs were reopening after the COVID lockdowns. It was described internally as the worst possible time for a shutdown, and the attackers likely knew it.

In May 2021, JBS, the world's largest meat processor, became a headline case. A ransomware attack disabled plants across the United States, Canada, and Australia.[4] In the United States, every beef facility went offline for days, eliminating nearly a quarter of national capacity. In Australia, thousands of workers were sent home. In Canada, one of the country's largest beef plants stopped receiving cattle. The FBI attributed the attack to a Russian-speaking ransomware group. JBS eventually paid $11 million in ransom to restore operations. What had once been a theoretical risk was now front-page news. If the world's largest meat processor could be brought down, who was safe?

The trend continued. In early 2023, Dole Food Company, one of the largest produce suppliers in the world, was forced to shut down operations in North America.[5] Grocery stores reported empty shelves in the salad section. Consumers who had never thought about cyberattacks related to food suddenly saw the result firsthand. For many, it was jarring: a hack had caused a shortage in their local grocery store, and the fear of pandemic-level outages shook the public.

In 2025, Dairy Farmers of America (DFA), the largest dairy cooperative in the United States, was also hit. The Play ransomware group disrupted multiple processing plants and threatened to release stolen financial and payroll data.[6] DFA managed to keep milk deliveries going with backups, but the attack underscored how even cooperatives owned by thousands of farmers are targets. Trust can be shaken as easily in a co-op as in a multinational corporation.

Together, these incidents painted a clear picture. Meat, dairy, produce, grain, and even candy were all on the list—and those are only the ones we know about. Many more have happened without major media coverage. No part of the food system has been spared. The illusion that "it won't happen to us" has crumbled. In boardrooms and barns alike, the conversation has shifted from "if" to "when."

Not Just Data at Stake: Food Safety

When attackers target the food system, they put food safety and quality at direct risk. Cybersecurity in food is not just about protecting computers but also about protecting the physical safeguards that keep food safe to eat.

The Dole incident in 2023 is a good example. Headlines focused on grocery shortages, but the real issue was safety. Produce depends on digital systems that track refrigeration, sanitation, and traceability. Without those systems, Dole could not prove food safety compliance. Out of caution, the company delayed shipments. It was the right decision, but it highlighted how fragile the system is. If you cannot prove food is safe, you cannot sell it.

This reality extends far beyond salads. In dairies, systems verify pasteurization cycles. In meat plants, they regulate freezer temperatures. In beverage facilities, they manage sanitation chemicals. When those systems fail, the line between a safe product and a dangerous or toxic one quickly becomes thin. A cyberattack does not have to poison food directly; all it has to do is disable the safeguards.

If the pasteurization process goes offline at a dairy, raw milk either spoils or reaches consumers unpasteurized. If refrigeration stops in a meat plant, the product can warm into the bacterial danger zone. These are scenarios food safety managers are now forced to consider in hazard planning. Regulators are also catching on; for example, the Food Safety Modernization Act (FSMA) does not specifically name cyber threats, but its requirement for preventive controls means that cyber risks are increasingly being treated as hazards.

And then there is trust. A cyberattack that disrupts supply or raises safety questions rattles confidence. Consumers wonder whether hackers have tampered with the food, even if they have not. Rebuilding that trust is not easy once it is lost. A recent non-cyber example illustrates this perfectly. In 2024, Boar's Head recalled more than 7 million pounds of ready-to-eat meats after a listeria outbreak linked to its Jarratt, Virginia, plant.[7] At least 60 people were hospitalized, and 10 died. Federal inspectors documented mold, pooling blood, and repeated sanitation violations inside the facility. The company responded by closing the plant indefinitely, discontinuing its liverwurst line, and offering transfers or severance to its 500-person workforce. For more than a year, the plant remained closed, its reopening contingent on strict U.S. Department of Agriculture (USDA) oversight and enhanced testing. Boar's Head paid a multimillion-dollar settlement, but the greater cost was public confidence in a brand long associated with premium quality. This was not a cyberattack, but the lesson is the same: once consumers start to wonder whether the food on their plate is safe, no amount of reassurance can easily restore trust.

From IT Problem to Food Safety Imperative

The good news is that the food and agricultural industry is waking up. Cybersecurity is finally being recognized as a food safety and business resilience issue (instead of just an IT problem). More companies are talking about "cyber hygiene" alongside sanitation schedules. Just as no plant would run without cleaning, companies are realizing they should not run without security controls and backups.

Regulators are beginning to slowly echo this. Although there is no single "food cyber law" in the United States, agencies like the Food and Drug Administration (FDA) and the USDA have acknowledged that cyber incidents can threaten food safety. FSMA is the key U.S. food law focused on preventing hazards before they occur. It requires preventative controls, and many now interpret that to include cyber risks. Auditors will begin inquiring about cybersecurity preparedness as well as safety logs.

In Europe, new requirements are already in place under the NIS2 Directive. NIS2 is the European Union's updated Network and Information Security law, which expands mandatory cybersecurity obligations to a wide range of industries, including food and agriculture. It requires companies to assess risks, put security measures in place, and report significant incidents quickly, sometimes within days. Crucially, it also pushes accountability down the supply chain, meaning food companies must ensure that their vendors meet cybersecurity standards as well. In short, the EU is treating cyber resilience in food as a regulated necessity, not a suggestion—a sharp contrast to the mostly voluntary approach in the United States.

Most importantly, the culture surrounding cybersecurity in food and agriculture is changing. It has become a common refrain: if you eat food, you should care about cybersecurity. It may sound cheeky, but it lands. Employees are learning to spot phishing attempts, and farmers are inquiring about protecting their precision agriculture software. Suppliers are being asked not only about quality standards but also about digital risks. The culture of safety that has always defined food is expanding to include cybersecurity. The new message is simple: cybersecurity is food safety.

A New Recipe for Resilience

The evolving role of cybersecurity in food systems can be summed up in one word: integration. Security is becoming part of operations, safety, and continuity planning. The old silos, with IT on one side and production on the other, are breaking down. Securing the digital means securing the physical. This is only the beginning, though. Attackers will continue to adapt, finding new weak points—AI attacks are next up to bat for sure. But defenders are adapting, too. Farmers, processors, regulators, and consumers are all part of the solution, and knowledge is power.

As we close this chapter, the next question is clear. What exactly are the systems we are protecting? We have talked about robots, PLCs, refrigeration units, and sanitation controls. But how do they actually work, and why are they so vulnerable? That is the subject of Chapter 5. If IT is the world of emails and spreadsheets, operational technology (OT) is the world of mixers, ovens, irrigation pumps, and packaging lines. It is where the digital meets the physical, transforming raw ingredients into food. Understanding OT is the key to understanding why cyberattacks on food systems are so disruptive and how we can defend against them.

Notes

1. Cybersecurity and Infrastructure Security Agency. (2021). "Cyber-attack against Ukrainian critical infrastructure." **https://www.cisa.gov/news-events/ics-alerts/ir-alert-h-16-056-01**
2. PBS News. (2023). "Russian drone strikes hit Ukrainian port key to grain exports." **https://www.pbs.org/newshour/world/russian-drone-strikes-hit-ukrainian-port-key-to-grain-exports**
3. Corfield, G. (2020). "Now you've done it: Cyber attack targeted Australian brewery 'n' dairy biz Lion." *The Register*. **https://www.theregister.com/2020/06/11/australia_brewery_lion_cyber_attack**
4. The Associated Press. (2021). "REvil, a notorious ransomware gang, was behind JBS cyberattack, the FBI says." *NPR*. **https://www.npr.org/2021/06/03/1002819883/revil-a-notorious-ransomware-gang-was-behind-jbs-cyberattack-the-fbi-says**
5. Lyngaas, S. (2023). "Cyberattack on food giant Dole temporarily shuts down North America production, company memo says." *CNN Business*. **https://www.cnn.com/2023/02/22/business/dole-cyberattack**
6. The Record. (2025). "Dairy Farmers of America confirms June cyberattack leaked personal data." **https://therecord.media/dairy-farm-leaked-info-ransomware**
7. Centers for Disease Control and Prevention. (2024). "Cases and additional death reported in listeria outbreak linked to deli meats." **https://www.cdc.gov/media/releases/2024/s0808-listeria-deli-meats.html**

CHAPTER 5

Understanding Operational Technology in Food Systems

Picture the journey of a freshly baked loaf of bread or a crisp gallon of milk, tracing every step from the sun-kissed fields to your kitchen table. Behind this process lies an intricate and dynamic network of machines, each silently communicating and working in harmony. This sophisticated process is known as *operational technology* (OT): the vital hardware and software that monitors and controls the physical components of food production and our modern ecosystem. Unlike the standard office computers used for sending emails or creating spreadsheets, OT serves as the unseen yet unwavering backbone that ensures food flows seamlessly from farm to consumer. It encompasses the precision of the pasteurizer, which dutifully maintains its rhythm, the irrigation systems that awaken fields with nourishing water at dawn, and the conveyor belts that tirelessly transport goods, never pausing or complaining, even during the longest shifts. What's truly remarkable is that OT is not confined to a single setting; it permeates every corner of food systems, whether nestled within a bustling factory, hidden inside a towering grain silo, or integrated into a rustic barn. It is ubiquitous, working diligently behind the scenes to ensure that the food we enjoy is produced, processed, and delivered with unmatched efficiency and reliability.

In earlier chapters, we chatted about how food shifted from manual labor to mechanization and then to digital control, shaping both culture and risk. Now it's time to pull back the curtain and be clear about what OT looks like. Not because I want you to become a control systems engineer, but because once you understand how these systems work, it becomes painfully obvious why they're vulnerable. OT isn't just computers sitting in a barn. It's the switches, sensors, and little robots that tell machines what to do, like holding milk at the right temperature, when to turn the fans in the chicken house, or when to move the conveyor belt. It's the behind-the-scenes helper that keeps food safe and moving, which also makes it the perfect thing for hackers to mess with, because when it breaks, the consequences can be deadly.

OT in Food Production

If you've ever walked into a modern food factory, you know it isn't quiet; you'll hear the rhythm of production. Conveyor belts hum, mixers churn, steam hisses, and packaging machines click in perfect time. It's like an orchestra that never stops playing, or, as I have always called it, "organized chaos radio." And the conductor isn't a human waving a baton; it's OT. The heartbeat of this system is the *programmable logic controllers* (PLCs), those rugged little industrial computers that act like recipe cards for the entire plant. A PLC doesn't improvise or take shortcuts. It follows instructions to the letter, just as you would when baking a loaf of bread: no imagination, just following the time and authentic recipe. If a temperature sensor indicates that the oven is a few degrees too cool, the PLC sends a command to fire up the burner. If a cap misses a bottle, the PLC halts the line before a hundred bottles of soda foam over. It's automation as precision cooking—reliable, repeatable, and built for scale, just like your grandma's cookie recipe. You don't mess with perfection.

Now, picture trying to keep track of hundreds of machines across a football-field-sized facility. That's where SCADA systems come in: *supervisory control and data acquisition*. Despite the bureaucratic-sounding name, SCADA is basically the head chef's clipboard for the entire operation. It pulls in all the information from sensors and PLCs and lays it out on screens in a control room. Operators can see whether the milk silos are chilled, if mixers are running too fast, or if a pasteurizer is falling behind schedule. Without SCADA, you'd need an army of people running from one end of the plant to the other with thermometers and clipboards. With it, a small team can supervise the entire factory floor, making sure the symphony plays on without a sour note.

The whole point of OT in production? Consistency. A loaf of bread baked in Boston should taste the same as one baked in Boise. OT keeps that promise by repeating the same processes at the same speeds, under the same conditions, on repeat, like your favorite song. And when these systems fail, it means food wasted, contracts broken, and consumers standing in front of empty shelves, wondering if they should look for another brand to buy.

Take Schreiber Foods in Wisconsin, one of the world's largest dairy processors. In October 2021, ransomware froze the company's systems, forcing shutdowns across facilities.[1] You might not recognize the name, but if you've had cream cheese on a bagel, you've probably eaten a Schreiber product. And when Schreiber stopped, bagel shops scrambled, grocery stores had holes in the dairy aisle (and not just in the Swiss cheese), and farmers sat with milk they couldn't move. Breakfast changed fast; cream cheese was out, butter was in, and nobody was happy about it.

In August 2023, Campbell Soup's Napoleon, Ohio, plant, the largest soup factory in the world, was forced to halt production for three days after a cyber intrusion.[2] On paper, the attackers hit IT. In reality, they froze OT, too. And here's the rule: if there's even a chance that the kettles or sterilizers are compromised, the lines stop. Nobody gambles with soup safety. Consumers saw it as a delay; inside the plant, workers stood idle while vats of soup sat unfinished, a reminder that when IT cracks, OT feels it.

Then there's Conagra Brands, a company behind dozens of household names in the frozen, packaged, and snack aisles. In fiscal year 2023, Conagra experienced supply chain disruptions caused by a third-party vendor's cybersecurity incident that affected its ability to fulfill customer orders.[3] On the surface, that may sound like a supply chain hiccup. But the reality was that production was interrupted because the OT systems couldn't run as planned. For a business that thrives on getting products to shelves predictably and reliably, disruption at that scale isn't just a technical problem; it's a breakdown in consumer trust. If the popcorn or peanut butter you count on has vanished, you don't care whether it's because of a server glitch or a cyberattack. All you know is that your snack aisle betrayed you, and now you're standing there hangry with an empty cart.

These examples highlight a fundamental aspect of OT in food production: when it fails, the consequences are immediate and visible. An IT system that's gone rogue might mean emails go unanswered or invoices get delayed: frustrating, yes, but survivable. An OT system that's affected means the bread doesn't bake, the soup doesn't cook, and the cream cheese spoils. Food doesn't wait. If OT goes offline, production halts. That means food sitting where it shouldn't, safety steps skipped, and workers left standing in front of equipment that wasn't designed for sudden halts. You don't make that up later; you lose the food, you risk the people, and you pay for it in real time.

And yet many of these systems are old, or *legacy*, to be more polite. Office laptops are replaced every few years. OT equipment, on the other hand, is built like tractors, designed to last decades. Walk into any food plant, and you'll find equipment from the 1990s still running production lines, sometimes with software that hasn't been updated since the turn of the millennium. (Windows 98 second edition is still a favorite in the wilds of production.) Why? Because if it isn't broken, nobody replaces it. The phrase "Don't mess with it" is practically a mantra on factory floors. There's a running joke between professionals who work in these types of plants: sneeze in the parking lot five miles away, and it might break production. In short, everything within the production space is sensitive (kind of like teenagers, but thankfully, no acne). However, this means many OT systems are still running on outdated code, default passwords, and sometimes no modern protections at all. They are the cast-iron skillets of food production: seasoned, dependable, and still cranking out product, but with no built-in lock to keep someone from misusing them.

I would be remiss if I didn't mention the cultural nuance; IT and OT don't have to be rivals. IT cares about confidentiality, data protection, and system recovery. OT prioritizes availability, safety, and keeping production running. In the food sector, availability is king. Stopping a line is the last resort because it means wasted ingredients, missed deliveries, employees sent home, and broken contracts. That's why OT security is such a balancing act: you must protect the process without jeopardizing it. And although IT teams might raise eyebrows about how old some of these OT systems are, they also recognize the stakes. We love our IT teams for guarding the data, but OT is the side of the house that makes sure dinner actually makes it to the table.

Ag-OT: Farming's Nervous System

When most people think about technology in farming, they still picture a tractor, maybe with GPS, rolling across a field. But the truth is that agriculture has quietly become one of the most technology-dependent industries on the planet. The same kind of automation that keeps soup kettles running in the Campbell's plant in Ohio now guides irrigation pivots in Nebraska and manages greenhouse climates in the Netherlands. This is why I use the term Ag-OT: operational technology explicitly tailored to agriculture. It's not a shiny new buzzword; it's a way to create a common language when talking about the technology that keeps modern agriculture moving right along.

Ag-OT is what keeps crops watered on schedule, balances fertilizer use, tracks livestock health, and controls storage conditions for harvests. It's the automated irrigation valve that opens at sunrise, the climate control system in a greenhouse that makes microadjustments to temperature and humidity, the sensors in a grain silo that trigger fans to prevent mold or keep chickens cool or warm in their houses, and the robots that milk cows in perfectly timed shifts. If OT in factories is about cranking out consistency and speed, making sure every loaf and every can comes out the same, Ag-OT has a tougher job. Out on the farm, it's about survival and timing. Miss the irrigation cycle, and the crop pays for it. Miss the feeding schedule, and the herd pays for it. Nature doesn't care if your system is down; it just keeps moving, with or without you.

That's what made the NEW Cooperative ransomware attack in Iowa in September 2021 so serious. The BlackMatter ransomware group demanded $5.9 million after locking up the co-op's IT backbone—the software used to manage feed formulations, soil data, and grain logistics for farmers across the Midwest.[4] Feed mills suddenly couldn't calculate balanced rations for livestock, grain elevators couldn't coordinate shipments, and farmers were left in limbo. If feed recipes are wrong, animals don't thrive. If elevators stall, harvests have nowhere to go. What started as an IT lockup cascaded into an operational breakdown, threatening meat, dairy, and grain producers downstream. One strike on IT reverberated through Ag-OT, proving just how interconnected and interdependent the whole system really is.

Around the same time, Crystal Valley Cooperative in Minnesota, a farmer-owned co-op that supplies feed, seed, fertilizer, and fuel, was also struck by ransomware.[5] Again, the attack targeted IT systems, networks, ordering, and logistics software, but the fallout was evident in Ag-OT. Orders slowed, communication lines jammed, and farmers who depended on the co-op as their one-stop shop were left wondering: Would the fuel delivery show up? Could they get the seed they had already booked? Even short delays throw off tightly planned schedules, and when margins are razor-thin, "wait and see" isn't an option. From the outside, it might look like a customer service hiccup. From the inside, it was Ag-OT grinding to a halt, and most of today's farms can't function on manual workarounds for long.

Then there's the equipment itself. In May 2022, AGCO, a global manufacturer of farm machinery that includes brands like Fendt and Massey Ferguson, was hit with ransomware.[6] The attack came at the worst possible time: planting season in the United States and Europe, when every day lost in the field means yield reductions months later. Farmers suddenly couldn't access the service tools or digital updates that keep their 20-ton machines running. Imagine standing in a field, ready to plant, but the digital systems that guide your planter are locked. The hackers had found a cruel kind of leverage: using time itself against farmers.

Yes, we eat in separate kitchens, but our food supply is global. When one link snaps, the effects ripple far beyond the farm or factory where it started. In February 2022, several Port of Antwerp companies, including Sea-Invest, were hit with a cyberattack that disrupted Zespri's kiwifruit shipments from New Zealand.[7] The orchards kept producing, the pickers kept picking, but the grading, packing, and shipping slowed to a crawl. For a company that lives on just-in-time exports to Asia and beyond, that's a nightmare. Fruit doesn't sit around waiting politely; it spoils.

We don't know every detail of how these logistics systems were impacted, but here's what a stall like that really means: growers don't get paid if shipments don't move. Employees on the packing line are stuck staring at crates of fruit they can't process, or, worse, hourly employees are sent home. Cold storage and shipping partners start bleeding money because the whole chain is off schedule. And eventually, the consumer notices; maybe the kiwifruit costs more that week, perhaps it's missing altogether. One clogged digital artery, and the ripple goes all the way to grocery stores across Asia, Europe, and yes, the United States, too.

Remember back in the first few pages of this book, when I told you about Swiss farmer Vital Bircher? In November 2023, he lost a cow and her calf after ransomware locked him out of his milking parlor.[8] Here's why that story matters here: it wasn't IT bleeding into operations, it was a straight OT hit. The parlor's control system itself was encrypted, and suddenly the farmer was blind. No corporate breach trickling down, no middleman, just Ag-OT cut off in the middle of daily chores, with consequences measured in animal lives.

And here's the bigger picture. A milking parlor doesn't run in isolation. It's tied into herd management software that tracks feeding, health, and milk production. It's linked to the cooling tanks that keep milk safe until pickup. Pull that thread, and the entire system starts to unravel: feeding schedules slip, health alerts vanish, and cooling tanks fall out of sync. What looks like "just one device" is actually the nervous system of the entire farm operation.

These examples show that Ag-OT failures have unique characteristics. Factories can sometimes restart production with overtime shifts. Sadly, agriculture sadly doesn't work like that; missed feeding can't be undone. A planting season doesn't get a do-over; fruit left unshipped spoils. Farming has always been tied to time and nature, but now it's tied to digital systems as well. When hackers target Ag-OT, they are essentially playing chicken with the natural clock, and farmers are the ones who pay the price.

Ag-OT also highlights the cultural differences in how technology is adopted. Many farmers are pragmatic. They don't chase every new gadget, but they've embraced precision agriculture tools that clearly improve yields or reduce costs. That pragmatism has created a patchwork of adoption: one farm might run GPS-guided tractors and Internet of Things (IoT) soil sensors, while their neighboring farm still relies on notebooks and muscle memory. This uneven adoption means some farmers are already exposed to cyber risks they don't fully recognize, and others will face them as soon as they adopt their first "smart" system.

At its core, Ag-OT shares the same vulnerabilities as factory OT: legacy systems, insecure remote connections, and a culture of availability over security. But the consequences are often harsher because agriculture is so unforgiving. That's why Ag-OT deserves its own spotlight. It isn't just another version of OT. It's the digital nervous system of food at its source, and when it falters, the entire food chain trembles.

When OT Fails, Food Safety Falters

If you've ever worked in food production, you know the golden rule: the food must be safe. You can miss a delivery window or send out a batch of cookies with crooked labels, but if what's inside makes someone sick, the whole operation grinds to a halt. This is where OT gets personal. It isn't just about running lines faster or keeping equipment humming. It's about making sure the food we eat won't land us in the hospital.

Take pasteurization. It looks like a simple heat setting, but in reality, it's a tightly controlled critical point. OT systems monitor flow, temperature, and time down to the second. Slip those settings, whether by accident or because someone decided to get cute with a cyberattack, and suddenly that carton of milk is a bacterial lottery ticket. The scary part? You wouldn't know until people started getting sick. That's why even the hint of tampering can shut a plant down overnight.

And for the lactose-intolerant crowd, don't think you're off the hook. Beer is just as vulnerable. Modern breweries run on OT, too: sensors call the shots on mash temps, fermentation times, pressure levels, even how long the hops steep. A few degrees too high or too low, and you don't get "craft," you get "toxic." Push it further, and the line between brewery and chemical plant starts to blur. That's why breweries trust OT to keep every pint safe. Remember the Lion ransomware attack in June 2020? One of Australia and New Zealand's biggest beverage companies had to shut down brewing altogether.[9] Beer drinkers weren't debating firewall rules; they were just mad their pint glasses were empty.

Allergens tell the same story. OT runs the recipe dosing, mixing, and labeling that keeps peanuts out of peanut-free products. Change one setting, and suddenly someone with a severe allergy is holding a death trap instead of a snack. Food companies treat this like life or death because it is. It used to be that allergens were a food safety problem, and cyber risks were a tech problem. Not anymore. Now, cyber risk includes peanut dust, because OT is the system making sure dosing, mixing, and labeling keep those allergens in check.

Cleaning or sanitation is also part of the OT and food safety process. The CIP, or *clean-in-place* process, is managed by OT: chemicals, rinse times, and the entire cycle. Skip a step, and you're breeding bacteria. And here's the thing: if there's even a suspicion that the CIP cycle didn't run right, the plant stops. Full stop. Because no brand wants to explain to regulators—or, worse, to parents—why contaminated food made it onto shelves.

This is why Campbell Soup shut down production in 2023 after the cyber incident. The incident started on the IT side, but the company couldn't guarantee that OT hadn't been touched. And when sterilization or kettles are in question, you don't gamble. The cost of downtime is ugly, but the cost of unsafe food in the market is catastrophic.

Food safety and cybersecurity used to be different conversations: one about microbes, the other about malware. Not anymore. OT married them. A cyber incident today can look precisely like a contamination event: unsafe food, recalls, broken trust. Regulators are starting to catch on, and some companies are even running cyberattacks as part of their food safety drills. It's not perfect, but it shows a recognition that food safety and cyber safety are now two sides of the same coin.

Looking Ahead

By now, the pattern should be clear: OT and Ag-OT are not optional add-ons to our food system; they are the nervous system itself. They keep silos breathing, kettles cooking, conveyors rolling, and barns temperature regulated. When they function, they fade into the background. When they fail, it's bare shelves, spoiled food, sick animals, and shaken trust.

In many ways, this is simply the next chapter in an ancient story. Back in Chapter 2, we explored how food shaped human culture from hearth fires to stored grain. In Chapter 3, we looked at how tools and machines transformed farming and production. And in Chapter 4, we began to see how cybersecurity became part of protecting food systems, not as an abstract IT concern but as a matter of public safety. This chapter adds the missing connective tissue: the invisible but very real systems that link technology to the food on our plates. OT and Ag-OT are the latest layer in humanity's long history of organizing, automating, and protecting the essentials of survival.

And this is where systems thinking really matters. You can't separate "technology" from "farming" or "production" anymore. They are woven together. A cyberattack on a co-op in Iowa can ripple down to chicken feed and ultimately affect grocery store prices in another state. A disruption at a soup plant in Ohio can raise questions about food safety that reach into homes across the world. A kiwi exporter in New Zealand can experience delays that alter availability in markets thousands of miles away. These aren't isolated IT issues; they're systemic shocks to our global food supply.

The lesson isn't to panic, but to recognize that securing what feeds us requires a new mindset. IT and OT teams need to see themselves as partners, not silos. Farmers and food processors need to see themselves as custodians of digital systems as much as barns and silos. Policymakers and regulators need to understand that cybersecurity guidance can be the difference between safe food and contaminated food. And the public needs to realize that when we talk about cybersecurity in food, we're talking about more than passwords and servers. We're talking about trust in what we eat.

As we move into Part II of this book, we'll zoom in on specific parts of the food chain: crops, livestock, processing, packaging, distribution, and more. Each sector has its own flavor of risk, pun fully intended. Irrigation controls on farms, refrigeration units in cold storage, high-speed blades in meat processing, and conveyor belts in warehouses all have unique vulnerabilities. But beneath those differences lies the same truth: each of them now depends on OT or Ag-OT to function.

Keep this foundation in mind as we step into the sector spotlights. What we've covered here—PLCs, SCADA, IoT, automation, and the roles of OT and Ag-OT in keeping food safe and flowing—is the language we'll use as we travel across the food system. We've set the table, so to speak. Next, we'll serve up the details sector by sector, exploring where the risks really live and how they touch all of us.

So, grab your boots, we're heading to the farm next.

Notes

1. Shepel, J. (2021). "Schreiber Foods hit with cyberattack; plants closed." *Wisconsin State Farmer*. **https://www.wisfarmer.com/story/news/2021/10/26/schreiber-foods-hit-cyberattack-plants-closed/8558252002**

2. Jones, D. (2023). "Campbell Soup says summer cyberattack caused limited business impact." *Cybersecurity Dive.* **https://www.cybersecuritydive.com/news/campbell-soup-cyberattack-limited-impact/694758**
3. Conagra 10-K (2023). Annual report filed with the SEC.
4. Starks, T. (2021). "Ransomware gang strikes Iowa agriculture business New Cooperative." *CyberScoop.* **https://cyberscoop.com/blackmatter-new-cooperative-ransomware-iowa**
5. Jones, D. (2021). "Ransomware compromises customer data in farm co-op attack." *Cybersecurity Dive.* **https://www.cybersecuritydive.com/news/agriculture-food-ransomware-coop/607080**
6. Toulas, B. (2022). "US agricultural machinery maker AGCO hit by ransomware attack." *Bleeping Computer.* **https://www.bleepingcomputer.com/news/security/us-agricultural-machinery-maker-agco-hit-by-ransomware-attack**
7. Flach, J. (2022). "Cyber attack at port companies affecting Zespri's kiwifruit shipments." *Fresh Plaza.* **https://www.freshplaza.com/article/9399298/cyber-attack-at-port-companies-affecting-zespri-s-kiwifruit-shipments**
8. Bayer, M. (2024). "Ransomware attack paralyzes milking robots — cow dead." *CSO Online.* **https://www.csoonline.com/article/3484349/ransomware-attack-paralyzes-milking-robots-cow-dead.html**
9. Corfield, G. (2020). "Now you've done it: Cyber attack targeted Australian brewery 'n' dairy biz Lion." *The Register.* **https://www.theregister.com/2020/06/11/australia_brewery_lion_cyber_attack**

PART II

Sector Spotlights—Where the Risks Live

CHAPTER 6

Primary Agriculture and Farming

Where the Ground Meets the Grid

In February 2022, as Russian forces crossed into Ukraine, farmers in the Zaporizhzhia region discovered something unsettling about their John Deere equipment. Tractors and combines that Russian soldiers had looted from a dealership in Melitopol were rendered inoperable, remotely disabled through the GPS and telemetry systems that normally helped optimize planting patterns.[1] The incident made headlines as a curious footnote in a brutal war, but for anyone watching the intersection of agriculture and technology, it revealed something significant: modern farming equipment is networked, trackable, and controllable from thousands of miles away. What protected Ukrainian assets in this instance could, under different circumstances, become a profound liability.

This chapter is about the foundation: the actual farming that happens before any algorithm processes a single data point. Primary agriculture is where seeds meet soil, where livestock are born and raised, and where the fundamental transformation of sunlight and water into food begins. It is also where cyber-physical risk has quietly embedded itself into operations that feed eight billion people daily.

Peeling Back the Curtain: Who Grows What We Eat

The global agricultural workforce defies easy categorization. The Food and Agriculture Organization estimates that agriculture employs roughly 866 million people worldwide, approximately 27 percent of the global workforce, although this figure masks enormous regional variation.[2] In sub-Saharan Africa, agriculture accounts for more than half of employment; in the European Union, the figure hovers around 4 percent.[3] These workers range from smallholder farmers cultivating single hectares in Vietnam to operators managing 10,000-acre spreads across the Canadian prairies.

A profound generational transition is underway across nearly every region. The average age of farmers in the United States is 57.5 years; in Japan, it exceeds 68.[4] Younger generations have migrated toward urban employment, leaving agricultural knowledge concentrated among aging populations. This demographic shift carries direct implications for cybersecurity: operators who learned their craft on analog equipment now manage systems with embedded processors, cellular modems, and cloud connectivity technologies. Additionally, this technology has arrived faster than training programs could adapt to it, and operators' aptitude to learn it varies.

Primary agriculture encompasses three broad production categories: crop cultivation, animal husbandry, and aquaculture, with many operations blending activities in mixed-farming systems. In terms of crop production, grains, oilseeds, fruits, and vegetables account for the largest share of cultivated land, with cereals alone covering more than 700 million hectares.[5] Livestock operations provide protein and dairy to global markets. Aquaculture, the fastest-growing food production sector, now supplies more than half the fish consumed by humans.[6] However, each sector carries distinct technology dependencies.

Geography shapes everything. Ukraine and the American Midwest produce grain that feeds populations continents away. The Netherlands, smaller than West Virginia, has become the world's second-largest agricultural exporter through intensive greenhouse cultivation.[7] Australia's wheat belt operates on margins defined by water scarcity; India's Punjab serves as the nation's breadbasket despite worsening groundwater depletion. These agricultural zones are connected by export flows and commodity markets that transmit disruptions across borders within hours.

The Quiet Digitization of the Field

The transformation from mechanical to networked agriculture happened incrementally. Tractors adopted GPS guidance in the 1990s, initially as expensive add-ons and eventually as standard equipment. Irrigation systems acquired automated controllers adjustable via cellular connections. Livestock operations installed sensors monitoring feed consumption and barn conditions. Milking parlors adopted robotic systems that identify individual animals using radio-frequency identification (RFID) tags.

What makes this digitization significant is its integration into operations that cannot pause while systems are patched. A dairy cow must be milked roughly every 12 hours; delays cause suffering and health complications.[8] Poultry houses require continuous ventilation; summer system failures can kill thousands of birds within hours.[9] The technology has delivered genuine efficiency gains, but it has created dependencies operators may not recognize until something fails.

The daily reality of modern farming involves constant interaction with digital systems, whether operators think of it that way or not. A farmer checks soil moisture on a smartphone; a ranch hand reviews cattle weights logged by automated scales: these mundane activities represent data flowing through networks, passing through vendor servers, and crossing borders to cloud infrastructure that could be located anywhere in the world. The farm is no longer an island; it is a node in a global network, even at the end of a gravel road with marginal cellular reception.

Where the Cracks Form

The vulnerabilities embedded in primary agriculture emerge from a collision between agricultural economics and technology lifecycles. Farm equipment represents substantial capital investment (a modern combine can exceed $500,000), and operators expect decades of useful life from it.[10] Yet embedded computer systems may become unsupported within five to seven years, long before mechanical systems wear out. The result is equipment running outdated operating systems and unpatched firmware designed before security was even a consideration.

Vendor dependence compounds these challenges. The agricultural machinery market has consolidated significantly, with a handful of manufacturers, such as John Deere, CNH Industrial, AGCO, and Kubota, dominating global sales.[11] When John Deere's systems experienced outages during the 2022 planting season, farmers were unable to access diagnostic tools or, in some cases, operate equipment.[12] The incident highlighted how deeply farming has become dependent on corporate digital infrastructure.

Remote access capabilities create attack surfaces inconceivable a generation ago. A 2022 University of Cambridge analysis identified vulnerabilities in agricultural telematics systems, including unencrypted data transmission and weak authentication.[13] Many agricultural control systems were designed for reliability rather than security, built to operate continuously in harsh conditions rather than to resist intrusion.

Satellite connectivity introduces dependencies that few farmers consciously consider. When Russia attacked Viasat's KA-SAT network on invasion day, the disruption extended beyond Ukraine. German wind turbines lost remote monitoring, and agricultural operations across Europe using the service experienced outages.[14] Farmers had no way to know their guidance systems (GPS) depended on infrastructure that would become a military target.

Human factors may be the most significant vulnerability. Agricultural operations run on thin margins; net farm income in the United States recently averaged just 8.6 percent of gross revenue, leaving little budget for security investments.[15] Operators serve as mechanics, agronomists, accountants, and now de facto IT and cybersecurity administrators. Training focuses on production techniques rather than phishing recognition. And farming's cultural ethos—trust, handshake agreements, community mutual aid—aligns poorly with zero-trust security frameworks.

When It Breaks

Documented cases likely represent a fraction of actual disruptions, as many farms lack monitoring capabilities to detect intrusions, and few have an incentive to publicize vulnerabilities. However, in September 2021, the NEW Cooperative, an Iowa grain cooperative serving over 60 farming communities, was struck by BlackMatter ransomware during harvest, a timing that suggests deliberate targeting for maximum leverage.[16] Systems for feed scheduling and grain logistics were disrupted, forcing the use of paper-based fallback procedures. BlackMatter reportedly demanded $5.9 million in ransom.[17] Weeks later,

Crystal Valley Cooperative in Minnesota experienced a similar attack.[18] These incidents demonstrated that agricultural operations meet ransomware groups' criteria: time-sensitive operations, thin margins making extended downtime existentially threatening, and limited security capabilities. Simply put, it means the agricultural sector is likely to provide threat actors with a payday due to the criteria.

Earlier that year, JBS S.A., the world's largest meat processor, paid an $11 million ransom following an attack that closed facilities in North America and Australia.[19] Although JBS operates at the processing level, the shutdown rippled upstream immediately; producers found themselves with animals ready for market and nowhere to send them, creating welfare and safety concerns and leading to financial losses that extended far beyond the company itself.

Quieter failures don't generate headlines. In 2023, Germany's Federal Office for Information Security documented vulnerabilities in agricultural machinery that could allow remote interference with steering and braking systems.[20] Australian farmers have reported GPS spoofing incidents affecting harvesting equipment.[21] These suggest ongoing low-level interference that rarely reaches formal incident reporting thresholds.

What It Teaches Us

Several patterns emerge across global contexts. First, there is a profound mismatch between the pace of technology adoption and the corresponding level of security awareness. Equipment manufacturers have moved aggressively to add connectivity features that generate ongoing data revenue, but investment in securing those features has lagged.

Second, regulatory fragmentation leaves agricultural cybersecurity in a gray zone. Agriculture focuses on production and food safety; cybersecurity agencies focus on critical infrastructure and government systems. The United States designated food and agriculture as critical infrastructure, but the actual requirements remain minimal relative to those for energy or finance.[22] The EU's NIS2 Directive will, for the first time, impose requirements on certain agricultural entities.[23]

Third, digital inequity tracks existing economic and geographic divides. Large commercial operations can afford dedicated IT staff; smallholder farmers in developing regions may lack reliable electricity or connectivity. Yet the global food system depends on both, and vulnerabilities anywhere can create cascading effects that cross borders.

Moving Forward: What Resilience Requires

Building resilience requires approaches acknowledging agriculture's distinctive characteristics and resource constraints. Farmers won't hire security operations centers; solutions must be embedded into equipment and everyday practices.

Integrating cybersecurity into existing farm safety culture offers a promising path. Operations already maintain protocols for equipment and chemicals; adding digital systems aligns with existing practice.[24] Vendor accountability matters: farmers deserve to know how long software will receive updates and what data will be transmitted. The right-to-repair movement has addressed farmer autonomy over equipment; security transparency should be part of that conversation.[25] Connectivity diversification, redundant providers, and offline-capable fallback modes would reduce single points of failure. Cross-border knowledge sharing remains underdeveloped; organizations like the International Society of Automation could facilitate sector-specific information sharing.[26]

Ultimately, protecting primary agriculture means recognizing that farms are no longer isolated operations. They are nodes in global networks, dependent on satellites overhead, servers in distant data centers, and software written by engineers who may never have set foot on a farm. The people who grow our food deserve security tools proportionate to the risks they currently face and have already faced.

If primary agriculture is the field, the physical place where food production begins, then precision agriculture represents the data layer emerging above it. That layer promises efficiency and sustainability gains that could help feed a growing population on a warming planet. It also introduces new categories of exposure that extend far beyond any individual operation, from the algorithms that decide when to plant to the genetic databases that shape what we grow. That's where we're headed next.

Notes

1. Faife, C. (2022). "Ukrainian farmers are using captured Russian tanks to tow equipment, but Russian soldiers can't use stolen John Deere tractors." *The Verge*.
2. FAO. (n.d.). "FAOSTAT: employment indicators." **https://www.fao.org/faostat/en/#data/OEA**
3. Eurostat. (2023). *Agriculture, Forestry and Fishery Statistics: 2023 Edition* (Publications Office of the European Union). **https://ec.europa.eu/eurostat/web/agriculture/publications**
4. Statistics Bureau of Japan, Ministry of Internal Affairs and Communications. (2021). *2020 Census of Agriculture and Forestry*. **https://www.e-stat.go.jp/en/stat-search/files?page=1&query=agriculture%20census**
5. FAO. (2023). *World Food and Agriculture – Statistical Yearbook 2023*. **https://doi.org/10.4060/cc8166en**
6. FAO. (2024). *The State of World Fisheries and Aquaculture 2024*. **https://doi.org/10.4060/cd0683en**
7. Viviano, F. (2017). "This tiny country feeds the world." *National Geographic*. **https://www.nationalgeographic.com/magazine/article/holland-agriculture-sustainable-farming**
8. Jacobs, J. A., & Siegford, J. M. (2012). "The impact of automatic milking systems on dairy cow management, behavior, health, and welfare." *Journal of Dairy Science*, 95(5), 2227–2247. **https://doi.org/10.3168/jds.2011-4943**

9. Oluwagbenga, E. M., & Fraley, G. S. (2023). "Heat stress and poultry production: A comprehensive review." *Poultry Science,* 102(12), 103141. doi: **https://doi.org/10.1016/j.psj.2023.103141**
10. Association of Equipment Manufacturers. (2023). *US Ag Tractor and Combine Reports.* **https://www.aem.org/market-share-statistics/us-ag-tractor-and-combine-reports**
11. Organisation for Economic Co-operation and Development. (2023). *Competition in the Agricultural Machinery Market.* **https://www.oecd.org/en/publications/2023/02/oecd-competition-trends-2023_4510a089.html**
12. Peters, M. A. (2022). "John Deere software outage leaves farmers in the lurch during planting." *Successful Farming.* **https://www.nbcnews.com/business/consumer/right-to-repair-farmers-challenge-john-deere-control-equipment-rcna199651**
13. Marshall, C., & Prior, M. (2023). "Cyber security: global food supply chain at risk from malicious hackers." *BBC News.* **https://www.bbc.com/news/science-environment-61336659**
14. Greenberg, A. (2022). "The untold story of the boldest supply-chain hack ever." *Wired.* **https://www.wired.com/story/the-untold-story-of-solarwinds-the-boldest-supply-chain-hack-ever**
15. USDA, Economic Research Service. (2024). "Farm income and wealth statistics." **https://www.ers.usda.gov/data-products/farm-income-and-wealth-statistics**
16. Cimpanu, C. (2021). "Iowa farm cooperative struck by ransomware attack during harvest season." *The Record.* **https://cyberscoop.com/blackmatter-new-cooperative-ransomware-iowa**
17. Greig, J. (2021). "BlackMatter ransomware hits U.S. farm Co-Op NEW Cooperative." *ZDNet.* **https://industrialcyber.co/news/blackmatter-ransomware-group-strikes-new-cooperative-makes-demand-of-5-9-million**
18. Bracken, B. (2021). "Crystal Valley cooperative hit with ransomware attack." *Threatpost.* **https://www.cybersecuritydive.com/news/agriculture-food-ransomware-coop/607080**
19. Turton, W., & Riley, M. (2021). "JBS Paid $11 million to resolve ransomware attack." *Bloomberg.* **https://www.bbc.com/news/business-57423008**
20. Bundesamt für Sicherheit in der Informationstechnik (BSI). (2023). IT-Sicherheit in der Landwirtschaft: Analyse und Handlungsempfehlungen. **https://www.tab-beim-bundestag.de/english/agriculture-and-food_cybersecurity-in-food-production.php**
21. Evans, M. (2023). "Farmers 'crippled' by satellite failure as GPS-guided tractors grind to a halt," *Sydney Morning Herald.* **https://www.smh.com.au/national/farmers-crippled-by-satellite-failure-as-gps-guided-tractors-grind-to-a-halt-20230418-p5d1de.html**
22. CISA. (2015). *Food and Agriculture Sector-Specific Plan.* **https://www.usda.gov/sites/default/files/documents/2015-food-and-agriculture-sector-specific-plan.pdf**
23. European Union. (2022). *Directive (EU) 2022/2555 on Measures for a High Common Level of Cybersecurity Across the Union (NIS2). EUR-Lex.* **https://eur-lex.europa.eu/eli/dir/2022/2555/oj**
24. American Farm Bureau Federation. (2025)."American Farm Bureau Federation partners with food and Ag-ISAC on cybersecurity." **https://www.fb.org/news-release/american-farm-bureau-federation-partners-with-food-and-ag-isac-on-cybersecurity**
25. Perzanowski, A., & Schultz, J. (2016). *The End of Ownership: Personal Property in the Digital Economy.* (MIT Press). **https://mitpress.mit.edu/books/end-ownership**
26. International Society of Automation. (2023). *ISA/IEC 62443 Series of Standards: Security for Industrial Automation and Control Systems.* **https://www.isa.org/standards-and-publications/isa-standards/isa-iec-62443-series-of-standards**

CHAPTER 7

Ag-Tech & Precision Agriculture

The Transformation Beneath Our Feet

During the harvest season of September 2021, combines rolled through Iowa cornfields, unaware of what was about to happen. As mentioned previously, a ransomware group called BlackMatter brought the NEW Cooperative to its knees and put that year's harvest in jeopardy. The NEW Cooperative, a Fort Dodge-based farmer cooperative that handles grain storage, feed distribution, and agronomic services across more than 50 locations in America's top corn-producing state, found its computer systems locked. The attackers demanded $5.9 million ransom, threatening to double the amount if payment wasn't received within five days.[1]

What made this attack particularly consequential wasn't just the ransom demand or the encrypted files. It was what went dark alongside them: the cooperative's proprietary precision agriculture platform that farmers relied on to guide irrigation decisions, fertilizer applications, and planting strategies. Within hours of the attack, farmers who had become dependent on digital soil mapping found themselves resorting to paper tickets to document grain hauls, reverting to practices their grandparents might have recognized.[2]

The incident illuminates a transformation that has reshaped agriculture over the past two decades. Precision agriculture, once a novelty adopted by early-technology enthusiasts, has become embedded in the operational fabric of modern farming. According to the USDA's Agricultural Resource Management Survey, automated guidance systems or GPS-enabled steering technology that allows tractors and combines to navigate fields with centimeter-level accuracy now operate on well over 50 percent of planted acreage for corn, cotton, rice, sorghum, soybeans, and winter wheat in the United States.[3] Among large-scale farms, adoption rates among those with gross cash farm income exceeding $1 million reach 70 percent or higher.[4]

This chapter examines what happens when an industry built on soil, weather, and biological cycles becomes inextricably linked to software, sensors, and cloud platforms. The vulnerabilities that emerge are not merely technical; they reflect the collision of agricultural rhythms with digital dependencies, the tension between equipment manufacturers' business models and farmers' operational autonomy, and the uncomfortable reality that the same connectivity enabling precision also enables intrusion.

The Architecture of Modern Precision Agriculture

To understand why precision agriculture creates cybersecurity challenges, we must first understand what it has become. The term *precision agriculture* encompasses an interconnected ecosystem of technologies that collect, transmit, analyze, and act on agricultural data. At the field level, sensors embedded in soil measure moisture content, nutrient levels, and temperature. Yield monitors mounted on combines record bushels per acre in real time, generating maps that reveal productivity variations across individual fields. Variable-rate technology systems adjust seed populations, fertilizer applications, and pesticide sprays based on prescription maps that account for soil type, topography, and historical yield data.

Above the fields, satellite imagery and drone surveillance capture multispectral data revealing plant health invisible to the naked eye. Weather stations deployed across agricultural operations feed local atmospheric conditions into predictive models. GPS receivers accurate to within 2 centimeters guide equipment along predetermined paths, eliminating overlaps and gaps that waste inputs and reduce yields.

What transforms these individual technologies into precision agriculture is their integration through farm management software platforms. Vendor services aggregate data from multiple sources, apply analytical algorithms, and generate recommendations that flow back to equipment in the field. A farmer using a smartphone application can identify which fields require attention, review yield comparisons across multiple seasons, and transmit updated prescriptions to machinery located miles away.

This integration creates value by enabling decisions that would be impossible without data synthesis. A variable-rate nitrogen prescription, for instance, might combine historical yield data, current soil test results, satellite imagery indicating crop vigor, and weather forecasts predicting rainfall that could lead to nutrient runoff. The prescription accounts for factors no farmer could simultaneously hold in working memory, optimizing applications that reduce costs while maintaining productivity.

Yet integration also creates dependency chains. During the NEW Cooperative attack, the co-op's service went offline. Farmers didn't simply lose access to a software application: they lost the accumulated analytical intelligence that had been guiding their decisions for the year, based on yield data, soil sampling results, and prescription histories that existed only in digital form. The attack revealed how thoroughly precision agriculture had become embedded in operational workflows and how vulnerable those workflows were to digital disruption.

The Human Dimension of Technology Adoption

Adoption of precision agriculture technologies has not been uniform, and understanding these patterns reveals much about the human factors shaping cybersecurity risk. USDA data consistently shows that farm size correlates strongly with technology adoption. In

2023, 70 percent of large-scale crop-producing farms used guidance autosteering systems, compared with far lower rates among small family farms.[5] The reasons extend beyond simple economics, although the capital costs of precision equipment certainly favor larger operations that can spread investments across more acres.

Larger farms tend to employ younger operators or hire staff with technical backgrounds who are more comfortable with digital tools. They often work with agronomic consultants and precision agriculture dealers who provide implementation support and ongoing troubleshooting. Smaller operations, particularly those in which the principal operator is nearing retirement, may lack both the financial resources and the technological fluency to fully adopt digital farming.

This adoption disparity creates a segmented risk landscape. Large, technologically sophisticated operations face greater exposure to cyber threats simply because they have a larger digital attack surface. Their equipment generates vast amounts of data, relies on cloud connectivity, and integrates multiple software platforms, each of which represents a potential entry point. Yet these same operations typically have access to IT support, whether through in-house staff or contracted services, that smaller farms lack.

Smaller farms present a different risk profile. Although they may have fewer digital systems to protect, those systems often lack even basic security measures. A yield monitor storing data on an unencrypted USB drive, a farm computer running outdated Windows software, and a wireless network secured with a default password are common across operations that adopted technology without adopting the security practices that should accompany it. When these farms do experience cyber incidents, they often lack both the technical capacity to respond and the insurance to recover.

The human psychology of farming compounds these challenges. Agriculture operates on biological timelines that don't accommodate system downtime. Seeds must be planted within optimal windows that may span only days. Crops must be harvested before the weather deteriorates. Animals require daily care regardless of whether computer systems are functioning. This operational urgency creates pressure to maintain uptime at any cost, which often means deferring software updates that require system reboots, maintaining legacy equipment because replacement would disrupt planting schedules, and paying ransoms to restore access to data needed for immediate decisions.

Ukraine and What It Revealed

In May 2022, a story emerged from occupied Ukraine that initially seemed like a feel-good narrative about technology thwarting Russian aggression. Russian military forces had looted the Agrotek dealership in Melitopol, making off with approximately $5 million worth of John Deere equipment: combines, tractors, and planters. The machinery was tracked via GPS as it traveled more than 700 miles to Chechnya, where the thieves discovered the equipment had been remotely disabled and would not start.[6] The incident was widely reported as a triumph of digital rights management, evidence that the same technologies farmers sometimes resent for limiting their repair options could serve legitimate security purposes. John Deere's ability to remotely disable stolen equipment seemed like a clear benefit of connected agriculture.

Security researchers saw something more troubling. The same capabilities that disabled stolen tractors in Chechnya could, in theory, disable legitimate tractors anywhere in the world. The researcher known as Sick Codes, who has spent years examining agricultural equipment security, observed that John Deere essentially maintains remote access to equipment it has sold. "They can access these devices at any time," he noted in an analysis of the Ukrainian incident.[7]

The implications extend beyond individual equipment. John Deere's Operations Center, the cloud platform that aggregates data from connected machinery, maintains a real-time map of late-model Deere equipment deployed globally. An attacker who compromised that platform wouldn't merely gain access to individual tractors; they would obtain visibility into the operational status of agricultural machinery across entire regions. During planting or harvest, such visibility, combined with disabling capabilities, could cause disruption at scale.

The Ukrainian incident also revealed a geopolitical dimension that is often overlooked in discussions of agricultural technology. Modern tractors are not standalone mechanical devices; they are nodes in global information networks that cross national boundaries. When Russian forces attempted to use equipment designed for connectivity, they discovered that connectivity works in both directions. The question of who controls agricultural equipment, and who can deny that control, has become a question of national security implications.

The Right-to-Repair Paradox

The tension between equipment manufacturers and farmers over repair rights has simmered for years, but the cybersecurity dimensions of this conflict deserve particular attention. At DEF CON 30 in August 2022, Sick Codes demonstrated a jailbreak of John Deere tractor touchscreen controllers, eventually running a modified, farm-themed game of *Doom* on equipment designed to guide agricultural operations.[8]

The demonstration was simultaneously a stunt, a proof of concept, and a warning. The jailbreak required physical access to circuit boards and months of reverse engineering—not a casual undertaking. Yet Sick Codes' work revealed that Deere's equipment ran outdated software, including Wind River Linux 8 and Windows CE, which no longer receive security updates.[9] As Kyle Wiens, CEO of repair advocacy organization iFixit, observed after the presentation, "Turns out our entire food system is built on outdated, unpatched Linux and Windows CE hardware with LTE modems."[10]

The right-to-repair conflict creates a cybersecurity paradox. Manufacturers argue that restricting access to equipment software protects both intellectual property and operational security. If farmers could modify tractor software, the argument goes, they might inadvertently introduce vulnerabilities or create pathways for malicious actors. Deere has repeatedly emphasized its commitment to safeguarding customers and the role they play in the global food supply chain.[11]

Repair advocates counter that manufacturer control creates its own vulnerabilities. When only authorized dealers can access diagnostic software, farmers become dependent on repair supply chains that may not respond quickly enough during time-critical agricultural operations. When manufacturers can remotely access equipment, that access becomes a potential attack vector. When equipment runs proprietary software that farmers cannot inspect, they cannot verify whether security patches have actually been applied or what data is being transmitted from their fields.

The paradox deepens when considering that some of the most sophisticated agricultural cybersecurity expertise exists outside manufacturer channels. Farmers have long demonstrated mechanical ingenuity, adapting equipment to local conditions and keeping machinery running beyond designed lifespans. That same ingenuity, applied to digital systems, could identify vulnerabilities that manufacturers miss. Yet current legal frameworks and technical restrictions treat farmer modification as a threat rather than a resource.

Data Sovereignty and the Question of Ownership

Beyond equipment security lies a more fundamental question: who owns the data generated by precision agriculture? A single modern farm can produce half a million data points per day, according to IBM research, a figure projected to grow to 4 million by 2036.[12] This data, which includes yield records, soil measurements, input applications, and equipment performance metrics, represents both operational intelligence and accumulated institutional knowledge about specific pieces of land.

The legal status of agricultural data remains surprisingly unsettled. Under U.S. law, raw data generally cannot be copyrighted because it consists of facts rather than creative expression. Farmers don't automatically own the data generated on their operations in the same way they own the crops they harvest or the equipment they purchase.[13] Instead, data ownership is typically determined by contracts between farmers and agricultural technology providers whose systems collect that data: contracts that farmers often sign without fully understanding their implications.

Recognizing these concerns, the American Farm Bureau Federation led an industry coalition that established the Privacy and Security Principles for Farm Data in 2014. These voluntary guidelines, updated in 2024, state that farmers should own information generated from their operations and control how that data is collected, used, and shared.[14] The Ag Data Transparent certification program, which emerged from these principles, evaluates technology providers' contracts against these standards and awards a seal of approval to companies meeting the criteria.[15]

Yet voluntary principles lack enforcement mechanisms. Research indicates that many farmers sign service agreements without fully understanding the data rights they are granting. Some agricultural technology providers' contracts contain broad provisions that permit data sharing with partners or the use of data in training artificial intelligence models, but these provisions may not be clearly disclosed in marketing materials.[16] The unevenness between large technology companies with substantial legal departments and individual farmers signing up for digital services creates negotiating conditions far from balanced.

The cybersecurity implications of data governance extend beyond privacy concerns. Aggregated agricultural data reveals patterns about productivity, practices, and vulnerabilities across regions. In the wrong hands, such data could inform commodity speculation, enable targeted attacks during vulnerable operational windows, or provide intelligence relevant to food security at national scales. Amid rising international scrutiny of AgTech ownership, questions persist about safeguarding national farm data from exposure to adversarial states, reflecting shared priorities in food security.[17]

A Global Perspective on Agricultural Cyber Risk

The vulnerabilities affecting American agriculture manifest differently but no less consequentially across the globe. In February 2020, a ransomware attack targeted Talman Software, an Australian company whose platforms handle more than 75 percent of wool trading in Australia and New Zealand. The attackers demanded 8 million Australian dollars to restore access to systems that managed auctions, sales records, and export documentation.[18] Talman refused to pay, instead rebuilding its systems on new infrastructure. But during the recovery period, wool sales were canceled across Australia, leaving an estimated 70,000 bales in limbo during a week when the industry typically processes up to $80 million in auction transactions.[19] Farmers already reeling from drought and bushfires found their market access cut off by an attack on software they had never directly touched.

The Talman incident illustrated a pattern increasingly visible across agricultural supply chains: concentration risk created by dominant platforms. A 2015 review by Australian Wool Innovation had warned that the industry's heavy reliance on a single software provider left it vulnerable, yet no alternatives had emerged before the attack proved those warnings prescient.[20] The same dynamic appears across agricultural sectors worldwide; critical functions depend on a limited number of software providers, creating single points of failure that threat actors are increasingly exploiting.

In Australia, where agriculture faces both drought and geographic isolation, the 2022 attack on GrainCorp targeted grain-handling and logistics systems, disrupting infrastructure that moves commodities from farms to export terminals.[21] The attack highlighted how agricultural cybersecurity extends beyond farm gates to encompass the entire chain connecting production to markets.

European agriculture confronts similar challenges within a regulatory environment increasingly focused on cybersecurity requirements. The European Union's NIS2 Directive, which took effect in October 2024, expands cybersecurity obligations to include food production and processing as essential sectors. Yet many European farms, like their American counterparts, operate with limited cybersecurity resources and aging technology infrastructure.[22]

The Behavioral Patterns That Create Vulnerability

Understanding precision agriculture cybersecurity requires examining not only technological vulnerabilities but also the human behaviors that create and perpetuate them. Farmers operate under conditions that systematically discourage security-conscious practices. Time pressure during planting and harvest creates incentives to defer maintenance, including software updates. Geographic isolation often means limited access to IT support. Thin profit margins prioritize spending on inputs that directly affect yield rather than security measures whose benefits are difficult to quantify.

The structure of agricultural technology markets reinforces these tendencies. Equipment dealers, who serve as the primary interface between manufacturers and farmers, are evaluated primarily by sales volume and service response times rather than security outcomes. The dealer who helps a farmer bypass a software requirement to get equipment running during harvest builds customer loyalty; the dealer who insists on security protocols that delay operations does not.

Generational transitions complicate the picture further. Many farm operations are transitioning across generations, with varying aptitudes for digital technology. An older generation that remembers farming before GPS may view precision agriculture as a useful tool but maintain paper backups and operational knowledge independent of digital systems. Younger operators who grew up with technology may assume connectivity and lack experience managing without it. The handoff between these generations often involves unstated assumptions about which practices are essential and which are merely habitual.

Trust relationships in agricultural communities also shape security behaviors. Farmers often share information freely with trusted neighbors, extension agents, and cooperative personnel: relationships built over generations that predate concerns about data security. The same openness that enables collaborative problem-solving can also serve as a vector for social engineering attacks. A phishing email impersonating a local cooperative or extension service exploits trust patterns established long before email existed.

The Sustainability Imperative

The adoption of precision agriculture is not merely a choice farmers make to improve efficiency; increasingly, it is a requirement embedded in environmental policy. The European Union's Common Agricultural Policy (CAP) for 2023–2027 introduced eco-schemes that explicitly support precision farming to reduce fertilizer use, minimize pesticide application, and meet climate targets.[23] The European Commission lists "precision crop farming to reduce inputs" among the sustainable agricultural practices that member states should incentivize.[24] Following its departure from the EU, the United Kingdom has pursued similar goals through its Environmental Land Management schemes. In January 2024, the Department for Environment, Food & Rural Affairs announced the introduction of "precision farming actions which use technology to target farming activities and reduce the use of costly pesticides or fertilizers."[25] Academic research has reinforced these policy directions. A 2017 study published in *Sustainability* by researchers at Wageningen University & Research in the Netherlands concluded that precision agriculture technologies can reduce greenhouse gas emissions while maintaining or improving farm productivity and economics.[26] The logic is sound: applying inputs precisely reduces waste, limits runoff into waterways, and lowers greenhouse gas emissions from excess fertilizer. Yet these well-intentioned policies push farmers toward technological dependencies that have not been designed with security as a foundational requirement. We are building the future of sustainable food production on systems whose vulnerabilities we have barely begun to assess. The road to a more resilient food system, it turns out, is being paved with good intentions and unsecured software.

What It Teaches Us: The Stakes Beyond the Farm Gate

Precision agriculture has delivered genuine benefits: reduced input waste, improved environmental outcomes, and enhanced productivity on constrained land resources. These benefits have enabled agricultural systems to feed growing populations while reducing per-unit environmental impacts. Rolling back digital adoption is neither feasible nor desirable.

Yet the same systems that optimize agricultural production have created dependencies whose risks are only beginning to be understood. When the NEW Cooperative's systems went dark, the BlackMatter ransomware group dismissed concerns about food supply disruption, claiming the attack would cause only financial losses. They were wrong about the scope: 40 percent of American grain production flows through software such as the systems they encrypted, but their dismissiveness reflected a broader failure to recognize agriculture as critical infrastructure that warrants serious security attention.[27]

The transformation of farming from analog to digital has outpaced the development of security practices appropriate to the "new reality." Closing that gap requires honest acknowledgment that precision agriculture's benefits come packaged with vulnerabilities that demand sustained attention. The choice is not between digital farming and analog farming; it is between digital farming done thoughtlessly and digital farming done with awareness of what is at stake when systems fail.

The NEW Cooperative attack ended without catastrophic consequences. The co-op implemented workarounds, farmers adapted, and the harvest continued. But the incident provided a preview of what more sophisticated attacks against more central targets could accomplish. The lesson should not be that agricultural systems proved resilient despite the attack. Rather, it should be that resilience requires deliberate cultivation, and that the time to cultivate it is before the next attack arrives.

Moving Forward: Resilient Precision Agriculture

Building cybersecurity resilience in precision agriculture requires approaches tailored to agricultural realities rather than generic security frameworks designed for corporate IT environments. Several principles should guide this work.

First, resilience must account for operational time constraints that are non-negotiable. Security measures that interfere with time-critical agricultural operations will be circumvented. Solutions must be designed for environments in which equipment may need to operate for weeks without interruption and in which an Internet connection cannot be assumed. This suggests a security architecture that can operate in degraded modes, maintain local functionality when cloud services are unavailable, and schedule maintenance windows to align with agricultural calendars rather than arbitrary update cycles.

Second, the agricultural community needs alternatives to third-party or vendor risk. The Talman attack demonstrated that industries dependent on single providers face systemic vulnerability. Encouraging interoperability standards, supporting alternative platforms, and maintaining offline backup capabilities all reduce the consequences of any single system failure. Data portability, the ability to move agricultural data between platforms, should be a baseline requirement, not a premium feature.

Third, security knowledge must be integrated into agricultural education and extension services. Land-grant universities, which have historically translated agricultural research into practical guidance for farmers, should expand their mission to include cybersecurity awareness. Extension agents, who maintain trusted relationships with farming communities, can convey security practices more effectively than IT professionals unfamiliar with agricultural contexts. I would also argue that cultivating professionals with hybrid knowledge of both industry operations and cybersecurity is the way of the future.

Fourth, equipment manufacturers bear responsibility for security outcomes proportional to the control they maintain over their products. If manufacturers insist on restricting farmers' access to equipment software in the name of security, they must demonstrate that such restrictions actually yield security benefits. Regular third-party security audits, transparent patch management, and clear communication about vulnerability remediation should become industry standards. Farmers should push back on contracts, including security riders, to make sure they can trust their equipment vendors.

Finally, policymakers and industry associations must recognize that agricultural cybersecurity is critical infrastructure, not merely private risk management. The food supply depends on agricultural operations functioning reliably. Attacks that disable farm equipment or corrupt agricultural data threaten public welfare, which justifies public investment in security resources, threat intelligence sharing, and incident response capabilities.

Notes

1. Ribeiro, A. (2021). "BlackMatter ransomware group strikes NEW Cooperative, makes demand of $5.9 million million." *Industrial Cyber.* **https://industrialcyber.co/news/blackmatter-ransomware-group-strikes-new-cooperative-makes-demand-of-5-9-million**
2. Bajak, F. (2021). "Iowa Farm cooperative hit by ransomware, systems go offline." Associated Press. **https://apnews.com/article/technology-business-iowa-malware-999b0f78fd5af4c5778d8ed1199f0fe2**
3. McFadden, J., Njuki, E., & Griffin, T. (2023). *Precision Agriculture in the Digital Era: Recent Adoption on U.S. Farms* (USDA Economic Research Service). **https://www.ers.usda.gov/publications/pub-details?pubid=105893.**
4. McFadden, J., & Lim, K. (2024). *Precision Agriculture Use Increases with Farm Size and Varies Widely by Technology* (USDA Economic Research Service). **http://www.ers.usda.gov/data-products/charts-of-note/chart-detail?chartId=110550**
5. Limm, K., McFadden, J., Miller, N., & Lacy, K. (2024). *America's Farms and Ranches at a Glance: 2024 Edition* (USDA Economic Research Service). **https://ers.usda.gov/sites/default/files/_laserfiche/publications/110560/EIB-283.pdf**
6. Lister, T., & Shukla, S. (2022). "Russian troops in Melitopol plunder $5m farm vehicles from Ukraine—to find they've been remotely disabled." *CNN.* **https://edition.cnn.com/2022/05/01/europe/russia-farm-vehicles-ukraine-disabled-melitopol-intl**
7. Roberts, P. F. (2022). "Feel good Ukraine tractor story highlights ag cyber risk." *The Security Ledger.* **https://securityledger.com/2022/05/feel-good-ukraine-tractor-story-highlights-ag-cyber-risk**
8. Calore, M., & Goode, L. (2022). "How a hacked tractor added fuel to the right-to-repair movement." *Wired (Gadget Lab Podcast).* **https://www.wired.com/story/gadget-lab-podcast-562**
9. Perlroth, N. (2022). "Q & A: sick codes talks tractor hacks." *The Record.* **https://therecord.media/q-a-sick-codes-talks-tractor-hacks**
10. Claburn, T. (2022). "John Deere jailbroken to run Doom at DEF CON." *The Register.* **https://www.theregister.com/2022/08/16/john_deere_doom**
11. Hazrati, M., Dara, R., & Kaur, J. (2022). "On-farm data security: practical recommendations for securing farm data," *Frontiers in Sustainable Food Systems,* 6, 884187. **https://www.frontiersin.org/journals/sustainable-food-systems/articles/10.3389/fsufs.2022.884187/full**
12. Ibid.
13. Yu, Z., De Vries, A., Ampatzidis, Y., & Sokol, D. D. (2021). "Who owns and controls farming data? AE564/AE564, 10/2021." *EDIS* 5. **https://doi.org/10.32473/edis-ae564-2021**
14. Janzen, T. (2024). "New concepts in the updated core principles for ag data." *Janzen Schroeder Ag Law.* **https://www.aglaw.us/janzenaglaw/2024/5/16/highlights-from-the-updated-ag-data-core-principles**
15. Ag Data Transparent. (n.d.). "Core principles."**https://www.agdatatransparent.com/principles**

16. Kaur, J., Hazrati Fard, S. M., Amiri-Zarandi, M., & Dara, R. (2022). "Protecting farmers' data privacy and confidentiality: recommendations and considerations." *Frontiers in Sustainable Food Systems,* 6. **https://www.frontiersin.org/journals/sustainable-food-systems/articles/10.3389/fsufs.2022.903230/full**
17. Pracın, B., & Schroeder, P. (2025). "Foreign ownership of agricultural land comes under new scrutiny." *Reuters.* **https://www.reuters.com/legal/legalindustry/foreign-ownership-agricultural-land-comes-under-new-scrutiny--pracin-2025-11-20**
18. Kirby, S. (2020). "Incident: Australian wool sales stopped by ransomware attack." *Kirby IDAU.* **https://kirbyidau.com/2020/02/27/incident-australian-wool-sales-stopped-by-ransomware-attack-itnews**
19. Hutchinson, T. & Glance, D. (2020). "Ransomware attack on sheep farmers shows there's no room for woolly thinking in cyber security." *The Conversation.* **https://theconversation.com/ransomware-attack-on-sheep-farmers-shows-theres-no-room-for-woolly-thinking-in-cyber-security-132882**
20. Cotton, J. (2020). "Wool exchange hit by ransomware attack: Talman Software." *iStart.* **https://istart.com.au/news-items/wool-exchange-hit-by-ransomware-attack-talman-software**
21. Kirby, S. (2022). "Incident: Ransomware attack on Australian agricultural company GrainCorp's payroll provider Kronos." *Kirby IDAU.* **https://kirbyidau.com/2022/02/23/incident-ransomware-attack-on-australian-agricultural-company-graincorps-payroll-provider-kronos**
22. European Commission. (2026). "NIS2 directive: securing network and information systems." *Shaping Europe's Digital Future.* **https://digital-strategy.ec.europa.eu/en/policies/nis2-directive**
23. European Commission. (2026). "Eco-schemes." *Agriculture and Rural Development.* **https://agriculture.ec.europa.eu/common-agricultural-policy/income-support/eco-schemes_en**
24. European Commission. (2026). "Sustainable agricultural practices and methods." *Agriculture and Rural Development.* **https://agriculture.ec.europa.eu/cap-my-country/sustainability/environmental-sustainability/sustainable-agricultural-practices-and-methods_en**
25. UK Department for Environment, Food & Rural Affairs. (2024). "Environmental land management in 2024: details of actions and payments." *Farming Blog.* **https://defrafarming.blog.gov.uk/2024/01/04/environmental-land-management-in-2024-details-of-actions-and-payments**
26. Balafoutis, A., Beck, B., Fountas, S., Vangeyte, J., Wal, T. V. d., Soto, I., Gómez-Barbero, M., Barnes, A., & Eory, V. (2017). "Precision agriculture technologies positively contributing to GHG emissions mitigation, farm productivity and economics." *Sustainability,* 9(8), 1339. **https://doi.org/10.3390/su9081339**
27. Kass, D. H. (2021). "New Cooperative ransomware attack timeline: status updates." *MSSP Alert.* **https://www.msspalert.com/news/new-cooperative-ransomware-attack-timeline-status-updates**

CHAPTER 8

Food Production and Processing

From Milling and Dairy to Ready-to-Eat Meals

On a Friday evening in early November 2022, Michael McCain faced the kind of call no CEO wants to receive. Maple Leaf Foods, Canada's largest prepared meats and poultry company, which his family had built over generations, was under cyberattack. Not by a competitor, not by regulators, but by criminals who had wriggled their way into the company's digital infrastructure and were now holding it hostage.

The Black Basta ransomware group, a sophisticated operation with suspected ties to Russian cybercrime networks, had encrypted critical systems across Maple Leaf's 21 manufacturing facilities. The company's highly automated operations, from order processing to manufacturing planning, from inventory management to payment systems, ground to a halt. Fourteen thousand employees across Canada, the United States, and Asia suddenly couldn't do their jobs the way they'd done them the day before.[1] What happened next would cost Maple Leaf Foods at least Can$23 million. But the decision McCain and his team made in those first hours would distinguish their response from nearly every other major food processor that had faced a similar crisis. They refused to pay.[2]

The Processing Floor

Food processing sits at the heart of the modern food system, the crucial middle ground where raw agricultural products become the packaged goods that fill grocery shelves and restaurant kitchens. It's a world most consumers never see, operating around the clock in facilities that transform milk into cheese, grain into flour, livestock into packaged meat, and produce into ready-to-eat salads or snacks.

The scale of food processing is staggering. In the United States alone, food manufacturing employs roughly 1.7 million workers across more than 35,000 establishments, according to the Bureau of Labor Statistics.[3] These facilities generate hundreds of billions of dollars in annual shipments and touch virtually every meal Americans consume. When you eat a sandwich, you're consuming the output of grain mills, bakeries, meat processors, and condiment manufacturers' machines, all running on their own complex systems.

For readers unfamiliar with the related terminology, *food processing* (sometimes called *food manufacturing*) refers to the transformation of raw agricultural ingredients into food products.[4] *Primary processing* involves initial steps like cleaning, sorting, and basic transformation: think wheat being milled into flour. *Secondary processing* combines ingredients and applies techniques like cooking, fermenting, or preserving: the flour becomes bread. *Tertiary processing* produces ready-to-eat or heat-and-serve products: the bread becomes a frozen breakfast sandwich.

What distinguishes food processing from other manufacturing is biology. Unlike a factory producing automobiles or electronics, food processing operates on biological timelines. Milk arriving at a dairy plant must be processed within hours, meat requires immediate refrigeration and careful temperature control, and fresh produce has days, sometimes hours, before it spoils. This perishability creates a constant pressure that shapes every operational decision, including decisions about technology and security.

The Technology That Runs Modern Food Plants

Walk into a modern food processing facility, and you'll find a landscape of interconnected technologies that would have been unrecognizable to workers from even 30 years ago. At the foundation sits operational technology (OT), the hardware and software that directly controls physical processes. As we discussed in Chapter 5, this includes programmable logic controllers (PLCs), which are specialized industrial computers that manage individual pieces of equipment. A PLC might control the temperature in a pasteurization system, the speed of a conveyor belt, or the timing of a packaging machine. These devices execute specific, repeatable tasks with precision that human operators couldn't match.

Overseeing these individual controllers are supervisory control and data acquisition (SCADA) systems: software platforms that collect data from multiple PLCs and present it to human operators through screens and dashboards. A SCADA system lets an operator in a control room monitor pasteurization temperatures across an entire dairy plant, adjust conveyor speeds on a meat processing line, or track the progress of a batch through a beverage bottling operation.

Between the plant floor and the corporate office sits the manufacturing execution system (MES), which translates business decisions into production instructions. When a sales order comes in, the MES helps schedule production, track ingredients, and document what was produced and when. Above that, enterprise resource planning (ERP) systems such as SAP and Oracle manage the business side: purchasing, accounting, customer orders, and supply chain logistics.

In practice, these systems increasingly talk to each other. The days when production floors operated in complete isolation—the "air gap" that security professionals once relied on—are largely over. ERP systems need real-time production data. Quality managers need remote access to monitoring systems. Equipment vendors need to troubleshoot problems without sending a technician to the site. Each connection adds capability, but it also adds potential vulnerability.

Let's consider what these systems control in a food production factory. In dairy processing, SCADA systems monitor pasteurization, the heat treatment that kills harmful bacteria in milk. The FDA requires that milk be held at 161 °F for at least 15 seconds, or at other approved time–temperature combinations.[5] If those parameters drift, the result isn't just an out-of-spec product; it's a potential public health hazard. The systems that maintain those parameters and document compliance are now digital.

In canning operations, retort systems control the pressure, temperature, and time curves that ensure shelf-stable products are safe from botulism and other hazards. In meat processing, refrigeration systems maintain the cold chain from slaughter through packaging. In bakeries, batch management systems control recipes, ensuring that the right ingredients are mixed in the right proportions. In ready-to-eat meal production, metal detectors and X-ray systems serve as final checkpoints before products ship.

The Legacy Problem

Here's an uncomfortable truth about food processing technology: much of it runs on software that its original developers no longer support. When a food company installs a new production line, it's typically making a 20- to 30-year investment. The equipment is expensive and the validation process is lengthy, and once a line is running and producing safe, consistent products, there's enormous reluctance to change anything. If a PLC running Windows XP has been reliably controlling a pasteurization system since 2008, the argument for replacing it with newer technology faces a steep climb. "If it ain't broke, don't fix it" isn't laziness in this context; it's a rational response to real operational risks.

Paper and Pencil

Back at Maple Leaf Foods, the decision not to pay the ransom meant finding another way forward. Within 48 hours of discovering the attack, McCain's team had accomplished something remarkable: they had pivoted their entire organization to manual operations. Order processing that had been automated now happened on paper. Manufacturing planning that had flowed through software now relied on spreadsheets and phone calls. Inventory tracking that had been real-time now required physical counts and handwritten logs. "In real time, we had to pivot the business from a highly efficient network of operations, one that is highly automated from orders to manufacturing planning, to inventory picking, receipts, and payment, to a business that had to run entirely on a manual basis," McCain later told market analysts. "Essentially going back to paper and pencils."[6] This wasn't elegant or efficient, but it worked. Maple Leaf's plants continued operating, animals continued to be processed, and products continued to ship to customers. The company's business continuity plans—the documented procedures for maintaining operations during disruptions—proved their value.

The Black Basta group, frustrated by Maple Leaf's refusal to negotiate, eventually published stolen data on its dark web leak site, including employee records, financial documents, and technical specifications—information that the criminals hoped would pressure the company into paying. Maple Leaf acknowledged the breach, offered affected employees two years of credit monitoring, and pressed forward with recovery.[7]

Four months later, McCain described the period from October through December as the "cyber quarter." The company reported a net loss of Can$41.5 million dollars for the quarter, compared to a small profit the year before. The direct costs of the attack: incident response, manual operations, and lost efficiency, totaling at least Can$23 million.[8] But Maple Leaf hadn't paid the ransom, it hadn't funded criminal enterprises, and it had demonstrated that a major food processor could survive a sophisticated ransomware attack without capitulating.

The contrast with other high-profile incidents is instructive. When JBS, the world's largest meat processor, faced a similar attack in June 2021, it ultimately paid $11 million in Bitcoin to the REvil ransomware group.[9] The company's rationale: protecting customers and preventing potential data exposure. The decision drew criticism from some, who argued that paying ransoms encourages more attacks, but JBS defended it as a business necessity given the circumstances.

Neither approach is obviously wrong. Both companies recovered. Both continued operating. But Maple Leaf's experience demonstrated that the "We had no choice but to pay" narrative isn't always accurate. The choice exists; it's just an extraordinarily difficult one.

When Salads Disappear

Three months after Maple Leaf's ordeal, in February 2023, shoppers in Texas and New Mexico began noticing something odd: the Dole salad kits they usually bought weren't on the shelves.[10] At Stewart's, a grocery store in Olney, Texas, customers started asking questions. The store's manager, Adam Wolfe, didn't have good answers until a memo arrived from Dole explaining the situation. The produce giant, one of the world's largest suppliers of fresh fruits and vegetables, was "in the midst of a cyberattack" and had shut down its systems throughout North America. Wolfe did something that would make the incident public: he posted Dole's internal memo on the store's Facebook page because he believed customers deserved to know why their salad options had vanished.[11]

Dole's attack followed a now-familiar pattern of ransomware encrypting critical systems. Production plants in North America temporarily shut down, and shipments to grocery stores halted. The company disclosed that half of its legacy servers and a quarter of its end-user computers were affected and that the personal information of nearly 4,000 U.S. employees had been accessed.[12] The direct costs reached $10.5 million in the first quarter alone. But what stood out in Dole's response was a candid admission from CEO Rory Byrne: the company didn't expect to recover those costs through insurance. "The simple answer," Byrne told analysts when asked about insurance coverage, "is no, we don't expect to recover on either of those categories" (referring to both the direct attack costs and the losses tied to the employee data breach). Getting cyber insurance in North America, he added, had become "prohibitive."[13] That comment reflected a dramatic shift in the cyber insurance market, rooted in an attack that predated both Dole and Maple Leaf by several years. The insurance implications of these incidents—war exclusions, the Mondelēz-Zurich litigation, and the hardening market that followed—are examined in depth in Chapter 23.

Milk, Cheese, and Friday Night Attacks

The timing of the Schreiber Foods attack was almost certainly not coincidental. It was Friday evening, October 22, 2021, when systems began failing at one of Wisconsin's largest dairy processors. By Saturday morning, plants and distribution centers across the company's network were offline. Milk trucks were arriving at facilities that couldn't process their cargo. The attackers, whoever they were, had chosen their moment well. Friday evening attacks are a deliberate ransomware tactic; security and IT staff are reduced over weekends, detection takes longer, and recovery loses a full business day before the work week resumes.[14]

Schreiber Foods is not a household name, but its products are everywhere. The company built itself as a customer-brand leader in cream cheese, natural cheese, processed cheese, and yogurt. By some estimates, Schreiber supplies roughly three-quarters of the yellow cheese that goes on fast-food hamburgers nationwide in the United States. When Schreiber stopped processing, the ripples spread far.[15] "Milk distribution was in disarray in Wisconsin this week," the *Wisconsin State Farmer* reported. Milk transporters had to be redirected to other plants. Some dairy farmers faced the agonizing prospect of dumping milk they couldn't deliver. The ransom demand, according to reports, was $2.5 million. Schreiber never confirmed whether it paid, but by late Monday, roughly 72 hours after the attack began, systems were coming back online. By Wednesday, the company was again accepting milk deliveries, producing dairy products, and shipping them to customers.[16] For Schreiber, those three days represented more than lost production. Dairy processing operates on biological terms, and milk arriving at a plant can't wait for IT systems to be restored; it must be processed or discarded. Every hour of downtime created cascading decisions about what to do with perishable inputs that kept arriving.

The attack came barely four months after JBS had paid its large ransom and just weeks after ransomware groups hit two grain cooperatives (NEW Cooperative in Iowa and Crystal Valley in Minnesota) during harvest season. A pattern was emerging: criminals had identified the food and agriculture sector as a lucrative target, and they understood how to maximize leverage by attacking when time pressure was greatest and to increase the likelihood of ransom payouts.

The Geography of Vulnerability

When cybersecurity professionals talk about the food processing sector's vulnerabilities, they often focus on technology: legacy systems, inadequate patching, and insufficient network segmentation. These factors matter, but the sector's challenges extend beyond technical debt.

Many food processing facilities are located in rural areas, close to the agricultural inputs they process. A cheese plant is located near dairy farms. A meat processor is located near cattle ranches. A grain mill is located near wheat fields. This geographic logic makes operational sense but creates staffing challenges that affect security.

Rural communities don't typically have deep pools of cybersecurity talent. The competition for qualified security professionals is fierce, and food processors, operating on thin margins in industries not known for lavish compensation, often can't match what technology companies or financial institutions offer. The result is security teams that may be stretched thin, understaffed, or simply nonexistent at smaller facilities.

Meals on Wheels and Hospital Trays

Not all food processing serves retail channels; some facilities serve populations who can't easily find alternatives when supply chains break. Apetito, a German frozen food manufacturer with operations across Europe, specializes in meals for institutional settings, including hospitals, care homes, schools, and meals-on-wheels programs that deliver hot food to elderly and disabled people in their homes. In late June 2022, the Hive ransomware group attacked Apetito's systems. The company's UK subsidiary, Wiltshire Farm Foods, found itself unable to process orders or coordinate deliveries. For more than a week, operations were severely disrupted.[17] "We currently have no access to our IT-supported systems because our servers have been attacked," Apetito announced. "For this reason, among other things, orders are currently not possible." The company's immediate priority was "to ensure supplies to clinics and retirement homes, as well as senior citizens at home." These weren't customers who could easily drive to another grocery store or order from a different delivery service. They were people who depended on Apetito's meals as a lifeline.[18] Apetito's crisis team worked around the clock to restore limited production capabilities and improvise delivery workarounds. The company warned customers that scheduled deliveries would be delayed and that meal selections might differ from what they had ordered. Recovery took over a week.

The Apetito incident illustrated a dimension of food processing vulnerability that extends beyond commercial disruption. When cyberattacks hit facilities serving vulnerable populations such as the elderly, the hospitalized, or children in schools, the stakes involve more than lost revenue. They affect people who may have few alternatives and limited ability to cope with disruption.

Crisps, Nuts, and Supply Chain Cascades

In late January 2022, shoppers in the United Kingdom began hearing an unusual warning: "Stockpile your crisps." Kevin Beaumont, a former Microsoft threat analyst, had noticed reports that KP Snacks, manufacturer of Hula Hoops, McCoy's, Tyrrells, KP Nuts, and other beloved British snacks, had been hit by ransomware.[19] The attack, attributed to the

Conti group, compromised the company's internal network and encrypted critical systems. "Through the weekend our IT Team and third-party experts have been assessing the scale of the intrusion and continue to do so," KP Snacks wrote to its retail partners. "As a result, at this stage we cannot safely process orders or dispatch goods."[20] The supply disruption was expected to last until the end of March: two full months. For UK retailers, this wasn't merely an inconvenience. KP Snacks supplies products to major supermarket chains across the country. When those products aren't available, shelves go empty, and consumers notice; also, competitors don't always have the capacity to fill the gap.

Conti's attack on KP Snacks followed a pattern the group had perfected. They gained access to the company's network, moved laterally through systems to maximize their footprint, and then encrypted everything they could reach while simultaneously stealing data for additional leverage. On their dark web blog, they posted samples of what they had taken: credit card statements, birth certificates, employee addresses, and phone numbers. The double-extortion tactic (encrypting systems *and* threatening to publish stolen data) has become standard among sophisticated ransomware groups.[21] It increases pressure on victims who might otherwise refuse to pay. Even if you can restore from backups, you can't un-steal data that's already been copied. KP Snacks never publicly confirmed whether it paid, but the extended timeline of the company's recovery, and the data that eventually appeared on Conti's leak site, suggested that negotiations either failed or never seriously began.

What It Teaches Us: Why Food Processing Gets Hit

The parade of incidents raises an obvious question: why does food processing keep getting attacked? Part of the answer is structural: food manufacturing operates on margins that leave little room for security investments that don't directly improve production. When every dollar matters and executives must choose between a new packaging line that increases output and security upgrades that prevent theoretical attacks, the packaging line often wins.

Another part of the answer is historical. Food safety regulation, which is extensive and sophisticated, is developed to address biological, chemical, and physical hazards. The FDA's Food Safety Modernization Act (FSMA), the most significant reform of food safety law in decades, requires facilities to have hazard analysis and preventive controls.[22] But those hazards are defined as things like pathogens, allergens, and foreign objects. Cyber risks aren't yet part of the framework.

A third part of the answer is operational. Food processing can't simply stop. Perishable inputs keep arriving. Customers keep ordering. Production schedules are built around biological timelines that don't pause for software updates. The 24-hour, 7-day-a-week nature of many operations means there's rarely a convenient time for security maintenance.

And part of the answer is that attackers have noticed all of the things just mentioned. Ransomware groups understand that food processors face enormous time pressure, operate on thin margins, and often lack sophisticated security capabilities. They understand that a dairy processor can't tell milk to wait while systems are restored. They understand that holiday cheese orders create leverage in October. They've done their homework and reminded us that they are going to ace their exams. The FBI has noticed the pattern, too. In a 2022 private industry notification, the bureau warned that ransomware groups had conducted attacks against six grain cooperatives during harvest season and two additional attacks during the spring planting season. The timing wasn't coincidental; it was strategic, designed to maximize pressure by striking when the industry was most vulnerable.[23]

What Food Processing Actually Cares About

Here's what food processing professionals know that cybersecurity professionals sometimes miss. The industry operates on three nonnegotiable priorities:

1. *Food safety:* Every decision in a food processing facility eventually connects to a single question: is this product safe to eat? Regulatory frameworks, quality systems, testing protocols, and operational procedures are all in place to ensure that what leaves the plant won't harm consumers. Food safety isn't just a legal requirement; it's an ethical obligation that defines the industry's identity.
2. *Production uptime:* Processing facilities exist to produce. When lines aren't running, products aren't made, customers aren't served, and money isn't earned. The financial pressure to maximize uptime shapes everything from maintenance schedules to staffing decisions. For facilities running on thin margins, every idle hour has a dollar figure attached to it.
3. *Employee safety:* Food processing involves heavy equipment, sharp implements, extreme temperatures, and various physical hazards. Safety culture in food processing is deeply embedded and fiercely protected. An injury on the floor means something in those protections broke down, and the industry treats that as seriously as a contamination event.

What makes cybersecurity challenging in this context isn't that these priorities conflict with security; it's that security professionals sometimes fail to connect their concerns to what the industry already cares about.

Consider how a cyber incident could compromise food safety. If an attacker manipulates pasteurization temperatures in a dairy plant, under-processed milk could reach consumers. If ransomware encrypts quality-control records during a contamination investigation, affected products can't be traced or recalled effectively. If production monitoring systems go dark, operators lose visibility into critical control points. Cybersecurity isn't separate from food safety; it's increasingly integral to it.

Think about how a cyber incident affects production uptime. Every case study in this chapter demonstrates the same pattern: ransomware attacks force production stoppages. Sometimes for hours. Sometimes for days. Sometimes with cascading effects that persist for months. The very thing food processing can't afford is downtime, and that's exactly what cyber attackers deliver.

Consider how a cyber incident could endanger employees. Industrial control systems manage equipment that can injure or kill workers if it operates incorrectly. Although no public case has documented a cyberattack causing physical harm to food processing workers, the possibility is real, and, sadly, it may have occurred without the public being aware. Systems that control heavy machinery, high temperatures, and pressurized equipment pose obvious risks to human health and safety if compromised.

When cybersecurity professionals talk to food processing audiences in terms of network segmentation and patch management, they're speaking a language that may not resonate. When they talk in terms of food safety, production uptime, and employee safety, they're speaking the industry's native tongue.

Moving Forward: Building Resilience Through Pivots

The incidents we've chatted about in this chapter share a common thread: the companies that recovered most effectively had prepared before the attack came. Maple Leaf Foods' ability to pivot to manual operations within 48 hours didn't happen by accident. The company had business continuity plans that it had practiced. It knew what to do when systems failed not because it expected a ransomware attack specifically but because it understood that any number of disruptions could force the company to operate without its normal technology. Preparation varies by organization, but certain elements recur across successful responses.

The practical frameworks behind that preparation—network architecture, backup strategies, incident response planning, vendor controls, and security culture—are examined in Part III of this book.

The digital transformation of food processing is not reversible. The efficiency gains, quality improvements, and competitive advantages that technology provides are too significant to abandon. No major food processor is going back to purely manual operations, but transformation brings risk. The same systems that optimize production can be encrypted by criminals. The same networks that enable real-time monitoring can serve as pathways for attackers. The same data that drives efficiency can be stolen and exploited. Food processing professionals didn't ask for this challenge; they entered an industry focused on transforming agricultural products into safe, nutritious food, not on defending against sophisticated cyber criminals. But the challenge has arrived regardless.

In the next chapter, we turn to food safety, quality, and compliance, the regulatory and operational frameworks designed to keep food safe. Those frameworks increasingly depend on digital systems for monitoring, documentation, and verification. When the systems that support food safety become targets themselves, the intersection of cybersecurity and food safety stops being theoretical and becomes urgent.

Notes

1. Constantin, L. (2023). "Ransomware gang takes credit for Maple Leaf Foods hack." *SecurityWeek.* **https://www.securityweek.com/ransomware-gang-takes-credit-maple-leaf-foods-hack**
2. Satariano, A. (2022). "Canadian food giant refuses to pay ransom after gang threatens data leak." *The Record.* **https://therecord.media/canadian-food-giant-refuses-to-pay-ransom-after-gang-threatens-data-leak**
3. U.S. Bureau of Labor Statistics. (2026). "Food manufacturing: NAICS 311." *Industries at a Glance.* **https://www.bls.gov/iag/tgs/iag311.htm**
4. FDA. (2018). "Food ingredient & packaging terms." **https://www.fda.gov/food/food-ingredients-packaging/food-ingredient-packaging-terms**
5. FDA. (2019). *"Grade "A" pasteurized milk ordinance."* **https://www.fda.gov/media/99451/download**
6. Maple Leaf Foods. (2023). *Maple Leaf Foods Inc.: Fourth Quarter and Full-Year 2022 Financial Results.* Conference call transcript. **https://www.mapleleaffoods.com/wp-content/uploads/sites/6/2023/03/Q4-2022-results-call-c.pdf**
7. Kovacs, E. (2023). "Ransomware gang takes credit for Maple Leaf Foods hack." *SecurityWeek.* **https://www.securityweek.com/ransomware-gang-takes-credit-maple-leaf-foods-hack**
8. Maple Leaf Foods. (2023). *Maple Leaf Foods Inc.: Fourth Quarter and Full-Year 2022 Financial Results.*
9. Reuters. (2021). "Meatpacker JBS says it paid equivalent of $11 mln in ransomware attack." **https://www.reuters.com/technology/jbs-paid-11-mln-response-ransomware-attack-2021-06-09**
10. Dole. (2023). "Form 6-K (Report of Foreign Private Issuer)." *U.S. Securities and Exchange Commission.* **https://www.sec.gov/Archives/edgar/data/1857475/000185747524000035/dole-20240331_d2.htm**
11. Kollewe, J. (2023). "Cyberattack on food giant Dole temporarily shuts down North American production." *CNN Business.* **https://www.cnn.com/2023/02/22/business/dole-cyberattack**
12. Dole. (2023). "Form 6-K (Report of Foreign Private Issuer)."
13. Heffernan, R. (2023). "Dole doesn't expect to recover full costs of ransomware attack." *Cybersecurity Dive.* **https://www.cybersecuritydive.com/news/dole-recovery-ransomware-attack/644445**
14. Rathje, J. (2021). "'Cyber event' knocks dairy giant Schreiber Foods offline amid ransomware wave." *CyberScoop.* **https://cyberscoop.com/schreiber-foods-cyber-event-ransomware-agriculture-food**
15. Elkin, E. (2021). "Cyberattack sends prices for 500-pound cheese barrels tumbling." *Bloomberg.* **https://www.bloomberg.com/news/articles/2021-11-08/cyberattack-sends-prices-for-500-pound-cheese-barrels-tumbling?embedded-checkout=true**
16. Wisconsin State Farmer. (2021). "Ransomware attack shuts down Schreiber cheese plants." **https://www.controleng.com/throwback-attack-dairy-giant-forced-to-halt-production-because-of-ransomware**
17. Glover, C. (2022). "Hive ransomware victims: Apetito, Exela and G4S among seven new alleged victims." *Tech Monitor.* **https://www.techmonitor.ai/technology/cybersecurity/hive-ransomware-victims-apetito-exale-g4s**
18. Best, D. (2022). "Frozen-food firm Apetito hit by cyber attack." *Just Food.* **https://www.just-food.com/news/frozen-food-firm-apetito-hit-by-cyber-attack/?cf-view**

19. Warren-Kachelein, D. (2022). "UK-based KP Snacks hit by ransomware in 'snack attack'." *Bank Information Security*. **https://www.bankinfosecurity.com/snack-attack-uk-based-kp-snacks-limited-hit-by-ransomware-a-18449**
20. Toulas, B. (2022). "KP Snacks giant hit by Conti ransomware, deliveries disrupted." *Bleeping-Computer*. **https://www.bleepingcomputer.com/news/security/kp-snacks-giant-hit-by-conti-ransomware-deliveries-disrupted**
21. FBI, Cybersecurity and Infrastructure Security Agency, and NSA. (2021). *Conti Ransomware*. Joint Cybersecurity Advisory AA21-265A. **https://media.defense.gov/2021/Sep/22/2002859507/-1/-1/0/CSA_Conti_Ransomware_20220309.PDF**
22. FDA. (2015). "FSMA final rule for preventive controls for human food." **https://www.fda.gov/food/food-safety-modernization-act-fsma/fsma-final-rule-preventive-controls-human-food**
23. FBI. (2022). *Ransomware Attacks on Agricultural Cooperatives Potentially Timed to Critical Seasons*. Private Industry Notification 20220420-001. **https://www.ic3.gov/CSA/2022/220420-2.pdf**

CHAPTER 9

Food Safety, Quality, and Compliance

In September 2024, Boar's Head recalled more than seven million pounds of deli meat after a Listeria outbreak killed 9 people and hospitalized 57 across 18 states.[1] The investigation revealed conditions that had persisted for years: black mold growing on walls and equipment, meat residue caked on surfaces, water dripping onto product, dead flies, and puddles of blood on the floor.[2] USDA inspectors had repeatedly documented these violations. Someone must have seen the mold or walked past the pooled blood, or at least someone in management should have reviewed inspection reports, yet production continued. This wasn't a cyberattack: no ransomware locked the doors, and no hacker altered temperature logs. This was a failure of people, process, and culture, resulting from the human decisions that determine whether food safety controls work or become theater. But it raises a question that sits at the heart of this chapter: if environmental monitoring systems had been tracking humidity levels, surface sanitation verification, and temperature stability with automated alerts and escalation protocols, would management have found it harder to ignore the deterioration? If digital systems had created immutable audit trails showing that corrective actions were overdue or incomplete, would accountability have forced intervention sooner?

The answer isn't simple; technology optimism and digital tools don't fix broken culture. A dashboard alert means nothing if the person who sees it lacks authority to stop production, fears retaliation for speaking up, or works for leadership that prioritizes throughput over safety. But this complexity, the intersection of digital systems, human behavior, organizational culture, and food safety consequences, is exactly why cybersecurity professionals and food safety professionals need to work together.

Food safety has entered the cyber-physical age. The sensors monitoring cook temperatures, the databases storing pathogen test results, the traceability platforms tracking which farm supplied which ingredient to which batch, the laboratory information systems generating certificates of analysis: all these digital tools have become embedded in how we prevent foodborne illness, verify safety controls, and prove compliance with regulations. When those systems fail or are compromised or ignored, the consequences aren't measured in lost data or business disruption; they're measured in hospitalizations, deaths, and shattered trust in the food supply.

Defining Food Protection: A Shared Framework

Before we can talk about how cybersecurity intersects with food protection, we need to define what we're protecting and why these terms matter. The vocabulary can be confusing, with overlapping concepts that mean different things in different contexts:

Food safety addresses the prevention of unintentional contamination and foodborne illness through proper handling, processing, and storage. It focuses on three categories of hazards: biological (bacteria like Salmonella and Listeria, viruses, parasites, allergens), chemical (pesticides, cleaning agents, naturally occurring toxins), and physical (glass fragments, metal shavings, bone fragments). Food safety controls answer questions like: Is the product cooked to a temperature that kills pathogens? Is raw product stored separately from ready-to-eat food to prevent cross-contamination? Are allergens properly labeled and managed to protect sensitive consumers?

Food defense addresses the protection of food from intentional contamination or sabotage by bad actors: terrorists, disgruntled employees, activists, or criminals seeking to cause harm, create fear, or extract a ransom. It focuses on identifying vulnerable points in the production process where someone with access and intent could deliberately introduce a contaminant. Food defense controls answer questions like: Who has access to ingredient storage areas? Are high-risk materials secured? Do background checks and personnel reliability programs reduce the risk of insider threats?

Closely related is *food fraud*, which involves intentional deception about food for economic gain. Examples include substituting cheaper ingredients, mislabeling origin or quality, and diluting products. Whereas food defense focuses on preventing intentional harm to public health, food fraud focuses on preventing economically motivated adulteration that may or may not pose safety risks but always undermines consumer trust and market integrity.

Food quality addresses whether food meets product specifications, customer expectations, and brand standards. It encompasses sensory attributes (taste, texture, appearance, aroma), nutritional content, shelf life, and consistency. Quality is what differentiates a premium product from an acceptable one, what builds brand loyalty, and what determines whether a customer buys again.

Food security is not to be confused with "security" in the cybersecurity context. It addresses the need to ensure consistent access to sufficient, safe, and nutritious food for all people. It encompasses availability (is enough food produced?), access (can people afford and obtain it?), utilization (can people use it safely and nutritiously?), and stability (will access remain consistent over time?).

Food protection serves as the umbrella term encompassing safety, defense, quality, and security. It represents the comprehensive approach to protecting food from harm and maintaining its integrity from farm to fork.

Why do these distinctions matter for cybersecurity? Because digital threats affect each domain differently. A compromised temperature sensor during cooking affects food safety by potentially allowing undercooked products to ship. A breach of access control systems could enable food defense threats by allowing unauthorized personnel into secure areas. Manipulating supplier certificates or altering product specifications in digital systems could facilitate food fraud by hiding ingredient substitutions or mislabeled origins. Manipulated quality-testing data undermines food quality by concealing off-spec products. And a ransomware attack that shuts down multiple processing facilities simultaneously can threaten regional food security by removing production capacity at scale.

The Global Regulatory Landscape

Food safety regulations vary by country and region, but the trend is universal: increasing reliance on digital verification, traceability, and real-time monitoring. Understanding this landscape matters because regulations drive technology adoption, and technology adoption creates cyber exposure.

In the United States, the Food Safety Modernization Act (FSMA), signed into law in 2011, shifted regulatory focus from responding to contamination to preventing it.[3] The law gave the Food and Drug Administration (FDA) new authority to require preventive controls at food facilities, set standards for produce safety, and oversee imported foods through the Foreign Supplier Verification Program (FSVP). The FDA regulates most food products, including seafood, produce, dairy, and packaged foods. The United States Department of Agriculture's Food Safety and Inspection Service (USDA-FSIS) regulates meat, poultry, processed egg products, and catfish (siluriform), maintaining a continuous presence of inspectors at slaughter and processing facilities. Both agencies increasingly expect digital record-keeping, electronic data submission, and technology-enabled traceability.

The European Union operates under the General Food Law (Regulation 178/2002), which established the European Food Safety Authority (EFSA) and created the principle of "one step back, one step forward" traceability: every food business must know where product came from and where it went.[4] The recent NIS2 Directive (Network and Information Security Directive) explicitly includes food production and distribution as critical infrastructure requiring cybersecurity risk management.[5]

The United Kingdom's Food Standards Agency (FSA) has emphasized the adoption of technology to enhance traceability and transparency, including blockchain pilots.[6] Canada's Safe Food for Canadians Regulations (SFCR), enforced by the Canadian Food Inspection Agency (CFIA), emphasize preventive controls and traceability similar to FSMA.[7] China has rapidly modernized food safety regulation through the State Administration for Market Regulation (SAMR), implementing mandatory traceability systems for many food categories, particularly infant formula.[8] Japan's Ministry of Health, Labour and Welfare (MHLW), India's Food Safety and Standards Authority of India (FSSAI), and Brazil's National Health Surveillance Agency (ANVISA) all increasingly incorporate digital requirements into their regulatory frameworks.[9,10,11]

Beyond national regulations, global standards facilitate international trade. The Codex Alimentarius, developed jointly by the World Health Organization (WHO) and the Food and Agriculture Organization (FAO), offers internationally recognized food safety guidance.[12] The Global Food Safety Initiative (GFSI) benchmarks certification schemes; these schemes increasingly expect digital documentation, automated monitoring, and electronic traceability. These schemes increasingly expect digital documentation, automated monitoring, and electronic traceability. In addition, ISO 22000 provides an international standard for food safety management systems.[13]

What do all these frameworks share? They were written before cyber threats to food systems became routine. They assume that digital records are trustworthy, that automated monitoring functions correctly, and that traceability systems remain accessible when needed. They require extensive documentation, now almost universally digital, without addressing how to protect that documentation from compromise, manipulation, or loss. As regulations evolve, food businesses face a paradox: they must digitize to meet compliance requirements, but digitization creates cybersecurity vulnerabilities that those same regulations rarely acknowledge or address.

HACCP in the Digital Age

Hazard Analysis and Critical Control Points (HACCP) is the systematic approach that underlies modern food safety worldwide. Developed in the 1960s for NASA's space program, HACCP has become the foundation for food safety regulations globally.[14] It's built on seven principles: conduct hazard analysis, determine critical control points (CCPs), establish critical limits, establish monitoring procedures, establish corrective actions, establish verification procedures, and establish documentation and recordkeeping. A CCP might be the cooking step for a ready-to-eat product, where heat must reach a specific temperature for a specific time to kill pathogens. The critical limit is the minimum temperature–time combination that achieves safety. Monitoring means continuously measuring temperature. Corrective action means holding or reworking product if temperature falls short. Verification means calibrating sensors and reviewing records. Documentation means records proving that all of this happened.

Thirty years ago, HACCP was largely manual. An operator watched a dial thermometer and recorded readings on a paper log every hour. A supervisor reviewed those logs daily, and records went into file cabinets. The system worked, but it was labor-intensive, was subject to human error, and provided only snapshot data. Today, HACCP continues digitally. HACCP plans exist in cloud-based platforms accessible across multiple facilities. Temperature sensors log data continuously and upload it to databases automatically. When a CCP goes out of specification, the system triggers automatic alerts. Some systems automatically divert products or stop lines when critical limits are breached. Corrective actions are documented digitally with timestamps, photos, and electronic signatures. Verification activities populate dashboards that management can view in real time from anywhere.

This transformation provides genuine benefits: faster detection of problems, more complete data capture, the ability to identify patterns across facilities, and proof of compliance that's more detailed than paper ever allowed. But it also creates dependencies and vulnerabilities that the original HACCP framework never contemplated. What happens when the temperature sensor fails and no one notices because they've stopped watching the physical thermometer? What happens when the database storing CCP records becomes inaccessible during a regulatory audit? What happens if automated alerts are routed to an email system that's been compromised, so no one receives the notification that the pasteurization temperature has dropped dangerously low? What happens if an employee with malicious intent alters critical limits in the system to make unsafe products appear acceptable? These aren't hypothetical questions. They represent scenarios that food safety professionals and cybersecurity professionals need to assess together.

HACCP plans have always required thinking through failure modes: what could go wrong at each step. Now those failure modes must include digital system failures and compromises. The most critical point: HACCP was designed with human oversight as the final control. Even with automated monitoring, someone must review the data, question anomalies, recognize when something seems wrong even if the system says everything is fine, and have the authority to stop production when necessary.

Risk Assessment: A Shared Language

One of the most promising opportunities for collaboration between food safety and cybersecurity professionals lies in a capability both disciplines already possess: risk assessment. Both groups think systematically about what could go wrong, how likely problems are to occur, what consequences might result, and how to prioritize limited resources. Food safety professionals conduct hazard analyses, identifying biological hazards (pathogens, allergens), chemical hazards (pesticides, cleaning compounds), and physical hazards (glass, metal). They assess likelihood and severity, prioritizing controls based on risk. They use SWOT analysis (strengths, weaknesses, opportunities, threats) to evaluate their programs comprehensively.

Cybersecurity professionals conduct parallel assessments using slightly different terminology. Threat modeling identifies adversaries and attack vectors. Vulnerability assessment examines system weaknesses. Impact analysis considers the consequences of exploitation. And risk prioritization combines likelihood and impact to determine where to invest in controls. They also use SWOT analysis to evaluate security posture.

The partnership opportunity emerges when these parallel risk assessments converge. Consider a facility where food safety identifies pasteurization as a CCP requiring continuous temperature monitoring. Cybersecurity examines the same control system and identifies threats: sensor calibration drift, network disruption, ransomware, and configuration changes that alter critical limits. The vulnerabilities include aging programmable logic controllers running outdated software, remote access for vendors without strong authentication, and a lack of network segmentation.

The combined assessment reveals that cyber vulnerabilities directly enable a food safety hazard. The food safety team understands the biological consequences and regulatory requirements. The cybersecurity team understands the technical attack vectors and defensive measures. Neither can adequately address the risk alone. The shared language of risk likelihood, consequence, control, and verification makes this collaboration natural rather than forced.

When Culture Fails: Lessons from Real Incidents

Two food safety catastrophes from recent history—one criminal fraud, the other operational neglect—offer critical lessons about how people, processes, and culture determine safety outcomes. Between 2008 and 2009, a *Salmonella Typhimurium* outbreak linked to peanut products from Peanut Corporation of America killed 9 people and sickened 714 across 46 states.[15] Investigation revealed systematic fraud: executives knowingly shipped contaminated product, falsified certificates of analysis, ordered retesting until negative results were obtained, and shipped product before test results returned. The company president was sentenced to 28 years in prison for conspiracy and intentionally distributing adulterated food.[16] Could stronger digital controls have prevented this? Perhaps partially. Certificate of Analysis systems with immutable audit trails could make falsification more visible. Automated flags for "retest" patterns might trigger an investigation. Laboratory information management systems with proper controls would timestamp when samples were received, tested, and reported, making it harder to ship before results are available.

But we must apply critical thinking; technology doesn't stop determined criminals. Peanut Corporation of America's leadership made deliberate choices to break the law and endanger consumers. Better digital controls might have made their fraud more difficult or helped detect it earlier, but fundamentally, this was a people problem: individuals lacking ethics, an organizational culture that valued money over lives, and insufficient regulatory oversight.

From a cyber-physical perspective, the lesson is this: cybersecurity controls such as data integrity verification, audit trails, and access management can make malfeasance harder and more detectable. But they're not substitutes for ethics, culture, and accountability. Organizations with weak ethical foundations will find ways around controls. Organizations with strong food safety culture use digital tools to reinforce an existing commitment to safety, not as the primary deterrent against wrongdoing.

The Boar's Head case represents a different failure mode: not criminal fraud but progressive deterioration. USDA inspectors documented violations repeatedly over the years, yet production continued. The violations weren't hidden; they were reported, documented, and seemingly ignored by management.

Could digital systems have helped? Possibly, if they were integrated with accountability mechanisms. Environmental monitoring systems tracking temperature, humidity, and sanitation verification could have provided early warnings. Automated alerts could have escalated problems that went unaddressed. Digital audit trails could have documented when problems were identified and what actions were or weren't taken. But critical thinking matters. Boar's Head received written USDA inspection reports documenting the violations. If management ignored those official government reports, would they have responded to digital dashboard alerts? The problem wasn't a lack of information; it was a lack of response. This was a culture failure, not an information failure.

Digital systems can't fix culture, but they can make deterioration harder to ignore. They create visibility, accountability, and documentation that make neglect harder to sustain over time. They force explicit decisions rather than allowing problems to be passively disregarded. They create evidence trials that regulators and investigators can follow. But technology alone doesn't compel action; people and organizational culture determine whether warnings lead to intervention.

What It Teaches Us: Cybersecurity as a Food Safety Control

The cases we've examined, from regulatory frameworks built before cyber threats existed (using HACCP systems that are now fully digitized) to food safety failures where better digital controls might have helped, all point to the same conclusion: cybersecurity is no longer separate from food safety. It's a control measure that protects the integrity of other control measures.

When we talk about CCPs, we mean steps where control can be applied to prevent, eliminate, or reduce food safety hazards. Traditionally, those controls were physical: heat to kill pathogens, refrigeration to prevent growth, and metal detectors to catch physical hazards. Now, many of those physical controls depend on digital systems to monitor, verify, and document their effectiveness. If the digital systems fail or are compromised, the physical controls may still function, but we lose the ability to prove they worked or, in some cases, to know whether they worked at all. This means cybersecurity controls—data integrity verification, access management, system monitoring, incident response, backup, and recovery—become part of the food safety control structure. They're not nice-to-have IT improvements; they're essential to maintaining trust and verification that underpin modern food safety.

Food safety professionals and cybersecurity professionals both understand risk assessment. Both think about what could go wrong, how to prevent it, and how to respond when prevention fails. Both deal with threats that evolve constantly and require vigilant monitoring. Both face the reality that perfect security is impossible and must make risk-based decisions about where to invest limited resources. Integration doesn't require either discipline to abandon its expertise. Rather, it requires both to recognize that in a digitized food system, biological hazards and cyber vulnerabilities intersect in ways that neither can address alone. A compromised temperature sensor isn't just a cybersecurity problem; it's a food safety problem. A pathogen-harboring site that goes undetected because environmental monitoring alerts fail isn't just a sanitation problem; it's a cyber problem.

Creating a cyber-physical food safety culture means breaking down the silos between these disciplines. It means food safety professionals learning enough about cybersecurity to ask the right questions about the systems they depend on. It means cybersecurity professionals learning enough about food safety to understand what's truly at stake when systems fail. It means organizations building teams, processes, and incident response plans that address both domains simultaneously rather than treating them as separate concerns.

Moving Forward: Food Safety Digital Resilience

Food safety has entered the cyber-physical age, whether we're prepared for it or not. The sensors, databases, and networks embedded in food systems aren't separate from food safety; they're integral to it. When those systems fail or are compromised, the consequences aren't measured in lost productivity or data breaches. They're measured in foodborne illness, deaths, shattered trust, and destroyed livelihoods.

The regulatory frameworks that govern food safety were written before digital transformations. They require extensive documentation and monitoring without adequately addressing how to protect those systems. They assume digital records are trustworthy without requiring the cybersecurity controls that make them so. As regulations evolve to acknowledge cyber risks, such as the EU's NIS2 Directive including food in critical infrastructure, the challenge will be providing practical guidance that helps food businesses, especially small and mid-size operations, implement both food safety and cybersecurity controls effectively.

The opportunity lies in the shared language both disciplines speak: the language of risk, consequence, control, and verification. Food safety professionals who conduct HACCP hazard analysis and cybersecurity professionals who conduct threat modeling are doing fundamentally similar work. Bringing them together to assess where cyber vulnerabilities enable food safety hazards creates more comprehensive protection than either discipline can achieve alone.

Food doesn't stop being physical because we monitor it digitally. Pathogens still grow, temperatures still matter, sanitation still requires physical cleaning, and product quality still depends on proper handling. But the systems verifying all of this have become digital, creating dependencies on technology that previous generations of food professionals never faced. In addition, this technology is near food or even touches food on its journey. Acknowledging this reality and building appropriate safeguards isn't optional; it's essential to maintaining food safety in a connected world.

As product leaves the production facility and enters distribution networks, all of these digital dependencies travel with it. The traceability systems tracking which batch went to which customer, the temperature sensors monitoring refrigerated transport, and the logistics platforms coordinating deliveries all become part of the extended food safety system. Every handoff creates new vulnerabilities, new responsibilities, and new opportunities for failure if the systems connecting those handoffs aren't appropriately secured. The next chapter follows that digital thread through distribution, where maintaining food safety depends as much on cybersecurity as on refrigeration.

Notes

1. CDC. (2024). "Investigation Update: Listeria Outbreak, Meats Sliced at Delis." **https://www.cdc.gov/listeria/outbreaks/delimeats-7-24/investigation.html**
2. USDA-FIS. (2024). "Boar's Head Provisions Co., Inc. expands recall of ready-to-eat meat and poultry products due to possible listeria monocytogenes contamination." FSIS notice. **https://www.fsis.usda.gov/recalls-alerts/boars-head-provisions-co--expands-recall-ready-eat-meat-and-poultry-products-due**
3. FDA. (2011). "Full text of the Food Safety Modernization Act (FSMA)." **https://www.fda.gov/food/food-safety-modernization-act-fsma/full-text-food-safety-modernization-act-fsma**
4. European Commission. (2002). "Regulation (EC) No. 178/2002 of the European Parliament and of the Council of 28 January 2002 laying down the general principles and requirements of food law, establishing the European Food Safety Authority and laying down procedures in matters of food safety." EUR-Lex. **https://eur-lex.europa.eu/legal-content/EN/TXT/?uri=CELEX%3A32002R0178**
5. European Commission. (2022). "Directive (EU) 2022/2555 on measures for a high common level of cybersecurity across the Union (NIS2 Directive)." EUR-Lex. **https://eur-lex.europa.eu/eli/dir/2022/2555/oj**
6. Food Standards Agency. (2026). **https://www.food.gov.uk**
7. Canadian Food Inspection Agency. (2019). Safe food for Canadians regulations (SOR/2018-108). **https://laws-lois.justice.gc.ca/eng/regulations/SOR-2018-108**
8. USDA FAS. (2021). *Food and Agricultural Import Regulations and Standards Country Report: China*. GAIN Report No. CH2021-0037. **https://apps.fas.usda.gov/newgainapi/api/Report/DownloadReportByFileName?fileName=Food+and+Agricultural+Import+Regulations+and+Standards+Country+Report_Beijing_China+-+People%27s+Republic+of_03-31-2021**
9. MHLW. n.d. "Food safety information." **https://www.mhlw.go.jp/english/topics/foodsafety**
10. FSSAI. n.d. "Food Safety and Standards Authority of India Website." **https://fssai.gov.in**
11. ANVISA. n.d. **https://www.gov.br/anvisa/pt-br**
12. FAO/WHO. (2018). *Understanding the Codex Alimentarius*. **https://www.fao.org/3/ca1176en/CA1176EN.pdf**
13. GFSI. n.d. "Global Food Safety Initiative (GFSI) Website." **https://mygfsi.com**; BRCGS. n.d. "Food safety." **https://www.brcgs.com/our-standards/food-safety**; SQF Institute. n.d. "What is SQF?" **https://www.sqfi.com/our-program/what-is-sqf**; Foundation FSSC 22000. n.d. "Food Safety System Certification 22000." **https://www.fssc.com/fssc-22000**; IFS. n.d. (International Featured Standards) Certification Database. **https://ifs-web-prod.ifs-certification.com/en**; International Organization for Standardization (ISO). n.d. "ISO 22000 Food Safety Management." **https://www.iso.org/iso-22000-food-safety-management.html**
14. Safe Food Alliance. (2020). "The history of HACCP." **https://safefoodalliance.com/haccp/the-history-of-haccp**
15. CDC. (2009). "2008-2009 salmonella outbreak linked to peanut butter." Investigation summary. **https://archive.cdc.gov/www_cdc_gov/salmonella/2009/peanut-butter-2008-2009.html**
16. United States Department of Justice. (2015). "Former peanut company officials sentenced to prison for their roles in salmonella-tainted peanut products outbreak." **https://www.justice.gov/archives/opa/pr/former-peanut-company-officials-sentenced-prison-their-roles-salmonella-tainted-peanut**

CHAPTER 10

Food Distribution and the Supply Chain

In April 2021, the Netherlands ran short of cheese, not because cows stopped producing milk or factories stopped making cheese, but because a logistics company that most Dutch consumers had never heard of could no longer find the cheese it had already stored. Bakker Logistiek operates six temperature-controlled warehouses across the Netherlands. The company coordinates food distribution to major retailers, including Albert Heijn, the country's largest supermarket chain, which has more than 1,000 locations. On Easter weekend in 2021, ransomware encrypted the company's systems. Within hours, Bakker Logistiek went from a sophisticated logistics operation to chaos. "We could no longer receive orders from customers," director Toon Verhoeven told Dutch broadcaster NOS. "And in our warehouses, we no longer knew where products were. These are very large warehouses; you don't just go looking for a pallet. We also couldn't plan our transport anymore. We have hundreds of trucks." Those too could no longer be scheduled or dispatched.[1]

The attackers, later linked to the Babuk ransomware group and Russian national Mikhail Matveev and now under U.S. indictment, had likely entered through a vulnerability in Microsoft Exchange servers.[2] Within days, Albert Heijn posted notices on its website apologizing for the limited availability of prepackaged cheese. The Netherlands, one of the world's top 10 cheese-consuming nations, was experiencing something that would have seemed absurd a generation earlier: a cheese shortage caused not by agricultural failure or a shipping disruption, but by encrypted computer files.[3]

This chapter follows food on its journey from production to plate through warehouses, across borders, in refrigerated trucks, and into the hands of the people who stock shelves and deliver groceries. Most consumers assume that their food comes directly from a farm, but in our modern food supply chain, it does not. Between harvest and table, food passes through a complex network of distributors, cold-storage facilities, freight forwarders, and logistics providers. Each handoff relies on digital systems, and each system represents a potential point of failure. When distribution fails, the problem isn't that food doesn't exist; it's that food can't reach the people who need it. Food protection professionals would call this a threat to food security, which, as you may remember from Chapter 9, ensures that all people always have physical, social, and economic access to sufficient, safe, and nutritious food.[4] Distribution disruptions are food security events even when they are temporary and the cause is a criminal with a keyboard rather than a drought or war.

Following the Food

To understand why a single logistics company matters so much, consider the journey a block of cheese takes from production to your refrigerator. A dairy processor in Wisconsin or the Netherlands turns milk into cheese. That cheese doesn't teleport to your grocery store; it enters a distribution network. First, the cheese moves from production to a warehouse, often operated by a third-party logistics (3PL) provider: a company that specializes in storing and moving goods for other businesses. Inside that warehouse, a warehouse management system (WMS) tracks exactly where each pallet sits among thousands of identical-looking shelves. Workers or, increasingly, automated systems locate products, pick them, and stage them for shipping. Then the cheese is loaded onto a refrigerated truck.

Throughout transport, the cold chain must remain unbroken. The *cold chain* is the continuous temperature-controlled storage and transportation that keeps perishable food safe. Internet of Things (IoT) sensors monitor temperature, humidity, and location. If the truck's refrigeration fails or the door stays open too long, alerts scream. The data logs, theoretically, provide proof that the product remained safe throughout its journey. The truck delivers to a distribution center, the regional hub where products from many sources converge, are sorted, and head to their final destinations. From there, the cheese might go to a grocery store's back room, a restaurant's walk-in cooler, or a food-service distributor, such as Sysco or U.S. Foods, that supplies hospitals, schools, and cafeterias.

Seafood often begins its cold-chain journey even earlier. Fish caught in Alaskan waters may be flash-frozen on the fishing vessel within hours of harvest, beginning temperature-controlled custody before the boat returns to port. From there, the frozen fish enters the same distribution infrastructure—warehouses, trucks, and distribution centers—before reaching a restaurant in Kansas City or a grocery store in Atlanta. At each stop, people make the system work. Warehouse workers in temperature-controlled environments can drop temperatures to −20 °F. Truck drivers maintain schedules while monitoring refrigeration equipment. Dock personnel coordinate arrivals and departures. Technology enables their work, but people execute it. When systems fail, these workers must improvise or watch food spoil while they wait for IT to restore access.

The North American Scale

What happened in the Netherlands in 2021 happened in North America in 2025, but at a far greater scale. United Natural Foods, Inc. (UNFI) is the largest publicly traded wholesale distributor of natural and organic food in North America. The company operates 52 distribution centers across the United States and Canada, serving approximately 30,000 customer locations. UNFI is the primary supplier to Whole Foods Market and serves independent natural food stores, conventional supermarkets, and institutional customers, including schools.[5] On June 5, 2025, UNFI detected unauthorized activity in its systems. By June 6, the company had shut down its entire network. Fifty-two distribution centers went dark simultaneously.[6] The impact was immediate and visible. Refrigerator cases at Whole Foods stores across America began to empty. Photos of bare shelves circulated on social media. The company that positions itself as the premium destination for fresh, natural food couldn't keep its shelves stocked.[7]

But the impact extended far beyond Whole Foods. The Community Food Co-op in Bellingham, Washington, a small cooperative grocery serving its local community, posted a notice to customers explaining that, because UNFI was its main supplier, some shelves might appear bare. In Connecticut, LaBonne's Markets, which sources specialty natural and gluten-free products from UNFI, found itself unable to place orders through normal channels.[8] Solo Italia Fine Pasta, a small pasta manufacturer in Ogema, Saskatchewan, couldn't get paid for products already delivered to Co-op stores, leaving a small business owner wondering whether to pull his products from shelves entirely.[9] UNFI reverted to manual operations; workers processed orders by phone, and distribution centers came back online at different rates, some within days, others taking longer. Electronic ordering wasn't restored until June 16, and the company didn't declare core systems "safely restored" until June 26, three weeks after the attack began.[10]

The financial damage was staggering. UNFI estimated that the attack would reduce fiscal 2025 net sales by $350 to $400 million, with a net income loss of $50 to $60 million. The company incurred $20 million in direct incident response costs, plus another $5 million for third-party cybersecurity, legal, and governance experts.[11]

The UNFI attack wasn't about a shortage of food. Farms were still producing. Processors were still processing. The food existed. But the digital systems that track inventory, process orders, coordinate shipments, and route trucks—the invisible infrastructure that moves food from where it is to where people need it—had been encrypted by criminals demanding ransom.

Cold Storage and the Canadian Dimension

Americold Realty Trust is the largest publicly traded real estate investment trust focused on temperature-controlled warehouses. The company operates approximately 250 facilities globally, including locations in Australia, New Zealand, Canada, and Argentina. For perspective, Americold holds the contract linking ConAgra, one of America's largest food-producing companies, to supermarkets and consumers.[12] In November 2020, ransomware hit Americold's network. The attack disrupted phone systems, email, inventory management, and order fulfillment across the company's global operations.[13] Customers reported being told to cancel or reschedule deliveries except for those considered critically time-sensitive and nearing expiration. One Reddit user described arriving at an Americold facility to find "their phones are down and they had the truck entrance barricaded off with the main entrance gates shut with no one manning the guard shack."[14] Then it happened again. In April 2023, Americold suffered a second cyberattack. This time, the Cactus ransomware group claimed responsibility and leaked a six-gigabyte archive of accounting and finance documents allegedly stolen from the company's network. The breach affected more than 129,000 current and former employees and their dependents, whose personal information was compromised.[15]

Cold storage is uniquely vulnerable because it runs on biological time. Temperature monitoring systems are often connected to networks so they can send alerts when refrigeration fails, but that connectivity creates an attack surface. When systems go down, operators face immediate questions with no good answers. Is the food still safe? Can we verify temperatures were preserved? If we can't find products in the warehouse, how do we prioritize what to ship first?

Temperature abuse is a leading cause of foodborne illness. The "danger zone" temperatures between 40 °F and 140 °F allow rapid pathogen growth. Products may look fine but harbor dangerous bacteria after even a few hours at improper temperatures. When monitoring systems are down, there's no way to prove that the cold chain remained intact. "We don't know the temperature" is sometimes worse than "The temperature is wrong," because at least a known excursion can be evaluated and decisions made about product disposition.

Meanwhile, in western Canada, Federated Co-operatives Limited serves a population that depends on cooperative grocery stores in ways that urban consumers rarely experience. The organization encompasses 160 co-ops across Saskatchewan, Alberta, British Columbia, and Manitoba. In remote and rural communities, the local Co-op may be the only grocery store within reasonable driving distance.[16] In June 2024, a cyberattack crippled Federated Co-op's systems: bread, frozen goods, produce, and dairy became scarce across western Canada. In Saskatoon, lettuce was difficult to find at one Co-op location. In Warman, shelves in the produce and frozen food sections sat bare. Signs apologized for "unforeseen IT disruption" and "temporary shortages in product supply."[17] The *Victoria Times Colonist* reported that Peninsula Co-op on Vancouver Island was struggling to restock because internal IT systems at Federated Co-operatives had been compromised.[18]

Three weeks passed before Federated Co-op declared all systems normal. During that time, attackers allegedly stole 80 gigabytes of data and threatened to release it on the dark web: human resources files, confidential business information, financial records, and client data.[19] For rural Canadian communities, this wasn't merely an inconvenience. When the Co-op is your primary grocery option and its systems are down, you're experiencing a food security event. The food exists somewhere. It just can't reach you.

The Regulatory Response

The European Union recognized the vulnerability of food distribution before many in the industry did. The Network and Information Security Directive 2 (NIS2), which took effect in October 2024, explicitly includes "food production, processing and distribution" among the sectors requiring cybersecurity measures.[20] Under NIS2, the food sectors are classified as "important entities," a designation that triggers specific obligations for medium and large companies. Organizations with 50 or more employees or an annual turnover exceeding €10 million must comply. The requirements are substantial: risk management measures for cybersecurity, supply chain security assessments, incident reporting within 24 hours of discovery, and, critically, management accountability. Executives are personally responsible for overseeing cybersecurity and can face temporary bans from management roles for serious compliance failures.[21] The penalties for noncompliance are significant. "Important entities" face fines of up to €7 million or 1.4 percent of global annual turnover, whichever is higher. "Essential entities," a category that includes energy and healthcare, face even steeper penalties of up to €10 million or 2 percent of turnover.[22]

What makes NIS2 particularly consequential for food distribution is its supply chain dimension. The directive requires companies to "address cybersecurity risks in the supply chains and supplier relationships." For a food distributor working in Europe, this means assessing the security posture of partners throughout the chain, from the farms that supply products to the technology vendors that provide warehouse management systems. A U.S. or Canadian company distributing food to European markets may need to demonstrate NIS2-equivalent security practices to keep business relationships with EU customers.[23]

In Chapter 28, we will examine NIS2 and other regulations in depth. Here, the key insight is that regulators have recognized what attackers already knew: food distribution is critical infrastructure, and its digital systems require protection. The incidents in the Netherlands, the United States, and Canada show why.

What It Teaches Us: The People Behind the Systems Matter

Every distribution system ultimately depends on people. Warehouse workers find and load products in environments cold enough to require insulated clothing and strict time limits for exposure. Truck drivers keep cold chain integrity during transport while navigating delivery schedules. Dock personnel coordinate the constant flow of arrivals and departures. Small business owners depend on distributor relationships for their livelihoods. When systems fail, these workers improvise. At UNFI, employees processed orders by phone when electronic systems went dark. At Bakker Logistiek, staff returned to pen and paper, trying to coordinate hundreds of trucks without the software that normally made such coordination possible. At Federated Co-op, store employees posted apologetic signs and fielded customer questions they couldn't answer.

Distribution workforces often have minimal cybersecurity training. The focus is on physical safety: how to work in cold environments, manage heavy loads, and avoid accidents on loading docks. Cybersecurity seems remote from the daily reality of moving pallets and driving trucks. High turnover in warehouses and the need to drive jobs make sustained training difficult. Seasonal workers and temporary staff cycle through without deep investment in any single company's security culture. Yet these workers are often the first to notice when something is wrong. A warehouse employee can't log into the inventory system. A driver's routing app won't load. A dock supervisor's radio goes silent. Social engineering attacks, such as fake delivery requests, spoofed invoices, and credential phishing, exploit the time pressure inherent in moving perishable goods. "We need to redirect this shipment immediately" is a compelling message when you know that delays mean spoiled food.

Moving Forward: Building Resilience in Food Distribution

The pattern across these incidents in the Netherlands, the United States, Canada, and beyond suggests that attackers have identified food distribution as a lucrative target. Multiple countries, multiple years, different ransomware groups, but the same playbook: encrypt the systems that track inventory and coordinate logistics, and then watch as perishable goods create time pressure that makes paying a ransom seem rational. For food companies working with distributors and logistics providers, this means asking tough questions. What is your distribution partner's cybersecurity posture? What happens to your products if their systems go down? Do your contracts include provisions for incident notification and business continuity? If your primary distributor is compromised, do you have contingency plans?

For distributors and cold-storage operators, resilience means planning for both digital and physical continuity. Business continuity plans must account for perishability, the ruthless biological clock that makes food distribution different from other logistics sectors. Manual fallback procedures need to function at scale, not just exist on paper. Communication plans must reach thousands of customers who are simultaneously panicking about empty shelves.

Cyber insurance has become part of the calculation, but insurance alone doesn't keep food moving. UNFI expects its cybersecurity insurance to cover incident costs, but insurance doesn't fill the refrigerator cases or pay the small suppliers waiting for invoices to clear.[24]

The organizations that weather these incidents are those that can operate, however imperfectly, without their digital systems. Bakker Logistiek recovered within a week by reverting to manual processes. UNFI took three weeks but kept distribution centers shipping. The question every food distribution company should ask is: if our systems go dark tomorrow, could we keep food moving at all?

From farm to your table, the journey of food depends on people and systems working together. Every handoff creates an opportunity for failure and responsibility for safety. The challenge for everyone who eats is recognizing that food security isn't just about production. It's about the invisible infrastructure that moves food from where it is to where we need it. That infrastructure is digital. It is vulnerable. And protecting it means protecting what feeds us.

Distribution delivers food to many destinations: grocery stores, convenience stores, and food service distributors who deliver to restaurants, hospitals, and schools—and, increasingly, direct to consumers via delivery apps and meal kits. The next chapter follows food into food service and hospitality, in restaurants, cafeterias, and institutional kitchens where it is prepared and served. Here, the cybersecurity landscape shifts again: point-of-sale systems, kitchen display screens, reservation platforms, and the vulnerability of an industry built on thin margins, high turnover, and customer-facing technology.

Notes

1. Abrams, L. (2021). "Dutch supermarkets run out of cheese after ransomware attack." *BleepingComputer.* **https://www.bleepingcomputer.com/news/security/dutch-supermarkets-run-out-of-cheese-after-ransomware-attack**
2. Arghire, I. (2021). "Babuk ransomware seen exploiting proxyshell vulnerabilities," *SecurityWeek.* **https://www.securityweek.com/babuk-ransomware-seen-exploiting-proxyshell-vulnerabilities**
3. Schwartz, M. J. (2026). "Gouda hacker charges tie to cheese-impacting ransomware hit." *BankInfoSecurity.* **https://www.bankinfosecurity.com/blogs/gouda-hacker-charges-tie-to-cheese-impacting-ransomware-hit-p-3449**
4. FAO. (2008). *An Introduction to the Basic Concepts of Food Security.* **https://openknowledge.fao.org/server/api/core/bitstreams/479ba332-4529-4c32-89ec-3912cd8b14ee/content**
5. United Natural Foods, Inc. (n.d.). **https://www.unfi.com/about-us.html**
6. United Natural Foods, Inc. (2025). "Form 8-K." *U.S. Securities and Exchange Commission.* **https://d18rn0p25nwr6d.cloudfront.net/CIK-0001020859/03a307f1-6da4-47b3-88ab-1a2a39d45b30.pdf**
7. Silverstein, S. (2025). "UNFI cyberattack disrupts grocery wholesale earnings for Q3 2025." *Cybersecurity Dive.* **https://www.cybersecuritydive.com/news/unfi-cyberattack-grocery-wholesale-earnings-q3-2025/750334**
8. Silverstein, S. (2025). "How the cyberattack against UNFI affected 4 independent grocers." *Cybersecurity Dive*, 2025. **https://www.cybersecuritydive.com/news/unfi-cyberattack-response-supply-chain-grocery-wholesale/750961**
9. Holowaychuk, M. (2024). "Shortage on shelves in midst of co-op cybersecurity incident." *CKOM.* **https://www.ckom.com/2024/07/02/shortage-on-shelves-in-midst-of-co-op-cybersecurity-incident**
10. United Natural Foods, Inc. (2025). "UNFI systems update." **https://ir.unfi.com/news/press-release-details/2025/statement/default.aspx**
11. Lennon, M. (2025). "United Natural Foods projects up to $400 m sales hit from June cyberattack." *SecurityWeek.* **https://www.securityweek.com/united-natural-foods-projects-up-to-400m-sales-hit-from-june-cyberattac**
12. Americold. (n.d.). **https://www.americold.com/about-americold**
13. Davis, J. (2020). "Hackers hit COVID-19 biotech firm, cold storage giant with cyberattacks." *TechTarget.* **https://www.techtarget.com/healthtechsecurity/news/366595492/Hackers-Hit-COVID-19-Biotech-Firm-Cold-Storage-Giant-with-Cyberattacks**
14. Greig, J. (2023). "Cold storage company Americold reports cyberattack." *The Record.* **https://therecord.media/cold-storage-company-americold-reports-cyberattack**
15. Gatlan, S. (2024). "Cold storage giant Americold discloses data breach after April malware attack." *BleepingComputer.* **https://www.bleepingcomputer.com/news/security/cold-storage-giant-americold-discloses-data-breach-after-april-malware-attack**

16. Federated Co-operatives Limited. (n.d.). **https://www.fcl.crs**
17. Patterson, D. (2024). "Food items remain sparse at co-op amid ongoing cyberattack." *CBC News.* **https://www.cbc.ca/news/canada/saskatchewan/food-items-remain-sparse-co-op-cyberattack-1.7253263**
18. Harnett, C. E. (2024). "Co-op struggles to re-stock in wake of cyberattack." *Victoria Times Colonist.* **https://www.timescolonist.com/local-news/co-op-struggles-to-re-stock-in-wake-of-cyberattack-9210692**
19. Walker, A. (2024). "Federated Co-operatives Limited announces all systems normal following cyberattack." *Moose Jaw Today.* **https://www.moosejawtoday.com/local-news/federated-co-operatives-limited-announces-all-systems-normal-following-cyberattack-9287743**
20. European Parliament and Council of the European Union. (2022). "Directive (EU) 2022/2555 of the European Parliament and of the Council of 14 December 2022 on measures for a high common level of cybersecurity across the Union, amending Regulation (EU) No 910/2014 and Directive (EU) 2018/1972, and repealing Directive (EU) 2016/1148 (NIS 2 Directive)." EUR-Lex. **https://eur-lex.europa.eu/eli/dir/2022/2555/oj**
21. European Commission. (n.d.). "NIS2 Directive: securing network and information systems." **https://digital-strategy.ec.europa.eu/en/policies/nis2-directive**
22. NIS2 Directive. (2023). "NIS2 fines." **https://nis2directive.eu/nis2-fines**
23. NIS2 Directive. (2023). "Food sector." **https://nis2directive.eu/foods**
24. Hîmpă, G., & RakovaI G. (2024). "Guide to good practice regarding the implementation of the NIS 2 Directive in the food sector." *Inforb.* **https://inforb.ro/wp-content/uploads/2025/05/D2.2-Good-practice-guide_EN_v10.pdf**

CHAPTER 11

Foodservice and Hospitality

Restaurants, Institutional Kitchens, and Vending Technology

The breach at Checkers and Rally's didn't make headlines the way ransomware attacks on meat processors would years later. No production lines went dark, and no grocery shelves were emptied, but for customers who swiped their payment cards at any of the 102 affected drive-thru locations between 2015 and 2019, the intrusion carried real consequences: stolen card numbers, expiration dates, and verification codes harvested by malware embedded in the restaurant chain's point-of-sale (POS) systems.[1] Checkers discovered the compromise in April 2019 and notified customers the following month. The malware had been present at some locations for nearly four years, quietly copying payment card data each time a customer ordered a burger and fries at the drive-thru window. This was not an isolated incident. The restaurant industry has become a favored target for financially motivated attackers, and the reasons aren't hard to understand: thin margins mean underinvestment in security, high employee turnover complicates training and access management, and the sheer volume of payment card transactions creates an attractive harvest.[2]

This chapter follows food to its destination before consumption in the restaurants, cafeterias, and institutional kitchens where meals are prepared and served. Here, cybersecurity concerns shift from production continuity and food safety to payment card protection, customer data privacy, and the operational systems that keep kitchens running safely. It's a sector where a ransomware attack might mean canceled reservations and spoiled inventory rather than contaminated product, but where the stakes for individual consumers, institutions, and the workers who feed them remain substantial.

The Restaurant Technology Stack

Walk into any modern restaurant, and you're surrounded by networked systems, most of them invisible to the guest experience. The POS terminal where servers enter orders and process payments sits at the center, but it connects to a constellation of other technologies:

kitchen display systems (KDSs) that route orders to the appropriate cooking stations, inventory management platforms that track ingredient levels, scheduling software that manages shift assignments, and, increasingly, third-party delivery integrations that pipe in orders from DoorDash, Uber Eats, and Grubhub. POS systems themselves have evolved considerably. Legacy systems often ran on Windows XP or specialized proprietary operating systems, storing payment card data locally and transmitting it through dedicated phone lines. Modern cloud-based platforms like Toast, Square, and Clover process transactions through Internet connections and may store data across distributed infrastructure.[3] Each of these architectures presents distinct security considerations.

The Landry's breach illustrated how complex restaurant technology environments can create multiple entry points for attackers. Landry's operates more than 600 locations under brands including Morton's The Steakhouse, Rainforest Cafe, Bubba Gump Shrimp Co., and Chart House. In January 2020, the company disclosed that malware had compromised payment card data at locations across its portfolio. The intrusion affected cards used between March 2019 and October 2019. What makes the Landry's incident interesting was how the malware gained initial access. Attackers compromised order-entry terminals—the devices servers use to input food and beverage selections—rather than the payment terminals themselves. In locations where staff occasionally swiped payment cards on the infected device, that data was captured. The company noted that chip-enabled transactions on payment terminals were not affected, a distinction that underscores how the transition to Europay, MasterCard, and Visa (EMV) chip technology has improved security for properly configured systems.[4]

Beyond payment processing, restaurants face data protection obligations for employee information and customer records. A breach at a restaurant chain can compromise Social Security numbers for thousands of employees, personally identifiable information from loyalty program databases, and reservation histories that reveal where customers dine and when.

The Franchise Security Challenge

McDonald's operates approximately 40,000 restaurants worldwide, and Subway has more than 37,000 locations. But most of these restaurants aren't owned by the corporate brand: they're independently owned franchises operating under license agreements.[5] This creates a security challenge that few other industries face at a comparable scale.

Franchise agreements typically impose operational standards on franchisees: food preparation procedures, ingredient sourcing requirements, and brand presentation guidelines. But cybersecurity mandates vary considerably. A franchisor can require franchisees to use approved POS vendors and maintain specific configurations, but enforcing those requirements across thousands of independent business owners, each with their own IT staff or contractors, networks, and risk tolerance, is difficult in practice.

When Wendy's disclosed a payment card breach in 2016, the company emphasized that fewer than 300 of its 5,700 North American locations were affected and that the compromised restaurants were all franchised locations, not company-operated stores.[6]

The malware had exploited a third-party POS service provider that served certain franchisees. Company-owned locations used different systems and were not compromised.

The National Restaurant Association has published cybersecurity guidance for its members, emphasizing that restaurants should work with payment-card-industry-qualified security assessors, maintain Payment Card Industry Data Security Standard (PCI DSS) compliance, and implement network segmentation to isolate payment systems from other business functions. But guidance is not mandated, and for small franchisees operating on razor-thin margins, security investments compete with equipment maintenance, labor costs, and rent.[7]

The franchise model also complicates incident response. When a breach occurs at a franchised location, who notifies customers? Who bears liability? Corporate brands understandably want to protect their reputation, but they may have limited visibility into the security practices of independent operators until something goes wrong.

When Hospital Cafeterias Go Dark

In September 2020, Universal Health Services (UHS), one of the largest hospital chains in the United States, with more than 400 facilities, experienced a ransomware attack that shut down computer systems across its network for weeks. The Ryuk ransomware strain encrypted servers and workstations throughout the organization, forcing clinical staff to revert to paper documentation and manual processes.[8]

The attention understandably focused on patient care: medication tracking, lab results, and imaging systems. But UHS facilities also include cafeterias that serve patients, staff, and visitors. Those cafeterias depend on the same network infrastructure that went offline. POS systems couldn't process transactions. Inventory management systems went dark. Dietary management software that tracks patient meal orders, nutritional requirements, and food allergies became inaccessible. Hospital food service operates in a regulatory environment distinct from that of commercial restaurants. Patient meals must accommodate documented allergies, religious dietary requirements, and medical restrictions, including diabetic and renal diets and texture-modified meals for patients with swallowing difficulties. The Joint Commission, which accredits healthcare organizations, requires documented processes for ensuring correct meal delivery.[9] When the systems supporting those processes fail, workarounds introduce safety risks.

The three dominant institutional foodservice companies—Compass Group, Sodexo, and Aramark—collectively serve billions of meals annually across hospitals, schools, corporate campuses, and other institutional settings.[10] They operate within client facilities, often using client network infrastructure, which means their operations are exposed to whatever cyber incidents affect their host institutions. A ransomware attack on a hospital doesn't just affect hospital-employed staff; it also affects the contracted foodservice workers trying to feed patients on time.

School Nutrition in the Crosshairs

The Los Angeles Unified School District (LAUSD) serves over 600,000 meals daily to students across more than 1,000 schools.[11] It's one of the largest school nutrition programs in the country, operating under the United States Department of Agriculture's (USDA's) National School Lunch Program and School Breakfast Program. Students qualifying for free or reduced-price meals (the majority of LAUSD students) depend on these programs as a primary source of nutrition. In September 2022, Vice Society ransomware operators breached LAUSD systems and eventually published stolen data after the district declined to pay a ransom. The attack disrupted district operations for weeks, affecting everything from administrative functions to classroom technology. What received less attention was the impact on school nutrition operations.[12]

School meal programs operate within a complex administrative framework. Students must be certified for free or reduced-price meal eligibility based on family income. Each meal served must be counted and categorized for federal reimbursement claims. Schools must demonstrate compliance with USDA nutritional standards. All of this generates data, and all of it depends on functioning information systems. When LAUSD's systems went down, staff had to improvise processes for meal counting and eligibility verification. The district ultimately recovered, but the incident illustrated how dependent school nutrition has become on the same network infrastructure targeted by ransomware operators.

LAUSD wasn't alone. Baltimore County Public Schools suffered a ransomware attack in November 2020 that shut down network systems for weeks.[13] Clark County School District in Nevada, the fifth-largest district in the nation, disclosed a data breach in August 2020 after ransomware operators published stolen files.[14] Each of these districts operates substantial nutrition programs affected by the broader system outages.

The Cybersecurity and Infrastructure Security Agency (CISA) has repeatedly warned about ransomware targeting the education sector, noting that K–12 institutions often lack dedicated security staff and operate legacy systems with known vulnerabilities.[15] School nutrition programs inherit whatever security posture their host districts maintain—or fail to maintain.

The Quiet Digitization of Vending

The vending machine has transformed from a coin-operated mechanical device into a networked computer that happens to dispense snacks. Modern vending equipment connects to management platforms that track inventory levels, report sales data, and process cashless payments. The same machine that sold candy bars for quarters in 1985 now accepts Apple Pay, maintains a cellular data connection, and may even incorporate a touchscreen for product selection. Canteen, a division of Compass Group, operates more than 200,000 vending machines across North America.[16] Five Star Food Service, Aramark's vending unit, maintains similar vast deployments. These aren't standalone devices; they're IoT endpoints in a distributed network that must be secured, patched, and monitored for food safety, such as temperature.

The micromarket concept has accelerated this transformation. Rather than traditional vending machines, corporate break rooms increasingly feature open shelving where employees select items; they then check out at a self-service kiosk. The kiosk tracks purchases, the shelving may incorporate weight sensors or cameras for inventory monitoring, and the entire system links to backend platforms for restocking and billing.[17]

From a cybersecurity perspective, vending infrastructure presents familiar challenges at an unfamiliar scale. Each connected device is a potential entry point. Many machines operate in physically accessible locations where USB ports or maintenance interfaces may be exposed. Payment processing must meet PCI DSS requirements even when deployed in environments with minimal physical security. And the low per-unit value of vending transactions can make security investments seem disproportionate until a breach compromises payment data across thousands of machines.

What It Teaches Us: Why Foodservice Security Struggles

The restaurant and foodservice industry faces structural challenges that make robust cybersecurity difficult to achieve. Profit margins in restaurants average between three and nine percent, leaving little room for security investments that don't show immediate operational returns.[18] Employee turnover in foodservice exceeds 70 percent annually, meaning that any training provided must be repeated constantly for new staff.[19] Many restaurant operators lack dedicated IT staff entirely, relying on point-of-sale vendors or managed service providers whose attention is divided across many clients.

The payment card industry has driven some improvement through PCI DSS compliance requirements. Merchants processing card payments must meet security standards or face penalties from card networks and potentially lose the ability to accept cards at all, but compliance is often treated as a checkbox exercise rather than a genuine security improvement, and the standards themselves may not address risks beyond payment card theft.

Ransomware targeting restaurants remains relatively uncommon compared to attacks on manufacturing or healthcare—the value extracted from encrypting a restaurant's systems is typically lower than what can be harvested from a hospital or meat processor—but restaurants do hold data that attackers find valuable: customer records, employee information, and payment credentials. And as restaurant operations become more dependent on digital systems for online ordering, delivery integration, dynamic pricing, and inventory optimization, the disruption potential of even a simple ransomware attack grows.

Moving Forward: Building Resilience Across the Service Counter

Improving cybersecurity in foodservice requires acknowledging the sector's constraints rather than prescribing enterprise security solutions that restaurants cannot afford or staff. Practical steps exist that respect the industry's economics while reducing risk. For restaurant operators, the most effective investment is often the simplest: working with POS vendors that take security seriously and maintaining configurations that follow best practices. Network segmentation can keep payment systems isolated from other business functions and costs little to implement but limits how far an attacker can move through compromised systems. Employee training on recognizing phishing and social engineering attempts reduces the most common initial access vector.

For franchise systems, corporate brands can provide security templates and approved vendor lists that make compliance easier for franchisees. Regular assessments, even simple ones, help identify locations that have drifted from security baselines. Shared incident response planning can ensure that when something goes wrong, the response isn't improvised.

For institutional foodservice, the challenge is often advocating within host organizations for security investments that protect food operations. When a hospital's IT department prioritizes clinical systems over cafeteria operations, foodservice leaders need to articulate why their systems matter for both operational continuity and patient safety.

As we'll see in the next chapter, the journey of food and agriculture extends far beyond land-based agriculture. The seafood industry, from commercial fishing vessels to aquaculture operations to processing facilities, faces its own cybersecurity challenges in environments that are often literally offshore, where connectivity is intermittent and physical isolation creates both security advantages and unique vulnerabilities.

Notes

1. Lemos, R. (2023). "Checkers breach underscores continued POS dangers." *Dark Reading*. **https://www.darkreading.com/cyberattacks-data-breaches/checkers-breach-underscores-continued-pos-dangers**
2. Kovacs, E. (2019). "Malware found on PoS systems at Checkers and Rally's restaurants." *SecurityWeek*. **https://www.securityweek.com/malware-found-pos-systems-checkers-and-rallys-restaurants**
3. MentorPOS. (2025). "The evolution of point of sale systems: from 1879 to the present." **https://www.mentorpos.com/the-evolution-of-point-of-sale-systems-from-1879-to-the-present**
4. Delaware, State of. Office of the Attorney General. (2020). "Landry's Inc. data breach notification." **https://attorneygeneral.delaware.gov/wp-content/uploads/sites/50/2020/06/Landrys-Inc.pdf**

5. McDonald's Corporation. (2023). "Restaurants by market 2023." **https://corporate.mcdonalds.com/content/dam/sites/corp/nfl/pdf/Restaurants%20by%20Market%202023.pdf**; Subway. (2023). "Subway® continues positive momentum with double-digit global sales growth." **https://newsroom.subway.com/2023-04-24-Subway-R-Continues-Positive-Momentum-with-Double-Digit-Global-Sales-Growth**
6. The Wendy's Company. (2016). "Update on payment card security investigation." **https://www.prnewswire.com/news-releases/wendys-update-on-payment-card-security-incident-300295206.html**
7. National Restaurant Association. (n.d). "How to protect cybersecurity at your restaurant." **https://restaurant.org/education-and-resources/resource-library/how-to-protect-cybersecurity-at-your-restaurant**
8. Cimpanu, C. (2020). "UHS hospital network hit by ransomware attack." *ZDNet*. **https://www.zdnet.com/article/uhs-hospital-network-hit-by-ransomware-attack**
9. The Joint Commission. (2024). *Hospital Accreditation Standards*.
10. Compass Group PLC. (2023). *Annual Report 2023*. **https://www.compass-group.com/content/dam/compass-group/corporate/oar-2023/annual-report-2023.pdf**
11. Los Angeles Unified School District. (n.d). "Everyone eats!" *YumYummi*. **https://lausd.yumyummi.com**
12. Sganga, N. (2022). "Ransomware group Vice Society targeted dozens of schools in 2022, Palo Alto network report." *CBS News*. **https://www.cbsnews.com/news/vice-society-ransomware-group-targeted-dozens-of-school-pala-alto-network-report**
13. Baltimore County Public Schools. (2020). Announcement on ransomware attack and recovery. Cited in Kost, E. "Ransomware attack forces Baltimore Public Schools to close." *UpGuard*. **https://www.upguard.com/news/ransomware-attack-forces-baltimore-public-schools-to-close-for-115-000-students**
14. Associated Press. (2020). "Clark County Schools reports computer ransomware attack." *SecurityWeek*. **https://www.securityweek.com/clark-county-schools-reports-computer-ransomware-attack**
15. Cybersecurity and Infrastructure Security Agency. (2020). "Cyber actors target K-12 distance learning and primary/secondary education to cause disruptions and steal data." **https://www.cisa.gov/news-events/cybersecurity-advisories/aa20-345a**
16. Compass Group USA. (n.d). **https://careers.compass-usa.com/who-we-are**
17. Vending Market Watch. (2024). *2023 State of the Vending and Micro Market Industry Report*. **https://www.vendingmarketwatch.com/reports/document/55056921/2023-state-of-the-vending-and-micro-market-industry-report**
18. National Restaurant Association. (2023). "National statistics." **https://restaurant.org/research-and-media/research/industry-statistics/national-statistics**
19. U.S. Bureau of Labor Statistics. (2023). "Job openings and labor turnover survey: September 2023." **https://www.bls.gov/news.release/archives/jolts_11012023.pdf**

CHAPTER 12

Seafood and Fisheries

Wild-Caught to Aquaculture and Everything in Between

Growing up in New England, I took seafood for granted in ways I only now appreciate. Summer meant lobster bakes on the beach, the smell of seaweed and salt air mixed with melted butter. Steamed clams and mussels were weeknight dinners. Fried scallops with shoestring fries and tartar sauce remain comfort foods that I actively crave. The fishing and lobster boats from Boston Harbor, Massachusetts, to Bar Harbor, Maine, were simply part of the landscape, their work invisible to a kid more interested in what ended up on the plate than how it got there.

That invisibility persists for most consumers. We see seafood at the grocery counter or on restaurant menus without considering the vast infrastructure required to harvest, process, and transport products that spoil faster than almost any other protein. And within that infrastructure lies a cybersecurity landscape unlike any we've examined so far in this book: one that operates across land and sea, spans the globe from Norwegian fjords to New Zealand's Southern Shores, and increasingly depends on sophisticated technology to function.

Here's the reality most people don't grasp: seafood represents the primary protein source for over 3 billion people worldwide. The global industry produces more than 200 million metric tons annually, and for the first time in history, farmed seafood (aquaculture) has surpassed wild capture. This isn't a niche market. For much of the world, disrupting seafood means disrupting survival.[1]

A Hidden World Beneath the Water

Aquaculture often operates out of sight and out of mind by necessity. Fish farms require clean, pristine waters away from human development, which means they tend to be in remote areas such as the lochs of Scotland, the fjords of Norway, the glacial lakes of New Zealand, the coastal regions of Chile, and island nations throughout Southeast Asia. China alone accounts for more than 50 percent of global aquaculture production, feeding a population that depends on species like tilapia and pangasius as dietary staples.[2]

The perception that these are simple operations couldn't be further from reality. Modern salmon farming, for instance, involves a level of technological sophistication that rivals any precision agriculture operation on land. Norwegian companies like Mowi, Cermaq, and SalMar have invested heavily in what the industry calls "smart farming," an approach that fundamentally transforms how fish are raised.[3] Consider the operational technology (OT) stack on a contemporary salmon farm. Pens containing hundreds of thousands of fish float in ocean waters 30 to 50 meters deep. Underwater cameras provide live 4K feeds of the fish population, connected via satellite links to monitoring centers on the mainland. AI systems analyze pellet detection and fish feeding behavior, determining when fish are satiated and adjusting feed rates in real time. Sensors continuously monitor water temperature, oxygen levels, salinity, and current patterns. Automated feeding systems deliver precisely calibrated nutrition through tubes connecting floating feed barges to individual pens.[4] These farmers know how their fish feel every day—data that would seem impossible to know when you're talking about populations too large to count by eye and at depths too murky to see through. That capability depends entirely on interconnected systems operating 24 hours a day in some of the harshest environments on Earth.

The Tech That Keeps It Running

Norway has become the epicenter of aquaculture technology development for practical reasons. The country produces approximately 1.5 million metric tons of salmon annually despite a population of only 5 million people. Salmon is Norway's second-largest export behind oil.[5] Economics demand efficiency, and efficiency demands automation. A single operator can now effectively manage 3,000 to 4,000 tons of salmon using centralized feeding systems and remote monitoring. Feed barges equipped with satellite Internet connections allow personnel sitting in control rooms to watch multiple pens simultaneously, reading fish behavior across screens that display conditions throughout a farming region. The feeding itself has become AI-controlled, with systems learning to recognize when fish are hungry, when they're satiated, and when something might be wrong.[6]

But feeding is only one piece. Net-cleaning robots, some the size of a small car, patrol the inside of pens, running on tank-like treads to remove biofouling that would otherwise restrict water flow. In Scotland, where farms may have more than 1,800 individual pens, each one requires cleaning roughly every two weeks during the summer months. These autonomous remotely operated vehicles (ROVs) carry out inspections, identify net tears, and even perform underwater repairs that would otherwise require commercial divers working in cold, low-visibility conditions.[7]

Shellfish operations have embraced similar technologies. Oyster and mussel farms increasingly use sensor arrays to monitor growth rates and environmental conditions. Scallop farms, particularly in Japan and Australia, deploy underwater monitoring systems to track populations across the seabed. The marine environment creates unique challenges for this technology. Equipment must withstand 100-mile-per-hour winds, 10-meter swells, constant saltwater corrosion, and biofouling that can render sensors useless within weeks. The industry has a saying about Silicon Valley solutions: they put it in their pen, and it immediately breaks and shatters, and the storms rip it away. Technology that survives in aquaculture environments must be built to withstand incredibly tough conditions.[8]

Wild Capture: Ships as Floating Data Centers

Commercial fishing vessels have transformed from diesel-powered workboats into floating technology platforms. Modern fishing operations depend on an array of electronic systems for navigation, fish finding, gear deployment, compliance monitoring, and catch documentation. Electronic chart display and information systems (ECDISs) have largely replaced paper navigation charts, integrating GPS data with radar and depth soundings to guide vessels through complex waters. Automatic identification systems (AISs) broadcast vessel position, speed, and heading to other ships and shore-based monitoring stations. Fish-finding sonar has evolved into sophisticated acoustic systems capable of estimating school size and species composition.

The International Maritime Organization (IMO) recognized the cybersecurity implications of this digitization, implementing requirements in 2021 that cyber risk management be addressed in vessel safety management systems.[9] But fishing vessels operate under different regulatory frameworks than cargo ships, and many fleets, particularly in the developing world, have been slow to adopt systematic cybersecurity practices.

Global Fishing Watch, an organization that monitors fishing activity worldwide using satellite imagery and AIS data, has documented how bad actors can exploit weaknesses in these systems. Vessels engaged in illegal, unreported, and unregulated (IUU) fishing routinely disable or spoof AIS transmitters to avoid detection.[10] The same techniques that hide illicit fishing could be used to falsify catch documentation, manipulate quota compliance systems, or interfere with fisheries enforcement.

GPS spoofing incidents have affected commercial fishing fleets in multiple regions. Vessels that depend on satellite positioning for navigation and gear deployment can be misled about their actual location, with implications ranging from inefficient operations to dangerous navigation errors near coastlines or other vessels.[11]

When the Systems Fail

The seafood industry's cybersecurity vulnerabilities came into sharp focus in January 2021 when AKVA Group, a Norwegian provider of aquaculture technology and services, suffered a ransomware attack. AKVA supplies the critical infrastructure that modern fish farms depend on for automated feeding systems, underwater cameras, net-cleaning equipment, and environmental monitoring sensors. The attack, which occurred over a weekend, took down several key systems and forced the company to work with Norwegian authorities to contain the damage.[12]

The financial impact was substantial: AKVA reported that the incident cost approximately $6 million in the first quarter alone.[13] But the broader implications extended beyond one company's balance sheet. When your technology provider goes dark, the farms relying on that technology face immediate operational challenges. AKVA wasn't the only aquaculture technology company targeted; its competitor Steinsvik (now ScaleAQ) was attacked the previous April.[14] Incidents revealed that attackers had identified the aquaculture supply chain as a viable target.

Similarly, a ransomware attack in late May and early June 2021 temporarily shut down or disrupted JBS operations in North America and Australia, affecting beef, pork, poultry, and prepared foods plants.[15] JBS is also active in farmed salmon through investments such as Huon Aquaculture in Australia,[16] so disruptions at the group level highlight cyber risk exposure that extends into seafood alongside traditional meat segments.

Seafood processing facilities face vulnerabilities similar to other food processing operations, but with a critical difference: seafood spoils faster. The window between harvest and processing must be measured in hours, not days. When a ransomware attack disables the systems that manage incoming products, maintain cold-chain documentation, and coordinate processing lines, the consequences materialize almost immediately.

Traceability and Its Double Edge

Seafood traceability has become increasingly important for both food safety and sustainability reasons. The U.S. Seafood Import Monitoring Program (SIMP), implemented by the National Oceanic and Atmospheric Administration (NOAA), requires electronic reporting of catch and chain-of-custody information for certain high-risk species imported into the United States.[17] The European Union maintains similar catch documentation requirements for its seafood imports.[18] These systems generate and rely on vast quantities of digital data. From the moment a fish is caught, it's documented through electronic logbooks, vessel monitoring systems, and landing declarations through each subsequent transfer of ownership and processing step, and the data trail must remain intact and accurate.

Blockchain-based traceability initiatives, including IBM Food Trust implementations by major seafood companies, have added additional layers of digital documentation.[19] Bumble Bee Foods, one of North America's largest shelf-stable seafood companies, implemented blockchain traceability for its yellowfin tuna in partnership with SAP, allowing consumers to trace their tuna from the Indonesian fishing vessel that caught it through processing and distribution.[20] That level of transparency depends on the underlying systems remaining secure and trustworthy.

The cybersecurity implications cut both ways. Robust traceability systems make it harder to launder illegally caught fish into legitimate supply chains. But those same systems, if compromised, could allow sophisticated actors to fabricate documentation for IUU-caught products, manipulate apparent catch origins to evade tariffs or sanctions, or corrupt food safety records that determine whether products are safe for consumption.

The Processing Bottleneck

Seafood processing is geographically concentrated in ways that create systemic vulnerabilities. In the United States, a substantial portion of domestic seafood processing occurs in Alaska, the Pacific Northwest, and along the Gulf Coast. Many wild-caught Alaskan species are processed in floating facilities or remote shore-based plants that operate only during fishing seasons.[21] These seasonal operations face unique challenges. Workers arrive for intensive periods of weeks or months, many unfamiliar with the specific systems and equipment at a given facility. Equipment may sit unused for extended periods, running outdated software that wasn't updated during the off-season. The remote locations often mean limited connectivity options and long lag times for technical support.

Hazard analysis and critical control points (HACCP) plans for the food safety management systems required for seafood processing increasingly depend on digital monitoring and documentation. Temperature logs, sanitation records, processing time documentation, and equipment calibration data must be captured and retained. When ransomware encrypts those systems, facilities face an impossible choice between operating without required documentation or halting operations.

The 48-hour problem haunts the industry. Fish farmers invest years and substantial capital in raising their stock—three years to bring salmon to market weight, for instance. All that time and money are concentrated into a narrow window after harvest. If systems fail during that window because of a ransomware attack in which operators halfway around the world have encrypted critical servers, the losses can be catastrophic.[22]

Lobstermen, Kelp, and Adaptation

The lobster industry in Maine offers an interesting case study in how fishing communities adapt to changing circumstances. Regulatory changes and warming waters have compressed traditional lobstering seasons, forcing many lobstermen to seek supplementary income. Some have become accidental stewards of marine ecosystems, implementing practices that sustain the resource they depend on.[23]

Others have turned to aquaculture. Lobstermen with boats already equipped for hauling pots have found that kelp farming provides a natural complement to their operations. The boats have cranes, deck space, and seawater plumbing already installed. Local kelp companies have established cooperative arrangements where lobstermen grow and harvest kelp during the off-season, diversifying their income while developing a new sector of the seafood industry.[24]

This small-scale, community-based aquaculture operates very differently from the industrial salmon farms of Norway or Scotland. The technology footprint is smaller, and the data requirements are less intensive. But as kelp farming scales—there are substantial efforts underway to expand seaweed cultivation for food, animal feed, and biofuel feedstock—the technology requirements will grow accordingly.

The transition also raises workforce cybersecurity considerations. Fishermen who spent careers relying on experience and instinct must now learn to operate and troubleshoot connected systems. Training programs that worked for vessel safety may not adequately address digital risks that seem abstract compared to the immediate physical dangers of working at sea. Also, who will train these fishermen on the dangers of connected systems?

New Zealand, Scotland, and Global Reach

Seafood cybersecurity concerns exist wherever the industry operates, but certain regions warrant particular attention given their role in global supply chains.

New Zealand's aquaculture industry, centered on King salmon and green-lipped mussels, operates in some of the world's most pristine waters.[25] The country's geographic isolation and strict marine monitoring regimes provide natural biosecurity advantages, but many farms are in remote regions where operations increasingly depend on digital monitoring systems and data links to on-shore facilities.[26] Freshwater King salmon raised in glacier-fed high-country systems beneath Mt Cook command premium prices in export markets, and physical remoteness amplifies reliance on automated monitoring, biomass cameras, and robotic processing systems that demand robust cybersecurity protections.[27]

Scotland's salmon farming industry, concentrated in the Western Isles and Highlands, operates in a regulatory environment shaped by the European Union's NIS2 Directive requirements for essential services.[28] The UK's departure from the EU hasn't eliminated the need to comply with EU cybersecurity requirements for companies that want to export to the European market. Scottish salmon farmers exporting to EU countries must demonstrate cybersecurity measures that meet EU standards, creating compliance obligations that exceed what domestic UK regulations currently require.

Norwegian aquaculture, as the world's technological leader, also faces the most sophisticated threats. The Norwegian National Security Authority has warned that critical infrastructure, including food production, faces elevated threat levels from state-sponsored actors.[29] For an industry so central to the Norwegian economy and so dependent on connected systems, those warnings carry weight.

What It Teaches Us: Vulnerabilities at Every Depth

Securing the seafood sector requires recognizing its unique characteristics. Unlike land-based food production, much of this industry operates in international waters or remote coastal zones beyond easy oversight. Equipment must function in harsh conditions with limited connectivity and even more limited on-site technical expertise. Supply chains span the globe, with fish caught in one hemisphere processed in another before reaching consumers in a third. The industry's growing technological sophistication has brought undeniable benefits: better animal welfare monitoring, more efficient feeding, improved environmental management, and enhanced traceability. But each sensor, each satellite link, and each automated system creates a potential attack surface that the industry has been slow to address.

Moving Forward: Charting a Course for Resilience

Three priorities emerge. First, the industry must recognize that IT and OT security in maritime and aquatic environments requires specialized expertise: not just cybersecurity professionals who understand food manufacturing, but those who understand the specific constraints of marine operations. Second, supply chain security deserves far more attention than it currently receives; the AKVA incident demonstrated how disrupting feed supply can threaten farms that themselves have never been directly attacked. Third, regulatory frameworks designed for terrestrial food production may not adequately address the unique circumstances of an industry that operates across jurisdictional boundaries in some of the world's most challenging environments.

The world depends on seafood in ways that few consumers appreciate. Those 3 billion people eating fish as their primary protein can't simply switch to alternatives if disruption occurs. And as warming waters push aquaculture further offshore and wild fisheries into new territories, the digital systems connecting this global industry become simultaneously more critical and more vulnerable.

The next chapter examines another liquid domain: the beverage industry, where many of the same process control and supply chain vulnerabilities apply, but the products on the line range from milk to wine to energy drinks.

Notes

1. White, C. (2024). "Global aquaculture surging, with production surpassing wild-catch fisheries—2024 UN FAO SOFIA Report Finds." *SeafoodSource.* **https://www.seafoodsource.com/news/supply-trade/global-aquaculture-surging-with-production-surpassing-wild-catch-fisheries-2024-un-fao-sofia-report-finds**
2. FAO. (2024). *The State of World Fisheries and Aquaculture 2024.* **https://openknowledge.fao.org/items/8ab20ccf-1e9d-4ae6-836c-ca770d16da01**
3. Mowi ASA. (2024). *Integrated Annual Report 2023.* **https://mowi.com/wp-content/uploads/2024/03/Mowi_Integrated_Annual_Report_2023.pdf**
4. Mowi ASA. (2025). "Pioneering the future of aquaculture." **https://mowi.com/blog/smart-farming-technologies**

5. The Fish Site. (2024). "Export boosts for Norwegian seafood." **https://thefishsite.com/articles/export-boosts-for-norwegian-seafood**
6. Cermaq. (2024). *Final Report for the iFarm Development License Project.* **https://www.cermaq.no/assets/Norway/PDF/Final-report-for-the-iFarm-development-license-project.pdf**
7. Marine Scotland. (2024). *Scottish Fish Farm Production Survey 2023.* **https://www.gov.scot/publications/scottish-fish-farm-production-survey-2023**
8. NOAA. (2024). "Marine aquaculture and the environment." **https://www.fisheries.noaa.gov/insight/marine-aquaculture-and-environment**
9. IMO. (2017). "Guidelines on maritime cyber risk management." MSC-FAL.1/Circ.3. (with subsequent revisions, including Rev.3 incorporating amendments effective January 1, 2021). **https://wwwcdn.imo.org/localresources/en/OurWork/Security/Documents/MSC-FAL.1-Circ.3-Rev.3.pdf**
10. Kroodsma, D. (2025). "Advances in satellite radar analysis help reveal fishing activity." *Global Fishing Watch.* **https://globalfishingwatch.org/article/advances-in-satellite-radar-analysis-help-reveal-fishing-activity**
11. U.S. Coast Guard. (2026). "Coast Guard maritime industry cybersecurity resource center." **https://www.uscg.mil/MaritimeCyber**
12. Keams, M. (2021). "Cyber-attackers target AKVA Group, down several key systems." *SeafoodSource.* **https://www.seafoodsource.com/news/business-finance/cyber-attackers-target-akva-group-downing-several-key-systems**
13. Hogan, H. (2025). "How vulnerable are seafood companies to cybercrime? More than they might think." *Global Seafood Alliance.* **https://www.globalseafood.org/advocate/how-vulnerable-are-seafood-companies-to-cybercrime-more-than-they-might-think**
14. Constantino, J. (2021). "Data sharing lowers the risk of cyber-attacks in aquaculture" *SalmonBusiness.* **https://www.salmonbusiness.com/data-sharing-lowers-the-risk-of-cyber-attacks-in-aquaculture**
15. Chertoff Group. (2021). "JBS SA ransomware attack security bulletin." **https://chertoffgroup.com/jbs-sa-ransomware-attack-security-bulletin**
16. White, C. (2021). "JBS moves to acquire 100 Percent of Australia's Huon Aquaculture." *SeafoodSource.* **https://www.seafoodsource.com/news/business-finance/jbs-acquires-100-percent-of-australia-s-huon-aquaculture**
17. NOAA. (2025). "Seafood import monitoring program." **https://www.fisheries.noaa.gov/international/international-affairs/seafood-import-monitoring-program**
18. EU IUU Fishing Coalition. (n.d.). "The catch certificate scheme." **https://www.iuuwatch.eu/the-iuu-regulation/catch-certification**
19. IBM Newsroom. (2021). "Nova Sea joins Norwegian Seafood Trust to track salmon from feed to food using IBM Food Trust technology." **https://newsroom.ibm.com/2021-06-01-Nova-Sea-Joins-Norwegian-Seafood-Trust-to-Track-Salmon-from-Feed-to-Food-Using-IBM-Food-Trust-Technology**
20. PRNewswire. (2019). "Bumble Bee Foods and SAP create blockchain to track fresh fish from ocean to table." **https://www.prnewswire.com/news-releases/bumble-bee-foods-and-sap-create-blockchain-to-track-fresh-fish-from-ocean-to-table-300808987.html**
21. NOAA Fisheries. (2024). *Annual Report to Congress on the Status of U.S. Fisheries.* **https://www.fisheries.noaa.gov/s3/2024-04/2023SOS-final.pdf**
22. Cook, J. (2012). "Seafood processing involves unique challenges." *Manufacturing.net.* **https://www.manufacturing.net/home/article/13149645/seafood-processing-involves-unique-challenges**

23. Maine Lobstermen's Association. **https://www.mainelobstermen.org**
24. NOAA Fisheries. (2021). "Tide to table profiles: Atlantic sea farms." **https://www.fisheries.noaa.gov/feature-story/tide-table-profiles-atlantic-sea-farms**
25. Crimp, D. (2007). "Aquaculture." *New Zealand Geographic.* **https://www.nzgeo.com/stories/aquaculture**; NIWA. "Greenshell™ Mussel." **https://niwa.co.nz/aquaculture/greenshelltm-mussel**
26. Cawthron Institute. (n.d.). "Deep diving into ocean data." **https://www.cawthron.org.nz/deep-diving-into-ocean-data**; New Zealand Primary Production Committee. (n.d.). "A future focus: science, technology and innovation." **https://www.pmcsa.ac.nz/topics/fish/a-future-focus-science-technology-and-innovation**
27. QTMagazine. (2019). "New partnership for Mt Cook Alpine Salmon, Scott Technology & Seafood Innovations." **https://qt.co.nz/new-partnership-for-mt-cook-alpine-salmon-scott-technology-seafood-innovations**; Ace Aquatec. (2024). "New Zealand's Mount Cook alpine salmon embraces Ace camera technology." **https://aceaquatec.com/news-and-resources/news/new-zealands-mount-cook-alpine-salmon-embraces-ace-camera-technology**; New Zealand King Salmon. *Annual Report 2022.* **https://www.kingsalmon.co.nz/wp-content/uploads/2022/05/NZKS-Annual-Report-FY22-WEB-FINAL-3.pdf**
28. European Union. (2022). "Directive (EU) 2022/2555 of the European Parliament and of the Council of 14 December 2022 on measures for a high common level of cybersecurity across the Union." EUR-Lex. **https://eur-lex.europa.eu/legal-content/EN/TXT/?uri=CELEX:32022L2555**
29. Norwegian National Security Authority. (2024). *Risiko 2024.* **https://nsm.no/getfile.php/1313477-1719434219/NSM/Filer/Dokumenter/Rapporter/Risiko%202024.pdf**

CHAPTER 13

The Beverage Industry: Alcoholic, Non-Alcoholic, Dairy, and Plant-Based

On March 11, 2021, Molson Coors Beverage Company filed an SEC Form 8-K disclosing what the document plainly described as a "cybersecurity incident" that caused a systems outage, disrupting brewery operations, production, and shipment of beer, wine, and other beverages.[1] The regulatory language masked operational chaos. Workers at brewing facilities in Chicago and Milwaukee were sent home. The company's SAP enterprise system, the software managing everything from ingredient orders to shipment scheduling, went dark. Molson Coors is the world's fifth-largest brewer.[2] Coors, Miller, Blue Moon, and Leinenkugel are brands that fill refrigerators across America, and production schedules are planned weeks in advance. Distribution networks move products daily. When systems fail, even briefly, the disruption cascades.[3] The company never disclosed whether ransomware was involved, whether ransom was demanded, or what the attack vector was. However, it did acknowledge significant recovery costs and warned that the incident might have a material impact on business operations. Full operational restoration took weeks.[4]

This chapter examines why beverage manufacturers from craft breweries to global spirits companies face cybersecurity challenges distinct from other food processing industries, and what these incidents reveal about protecting systems that most consumers never think about—until their favorite drink isn't on the shelf.

How We Got Here

The history of beverages is, in many ways, the history of human survival. For most of recorded history, water was dangerous. Contaminated wells and rivers spread cholera, typhoid, and dysentery. Fermentation solved this problem: the alcohol in wine and beer killed pathogens, making these drinks safer than the water available to most people.

Monasteries became brewing centers not just for spiritual reasons but because monks needed safe hydration. The phrase "small beer" referred to low-alcohol brews consumed daily by everyone, including children, because the alternative included the risk of disease.[5] This isn't ancient history, as we might assume. Municipal water treatment that is reliable enough to make tap water universally safe is barely a century old in developed nations and still incomplete globally.[6] The bottled water industry exists partly because trust in tap water varies by geography and generation.[7]

Wine appears in the foundational texts of multiple religions. Catholic Mass requires wine for communion. Jewish Shabbat and holidays begin with Kiddush over wine. And ancient Greek and Roman religious ceremonies incorporated wine offerings. Regional identity also intertwines with beverage production in ways that transcend commerce. Champagne isn't just sparkling wine, but a protected designation that defines a French region's economy and heritage. Scotch whisky, Kentucky bourbon, Mexican tequila, and Portuguese port carry legal protections because they represent cultural patrimony.[8] An attack on these industries threatens something deeper than quarterly earnings; it's intimate in more human ways.

Walk through a grocery store beverage aisle today and try to count the options. Fifty years ago, you'd have found a fraction of what's there now. Soda formulations have evolved. Bottled water brands have multiplied from a novelty into a $350 billion global industry.[9] Energy drinks didn't exist as a category until the 1980s. Hard seltzers appeared in the 2010s and captured significant market share almost overnight. And plant-based beverages—oat, almond, soy, and coconut—now compete for refrigerator space with dairy.

How did we get here? Like most things, it was thanks to technology. Pasteurization extends shelf life. Refrigerated distribution made fresh dairy beverages commercially viable beyond local markets. Aseptic packaging allowed shelf-stable products without refrigeration. Automated bottling lines made small-batch production economically feasible, and digital recipe management enabled companies to produce dozens of variations efficiently. But the same technology that created this abundance now creates vulnerability.

Wine: Where Agriculture Meets Beverage Production

A bottle of wine from a particular vintage is a time capsule. It tells the story of that year: the soil conditions, the weather patterns, the insect pressures, and the decisions made by the winemaker. Each bottle captures something unrepeatable, and you won't get that exact glass again unless you've bought more of that vintage; even then, the wine continues evolving. It's not sentimental; it's an agricultural and biological reality. It's beautiful, really, right?

When asked about favorite wines, enthusiasts often answer by grape variety rather than brand—Pinot Noir from a cooler year, Cabernet Sauvignon from a warm one—because vintages differ so dramatically that brand loyalty means less than understanding what conditions shaped the grapes. The distinction between grower champagne (small producers who grow their own grapes) and house champagne (large négociants who purchase grapes from the same vintage year) reflects different philosophies and scales and, from a cybersecurity perspective, fundamentally different risk profiles.

Unlike brewing or soft drink production, where raw materials arrive at a facility, wine begins in the vineyard. A winery is simultaneously a farm and a manufacturing operation, and both halves depend on digital technology. Climate change has fundamentally reshaped wine production, driving the adoption of precision viticulture technologies: weather stations tracking microclimate conditions, soil moisture sensors, drone imagery mapping vine vigor, and predictive analytics forecasting harvest timing.[10] The precision viticulture market reached $3.24 billion in 2024 and is projected to exceed $9.5 billion by 2032.[11]

In small and medium wineries, the winemaker wears many hats, including viticulturist, cellar master, sales manager, and, increasingly, IT administrator and de facto cybersecurity officer. Winemakers install sensors, configure weather stations, manage guest Wi-Fi for the tasting room, troubleshoot the point-of-sale system, and somehow still find time to make award-winning wine. This concentration of responsibilities creates obvious vulnerabilities. Security configurations may be set once and forgotten. And network segmentation between vineyard Internet of Things devices, winery control systems, tasting room point-of-sale systems, and guest Wi-Fi may not exist at all.

When Crises Compound

For U.S. wine regions in California, Oregon, and Washington, wildfire has become an existential threat. But flames are only part of the danger. *Smoke taint*, which occurs when volatile phenol compounds from smoke are absorbed by grape skins and released during fermentation, creating unpleasant flavors, can destroy an entire vintage even when fire never touches the vineyard.[12] The 2020 wildfire season cost the U.S. wine industry an estimated $3.7 billion.[13]

The wine industry is responding with technology: AI-powered smoke detection sensors deployed across vineyards in Sonoma and Willamette Valley, atmospheric modeling predicting smoke movement, and rapid testing protocols to detect taint compounds before harvest.[14] Consider what happens when a cyberattack coincides with wildfire conditions: smoke detection sensors go offline, weather station data stops flowing, irrigation systems malfunction, and communication with harvest crews breaks down. A cyberattack during wildfire season doesn't just disrupt operations; it blinds the winery at precisely the moment when real-time information matters most. This type of disruption would affect not only the vineyard but also the surrounding community.

Brewing, Spirits, and the Technology That Runs Them

The United States alone has more than 9,500 breweries, ranging from massive operations like Anheuser-Busch InBev's facility in St. Louis to neighborhood craft breweries producing a few hundred barrels annually. Global beer production exceeds 185 billion liters per year. What unites operations across this vast scale is increasing dependence on digital systems.[15]

Modern breweries rely on supervisory control and data acquisition (SCADA) systems and programmable logic controllers (PLCs) to manage the brewing process. Mashing converts grain starches to sugars, and temperature control is critical, as a few degrees can change the final product's character. Boiling and hopping add bitterness and flavor, and timing matters for extraction. Fermentation is where biology sets the timeline: yeast converting sugars to alcohol takes days to weeks, depending on the beer style, and interruptions can ruin entire batches.[16]

Recipe management systems store proprietary formulations digitally—not just ingredient lists but precise process parameters. For craft breweries, these recipes represent competitive differentiation. For major brewers, formula databases cover hundreds of products across global facilities. Clean-in-place (CIP) systems that automate cleaning cycles between batches depend on digital controls to ensure equipment is sanitized without manual intervention.

Spirits production involves different timelines and vulnerabilities. Distillation requires precise temperature control: too hot, and you collect undesirable compounds; too cool, and efficiency drops. But the greater exposure may be in the aging process: a bourbon distillery might have millions of dollars' worth of whiskey sitting in barrel warehouses for years. Digital inventory systems track barrel locations, age, and projected quality. Lose that data, and you face chaos trying to identify which barrels are ready for bottling.[17]

A Global Pattern of Attacks

Molson Coors wasn't an isolated incident. Beverage companies across multiple continents have faced attacks that reveal both the industry's vulnerabilities and the varied approaches to response.

Arizona Beverages (United States, March 2019)

Before the surge of attacks in 2020, Arizona Beverages, known for its iconic iced tea, suffered an iEncrypt ransomware attack in March 2019. Approximately 200 servers and networked computers were affected. The company operated with pen and paper for days.[18] The attack occurred shortly after Keurig Dr Pepper acquired Arizona, timing that highlights how corporate transitions can create vulnerability. Systems integration, credential changes, and network reorganization can open windows that attackers exploit. Full recovery took approximately two weeks.[19]

Lion Dairy and Drinks (Australia, June 2020)

Lion is Australia's largest beverage company, producing both beer (Tooheys, XXXX Gold, James Squire, and New Zealand's Speight's and Steinlager) and dairy products (Dairy Farmers milk, Pura, Dare iced coffee). In June 2020, a REvil ransomware attack took systems offline for multiple days.[20]

The dual disruption is particularly instructive. Beer production could be partially paused, although fermentation already underway had to be monitored manually. But dairy operations faced different pressures: milk from farms still arriving daily required processing regardless of IT status. Lion operated in a degraded state for approximately two weeks. Some systems took months to fully restore.[21]

Brown-Forman (United States, August 2020)

Jack Daniel's parent company, Brown-Forman, faced a REvil attack in August 2020. Attackers claimed to have stolen 1 terabyte of data. The company confirmed unauthorized access but stated that it successfully prevented its systems from being encrypted. It refused to pay a ransom, and the attackers published data samples. Brown-Forman illustrates an important distinction: preventing ransomware isn't binary. Data theft may occur even when operational systems survive encryption. The company stopped the worst outcome but still faced data exfiltration and the ongoing threat of public disclosure.[22]

Campari Group (Italy, November 2020)

Campari Group, headquartered in Milan, produces Campari, Aperol, Wild Turkey bourbon, Grand Marnier, Skyy Vodka, and Appleton Estate rum. The aperitivo tradition of drinking Campari or Aperol before dinner is woven into Italian cultural identity. An attack on Campari resonated beyond business disruption—it felt personal. RagnarLocker ransomware struck on November 1, 2020, a Sunday, typical timing for attackers seeking to exploit weekend skeleton crews. IT systems shut down across global operations. The attackers claimed to have stolen 2 terabytes of data and demanded $15 million. Campari refused to pay the ransom.[23] The RagnarLocker operators published samples of allegedly stolen data, including banking information, employee details, and business agreements. This again illustrates the double-extortion model now common in ransomware: even if you refuse to pay for decryption, attackers threaten to release sensitive data. Campari took approximately one week to restore critical systems.[24]

Duvel Moortgat (Belgium, March 2024)

Belgian brewing company Duvel Moortgat, known for Duvel golden ale and craft brands including Boulevard Brewing (Kansas City) and Firestone Walker (California), suffered a ransomware attack in March 2024 that caused bottling operations to halt at Belgian facilities.[25] Belgium has unique brewing traditions: Trappist ales and lambics that cannot be replicated elsewhere. Like the one against Campari Group in 2020, an attack targeting Belgian brewing heritage carried cultural weight beyond commercial impact. Duvel Moortgat also represents the craft segment: smaller production volumes but higher margins and intense brand loyalty.[26] The attack demonstrated that midsized specialty brewers face the same threats as industry giants.

Water: The Invisible Ingredient

Water isn't just an ingredient in beverages; it's often the primary ingredient. A can of soda is roughly 90 percent water.[27] Beer is approximately 95 percent water.[28] The quality, treatment, and availability of water directly affect what ends up in the bottle. Beverage producers either draw from municipal supplies or operate their own treatment systems. Either approach creates cybersecurity dependencies connecting beverage manufacturing to broader water infrastructure.

In November 2023, the Municipal Water Authority of Aliquippa, Pennsylvania, which serves roughly 15,000 customers, discovered that attackers had accessed an Internet-connected controller at a booster station. The intrusion targeted a Unitronics PLC and displayed a message claiming responsibility by CyberAv3ngers, an Iranian-backed hacktivist group. The attack forced the utility to switch to manual operations.[29] The U.S. Treasury Department subsequently sanctioned Iranian actors connected to the attack. Federal investigators confirmed that fewer than 10 other U.S. water facilities were breached through the same vulnerability: default passwords on Internet-exposed equipment. The same Unitronics controllers compromised at Aliquippa are used in industrial settings beyond water utilities, including beverage manufacturing.[30]

In October 2024, American Water, the largest regulated water utility in the United States, serving more than 14 million people across 14 states and 18 military installations, disclosed a cyberattack that forced a shutdown of its MyWater customer portal and paused billing systems for approximately one week. Water quality and core operations were not compromised, but the incident illustrated that even utilities with significant resources face successful intrusions.[31]

The systemic connection matters: a regional attack on water infrastructure doesn't just affect residential customers. Food and beverage manufacturers drawing from that supply face cascading disruptions, water quality uncertainties, production shutdowns, and potential product safety concerns.

What It Teaches Us: What Makes Beverages Different

Unlike food processing, where production can often pause and resume, fermentation operates on a biological clock. Without monitoring and temperature control, batches can spoil, produce off-flavors, or fail entirely. A brewery that is mid-fermentation when systems go down faces real decisions about product loss. Wine fermentation during harvest season is particularly vulnerable, as grapes arrive daily for a few weeks, and juice must begin fermentation promptly. An attack during crush season could cost an entire vintage year.

Beverage formulations represent core business assets. Coca-Cola's formula famously resides in a vault, but most beverage companies store formulations digitally. Recipe theft or corruption could enable counterfeiting or destroy competitive advantage. How a brewery ferments isn't just what yeast strain it uses, but specific temperatures, timing, and techniques developed over years. Spirits companies blend products from different sources, and master blenders' decisions about which barrels to use in what proportions are proprietary and irreplaceable.

Beverage brands also enjoy intense consumer loyalty. A security incident raising questions about product integrity, even without actual contamination, can damage brand value disproportionately. Craft beers, fine wines, and premium spirits command prices based on authenticity and quality perception, so these segments are especially vulnerable to reputational damage.

The patterns from the 2019–2024 attacks offer some lessons:

1. **Weekend timing is deliberate.** Lion and Campari were hit on days when skeleton crews staffed operations.
2. **Data exfiltration accompanies encryption in most modern attacks.** Brown-Forman and Campari both faced stolen data regardless of whether they paid the ransom.
3. **Ransom refusal is viable but not painless.** Both companies demonstrated that it's possible to refuse and recover, although at significant cost. Recovery takes weeks, not days, across all documented incidents.

Moving Forward: Building Resilience in the Brew

Small and midsized operations face difficult challenges. Craft breweries often have zero dedicated security personnel, rely on consultant-installed control systems, use consumer-grade networking equipment, and manage recipes on general-purpose computers. Low margins limit security investment. Trade organizations like the Brewers Association provide guidance, but implementation remains uneven across thousands of small operations.

Industry-specific considerations include separating fermentation control systems from enterprise IT, isolating recipe management databases, and protecting quality testing systems as critical infrastructure. The ability to monitor fermentation manually, paper-based recipe backups for essential products, and cold storage contingency plans for interrupted dairy processing can mean the difference between a disruption and a disaster.

The beverages examined in this chapter depend on ingredients that must be produced and kept secure. Beer requires malted barley and hops. Dairy beverages begin with feed consumed by cattle. The supply chain extends backward from the bottling line to farms and feed mills. The next chapter examines a sector most consumers never consider: animal feed and nutrition. The companies formulating and producing feed for livestock, poultry, and aquaculture represent a critical and increasing digital link in the food chain. When feed production falters, the animals that become our food and beverages suffer first. From grain to glass, the systems are connected. Protecting what we drink requires protecting every link in that chain.

Notes

1. Molson Coors Beverage Company. (2021). "Form 8-K." *SEC*. **https://www.sec.gov/Archives/edgar/data/24545/000110465921034789/tm219417d1_8k.htm**
2. VinePair. (2025). "The 40 biggest beer companies in the world in 2025." **https://vinepair.com/booze-news/40-biggest-beer-producers-2025**
3. Greig, J. (2021). "Molson Coors says cyberattack disrupted beer brewing operations." *ZDNet*. **https://www.zdnet.com/article/molson-coors-discloses-cyberattack-disrupting-its-brewery-operations**
4. Molson Coors Beverage Company. (2021). Q1 2021 earnings conference call transcript. **https://ir.molsoncoors.com/events-and-presentations/event-details/2021/Q1-2021-Molson-Coors-Beverage-Company-Earnings-Conference-Call/default.aspx**
5. Hornsey, I. S. (2003). *A History of Beer and Brewing* (Royal Society of Chemistry). **https://www.academia.edu/22980855/A_History_of_Beer_and_Brewing**
6. Bosaq. (2020). "The history of water treatment." **https://bosaq.com/water-treatment-history**
7. UltraPure H2O Tech. (2024). "A journey through the history of water purification." **https://ultrapureh2otech.com/2024/04/a-journey-through-the-history-of-water-purification**
8. World Intellectual Property Organization. (2017). *Geographical Indications: An Introduction*. **https://www.wipo.int/publications/en/details.jsp?id=4562**
9. Grand View Research. (2024). *Bottled Water Market*. **https://www.grandviewresearch.com/industry-analysis/bottled-water-market**
10. Wine & Spirit Education Trust. (2025). "Drones, robots and how vineyards are getting smarter." **https://www.wsetglobal.com/knowledge-centre/blog/2025/drones-robots-and-how-vineyard-technology-is-getting-smarter**
11. DataM Intelligence. (2025). *Precision Viticulture Market Report*. **https://www.datamintelligence.com/research-report/precision-viticulture-market**
12. Wilkinson, K. (2022). "Detection of smoke taint in grapes and wine." *HORIBA*. **https://www.horiba.com/usa/scientific/resources/science-in-action/detection-of-smoke-taint-in-grapes-and-wine**
13. Oregon State University, Extension Service. (2021). *Impacts to Oregon's Wine Industry: Covid-19 and the 2020 Wildfires*. **https://industry.oregonwine.org/wp-content/uploads/2020-Vineyard-and-Winery-Report-COVID-and-Wildfire-Impacts-09-07-21.pdf**
14. Rumbaugh, A. (2025). "Safeguarding US vines and wines from wildfire smoke." *National Grape Research Alliance*. **https://graperesearch.org/2025/07/29/safeguarding-us-vines-and-wines-from-wildfire-smoke**
15. Brewers Association. (2023). 2023 Annual Report. **https://www.brewersassociation.org/annual-report**
16. Pisney, C. (n.d.). "How is beer made? - The basics." *Cicerone Certification Program*. **https://www.cicerone.org/us-en/blog/how-is-beer-made-the-basics**
17. Bumpkins Stills. (2025). "Precision in distilling: why temperature control is everything." **https://bumpkinstills.co.uk/why-temperature-control-is-everything**
18. Whittaker, Z. (2019). "Arizona Beverages knocked offline by ransomware attack." *TechCrunch*. **https://techcrunch.com/2019/04/02/arizona-beverages-ransomware**
19. Keurig Dr Pepper Inc. (2024). "Keurig Dr Pepper Strengthens National Direct-Store-Delivery Operations with Acquisition of Strategic Assets from Kalil Bottling Company." **https://www.keurigdrpepper.com/keurig-dr-pepper-strengthens-national-direct-store-delivery-operations-with-acquisition-of-strategic-assets-from-kalil-bottling-company**

20. Becker, J. (2020). "Cyber attack halts Lion production of milk and beer." *ABC News Australia*. **https://www.abc.net.au/news/rural/2020-06-11/lion-dairy-drinks-cyber-attack/12343666**
21. Paganini, P. (2020). "REvil operators threaten to leak files stolen from Australian firm Lion." *SecurityAffairs*. **https://securityaffairs.com/105317/uncategorized/revil-operators-threaten-lion.html**
22. Ilascu, I. (2020). "U.S. spirits and wine giant hit by cyberattack, 1TB of data stolen." *BleepingComputer*. **https://www.bleepingcomputer.com/news/security/us-spirits-and-wine-giant-hit-by-cyberattack-1tb-of-data-stolen**
23. Abrams, L. (2020). "Campari hit by Ragnar Locker ransomware, $15 million demanded." *BleepingComputer*. **https://www.bleepingcomputer.com/news/security/campari-hit-by-ragnar-locker-ransomware-15-million-demanded**
24. Bîzgă, A. (2020). "Ragnar Locker group claim to have stolen 2 terabytes of data in Campari Group cyberattack." *Bitdefender*. **https://www.bitdefender.com/en-us/blog/hotforsecurity/ragnar-locker-group-claim-to-have-stolen-2-terabytes-of-data-in-campari-group-cyberattack**
25. Brauwelt International. (2024). *Brewer Duvel Moortgat Becomes Victim of Cyber Attacks*. **https://brauwelt.com/en/international-report/europe-russia/646761-brewer-duvel-moortgat-becomes-victim-of-cyber-attacks**
26. Duvel Moortgat. (2026). "Who we are." **https://www.duvelmoortgat.com/en/who-we-are**
27. Coca-Cola Company. (n.d.). "What are the ingredients of Coca-Cola Classic?" **https://www.coca-cola.com/ph/en/about-us/faq/what-are-the-ingredients-of-coca-cola-classic**
28. Hukins Hops. (n.d.). "What percentage of beer is water?" **https://www.hukins-hops.co.uk/our-story/news-press/what-percentage-of-beer-is-water**
29. Industrial Cyber. (2023). "Iranian hacker group CyberAv3ngers allegedly breach municipal water authority of Aliquippa." **https://industrialcyber.co/industrial-cyber-attacks/iranian-hacker-group-cyberav3ngers-allegedly-breach-municipal-water-authority-of-aliquippa**
30. DiMolfetta, D. (2024). "Treasury sanctions Iranian cyber officials tied to 2023 water system hacks." *Nextgov/FCW*. **https://www.nextgov.com/cybersecurity/2024/02/treasury-sanctions-iranian-cyber-officials-tied-2023-water-system-hacks/393877**
31. Industrial Cyber. (2024). *American Water Works Reports Cybersecurity Incident Following Unauthorized Hacker Activity*. **https://industrialcyber.co/utilities-energy-power-water-waste/american-water-works-reports-cybersecurity-incident-following-unauthorized-hacker-activity**

CHAPTER 14

Animal Feed and Nutrition

Feed Production, Additives, and Diet Formulation

Ever wondered what goes into the bag of kibble you pour into your dog's bowl each morning? Or what the chickens that laid your breakfast eggs were fed yesterday? What about the cattle that became last night's steak, or the salmon in your lunchtime salad? These questions lead to the same hidden industry: animal feed and nutrition. It's a sector worth more than $500 billion globally, responsible for nourishing the livestock, poultry, and aquaculture that billions of humans depend on daily.[1] It's also becoming digitized at a pace that's outrunning security.

The numbers challenge common assumptions. Less than half of the world's cereal crops feed people directly; the majority goes to animals.[2] In the European Union, nearly two-thirds of cereals become livestock feed. Globally, more than three-quarters of soy production ends up as animal feed; only a fraction becomes tofu or soy milk.[3] More than 80 percent of agricultural land worldwide supports livestock production, either as pasture or as crops grown specifically for feed.[4]

This makes animal feed not a niche sector but the foundation on which much of human nutrition ultimately rests. Feed typically accounts for 60 to 70 percent of livestock production costs, the single largest expense for anyone raising animals for food.[5] If you wanted to disrupt food systems with maximum leverage, animal feed offers a compelling target. Attack the feed, and you affect every animal-dependent product in that supply.

More Than Calories

Feed isn't just about filling stomachs. It's medicine, disease prevention, and precise nutritional science delivered through every meal. Put on your lab coat and consider coccidiosis, a parasitic disease that devastates poultry operations worldwide. Caused by protozoan parasites that colonize the intestinal tract, coccidiosis impairs nutrient absorption, reduces growth, and kills birds outright in severe cases.[6] Annual losses in the United States alone run into the hundreds of millions of dollars.[7] The solution: coccidiostats, compounds added to feed that prevent the parasites from reproducing. Ionophore coccidiostats such as monensin and salinomycin work by disrupting ion transport across cell membranes; synthetic compounds like diclazuril attack different biochemical pathways. The European Union has authorized 11 different coccidiostats for use in poultry and rabbits, each with specific dosing requirements and withdrawal periods before slaughter.[8] Okay, lab coats off. Here's what that means: chickens get gut parasites that make them sick, and, sadly, they die. We put medicine directly in their feed to stop those parasites from spreading. Chemistry is complicated, but the goal is simple: healthy birds. And because what animals eat eventually becomes what we eat, there are strict rules about dosing and timing.

Cattle receive their own feed additives. Ionophores improve feed efficiency by altering rumen microbiology. The same compounds that kill poultry parasites help cattle extract more nutrition from the same amount of feed. Vitamins, minerals, amino acids, and trace elements are blended into rations calculated to support specific production goals: growth in beef cattle, milk production in dairy cows, and egg laying in hens.[9]

Working animals depend on this precision, too. Farm dogs—the border collies and Australian shepherds that work livestock from dawn to dusk—require diets with higher protein and fat content than typical pets.[10] Horses that work cattle operations need rations balanced for sustained energy output. These animals aren't just companions; they're essential to agricultural operations, and their nutritional requirements are every bit as calculated as the livestock they help manage. Getting this wrong has consequences: ionophore toxicity in horses can be fatal, and horses are extremely sensitive to compounds that cattle tolerate well.[11] Cross-contamination at a feed mill, where horse feed picks up residues from cattle feed produced on the same equipment, has caused deaths. The precision required isn't optional: it's a matter of animal welfare and, ultimately, human food safety.

Your Dog's Food: A Window into the Industry

If the feed industry seems remote, consider that you may interact with it every day. Pet food is animal feed, manufactured using the same fundamental systems, drawing from the same ingredient supply chains, and facing the same vulnerabilities.

The American pet food market exceeds $60 billion annually. Approximately two-thirds of households own at least one pet.[12] The industry has professionalized dramatically over the past decades, with specialized formulations for different life stages, breeds, and health conditions. Behind that variety lies manufacturing infrastructure: batching systems that measure and mix ingredients, quality control processes that verify composition, packaging lines that fill bags and cans, and inventory systems that track raw materials and finished products.

The 2007 melamine crisis demonstrated what happens when this infrastructure fails. Wheat gluten and rice protein concentrate from Chinese suppliers had been adulterated with melamine, a nitrogen-rich industrial chemical that made the ingredients appear higher in protein than they were. The contaminated ingredients entered pet food supply chains and killed at least 4,000 dogs and cats, with thousands more sickened. Recalls ultimately covered more than 60 million containers across more than 100 brands.[13]

That crisis was physical contamination, not a cyberattack. But it illustrates the control points where digital systems now operate. Ingredient verification, formulation management, batch tracking, and quality testing are all digitized in modern facilities. Compromise those systems, and you can alter what goes into the food. Disable quality control, and contamination may pass undetected. Attack supply chain coordination, and products may not reach shelves or may arrive without proper traceability.

More recent incidents underscore the ongoing risks. In 2021, aflatoxin contamination—a mycotoxin from moldy corn—killed at least 70 dogs after passing through manufacturing processes that should have caught it.[14] In late 2024, a cat died after consuming raw frozen pet food contaminated with H5N1 avian influenza, with genetic sequencing confirming the match.[15] Each failure erodes the trust that pet owners place in commercial food systems.

I think about this every time I open a can of food for my kittens. Some of their wet food comes from New Zealand, and I've developed a new appreciation for what that means after researching this book. The ingredient list includes green-lipped mussels, something I'd never heard of until an aquaculture professional mentioned them during an interview. Now my cats eat them daily. My two prefer beef or chicken, but I introduce novel proteins occasionally: rabbit, boar, things that push beyond the usual to keep them from becoming picky eaters. Every time I crack open that can, I'm reminded of all the people, processes, and technology that brought this moment together: the mussels harvested off New Zealand's coast, the cattle raised somewhere I might never visit, the formulation software and quality systems, operational technology teams, and cold-chain logistics that delivered it safely to my kitchen. That might sound dramatic for cat food, but pets are family. Understanding where their food comes from matters to me, and it should matter to anyone who cares about the animals in their lives, whether in your home or on your plate.

From Field to Factory to Bowl

Understanding feed-industry cybersecurity requires understanding how feed is made. The process spans three distinct stages, each with its own technology and vulnerabilities.

It starts in fields: the corn, soybeans, wheat, and other crops that become feed ingredients are grown, harvested, and stored using the agricultural technologies explored in Chapters 6 and 7. From there, commodities move to feed mills, which are facilities that transform raw agricultural products into formulated feeds.

A modern feed mill is an industrial operation. Ingredients arrive by truck or rail and move into storage silos. Batching systems weigh and combine ingredients according to digital recipes. Mixers blend the components. Pelleting equipment compresses the mixture into dense, durable pellets that resist spoilage and are easier to handle. Quality control systems verify that the finished product meets specifications. Inventory management tracks everything from incoming ingredients to outgoing shipments.[16]

The heart of mill operations is the programmable logic controller (PLC): the same industrial computer that runs the food processing plants we examined in Chapter 5. PLCs control the batching sequences, manage the mixers, and regulate the pelleting equipment. Supervisory control and data acquisition systems provide operators with visibility across the entire operation: silo levels, equipment status, production rates, and quality metrics.

For pet food and some specialty feeds, there's an additional stage: production facilities that transform basic feed into finished consumer products. Here, ingredients undergo cooking, extrusion, canning, or other processes that create the kibble, wet food, or treats that end up on store shelves. These facilities look more like food-processing plants than agricultural operations, with all the technological dependencies that implies.

At each stage, formulation software guides operations. Feed formulation is a computationally intensive optimization problem: balancing nutritional requirements against ingredient costs, availability, and quality variability. The software calculates the least-cost combination that meets all specifications, and those digital recipes represent significant intellectual property. A company's competitive advantage may depend on formulation expertise that exists only in digital form.[17]

When the Systems Break

In March 2025, Astral Foods, South Africa's largest chicken producer and largest animal feed company, experienced a cyberattack that cascaded across its integrated operations.[18] Astral isn't just a poultry producer. Through its Tiger Animal Feeds subsidiary, the company manufactures feed for its own operations and external customers. When attackers struck, they didn't hit one function; they disrupted the entire chain. Feed mills couldn't operate normally. Processing lines halted. Deliveries failed. The company disclosed losses exceeding $1 million and issued a profit warning to investors.[19] No ransomware group claimed responsibility, and Astral confirmed that no customer or supplier data was compromised. But the incident demonstrated how integrated agribusiness creates concentrated risk. A single intrusion multiplied across business functions from feed manufacturing to livestock production to food processing.

Similar patterns have emerged globally. Between 2021 and 2022, five major animal nutrition companies in northwestern France fell victim to cyberattacks, a pattern suggesting systematic targeting of a regional industry cluster rather than random opportunity.[20] In March 2022, a cyber event shut down 13 dairy processing plants across the United States, forcing milk disposal and threatening school lunch supplies.[21] In June 2025, the largest dairy cooperative in the United States disclosed that multiple manufacturing facilities had experienced ransomware attacks.[22]

The agricultural sector has been described as vulnerable precisely because protection has lagged behind other industries. Cybersecurity spending in North American agriculture accounts for roughly 3 percent of total cybersecurity investment, a fraction of what banks, hospitals, and energy companies invest. The defenses that exist elsewhere haven't materialized for the systems that feed us.[23]

Feed Safety, Quality, and Compliance

Feed safety isn't separate from food safety; it's the foundation. What animals eat determines the safety of what humans ultimately consume. This principle drives regulatory frameworks worldwide.

The Codex Alimentarius—the international food standards body established by the Food and Agriculture Organization of the United Nations (FAO) and the WHO—developed specific guidance for animal feeding called the Code of Practice on Good Animal Feeding that covers prioritizing hazards in feed and provides guidelines on risk assessment.[24] These documents establish the principle that feed safety systems must cover the entire chain from ingredient sourcing through manufacturing to feeding.

The European Union regulates feed through multiple frameworks. The Feed Hygiene Regulation requires hazard analysis and critical control points (HACCP) for feed manufacturers.[25] Maximum limits are in place for contaminants, including heavy metals, dioxins, mycotoxins, and pesticide residues. The Rapid Alert System for Food and Feed tracks notifications across member states; three out of four notifications concern microbial contamination.[26]

Coccidiostats occupy a peculiar regulatory position. In most contexts, compounds that affect living organisms in feed would be classified as veterinary medicines. But in the EU, coccidiostats are authorized as feed additives, with their safety evaluated by the European Food Safety Authority rather than the European Medicines Agency. Each authorization specifies maximum and minimum dosages, animal categories, maximum residue limits in food products, and withdrawal periods where necessary.[27]

In the United States, medicated feeds require compliance with the Veterinary Feed Directive, which mandates veterinary oversight for certain antimicrobials. Feed mills producing medicated feeds must maintain detailed records, verify prescriptions, and ensure proper labeling.[28]

Quality systems overlay these regulatory requirements. The Safe Feed/Safe Food certification program establishes standards for feed manufacturing, with third-party audits verifying compliance. Documentation requirements create paper trails, increasingly digital trails, that enable traceability when problems emerge.[29]

All these compliance functions depend on information systems. Formulation records, batch documentation, quality test results, and shipping manifests are the data that demonstrate regulatory compliance in databases and software systems. Compromise those systems, and you compromise the ability to demonstrate that feed was manufactured correctly and safely.

The Insects Are Changing Everything

Three insects are transforming the feed industry: black soldier flies, crickets, and mealworms. Reared at commercial scale, they convert organic waste into protein-rich feed ingredients, creating circular economy loops that address both waste disposal and ingredient sourcing.[30]

The nutritional case is compelling. Black soldier fly larvae offer amino acid profiles similar to fish meal, making them valuable for aquaculture feeds. They contain lauric acid, which supports immune function in poultry. Cricket meals provide complete protein. Mealworms offer high fat content suitable for certain applications.[31] For an industry constantly seeking protein sources that don't compete with human food, insects represent a genuinely novel solution.

But insect agriculture is intensely dependent on technology. These aren't outdoor operations—they occur in controlled-environment facilities where temperature, humidity, and airflow must remain within narrow parameters. Black soldier flies require specific conditions for each life stage. Too hot, and colonies collapse. Too humid, and disease spreads. The parallels to vertical farming are striking: both depend on digital control systems to manage environments that are too precise for manual operation.

The expertise required to rear insects at a commercial scale has unexpected origins.

Screwworms: From Eradication to Innovation

In a facility just east of the Panama Canal, workers breed flies. Not a few flies, but approximately 100 million per week. More than a billion per year.[32] The flies are New World Screwworms, a parasite whose larvae burrow into the flesh of living animals. Before eradication efforts began, screwworms caused devastating losses across the Americas. Cattle with open wounds were literally eaten alive. Newborn animals were particularly vulnerable because the umbilical cord provided an entry point. Wildlife suffered alongside livestock. Economic losses ran into hundreds of millions of dollars annually.[33]

The solution was elegant and counterintuitive: breed massive numbers of screwworms, sterilize the males with radiation, and release them to mate with wild females. Because female screwworms mate only once in their three-week lifespan, matings with sterile males produce no offspring. Saturate an area with enough sterile males, and the population collapses.[34]

The United States was declared screwworm-free in 1966. The barrier has since been pushed south through Mexico and Central America.[35] Today, the Panama facility, jointly operated by the U.S. and Panamanian governments through the Commission for the Eradication and Prevention of Screwworm (COPEG), releases sterile flies at the Colombian border to prevent northward spread.[36]

This sterile insect technique represents one of the most successful pest eradication programs in agricultural history. And the infrastructure required precisely controlled environments for mass insect rearing, careful management of temperature and humidity, and quality systems ensuring consistent production directly parallels what commercial insect agriculture requires.[37]

The techniques developed to eliminate a devastating livestock pest now inform an industry producing feed ingredients. The controlled environment expertise, the mass-rearing protocols, and the quality control systems all transfer from eradication to production. But recent events underscore the vulnerability of both applications. In 2023, screwworm detections in Panama spiked from an average of 25 cases per year to more than 6,500. The pest spread northward into Mexico, prompting emergency measures, including new sterile fly-production facilities in Texas.[38] If the barrier fails, screwworms could reestablish across North America, with catastrophic consequences for livestock producers and wildlife alike.

The same digital monitoring and coordination systems that maintain the screwworm barrier support emerging insect agriculture. Compromise either, and the deadly consequences extend far beyond any individual facility.

Supply Chain Fragility

Feed doesn't help animals if it's sitting in a mill. It must reach farms, and that logistics chain introduces its own digital dependencies. Modern feed delivery relies on routing and scheduling software that optimizes truck movements across delivery networks: GPS tracking monitors fleets. Weighing systems verify loads. Digital records enable regulatory compliance and traceability.[39]

Cross-contamination represents a particular concern. Medicated feeds must be segregated from nonmedicated feeds. Flushing protocols between loads prevent residue transfer; remember the ionophore toxicity risk for horses. Delivery errors have caused illness and death among livestock. When scheduling systems go offline, the coordination that prevents these errors becomes much harder to maintain.

When the BlackMatter ransomware group hit the NEW Cooperative in September 2021, an incident we examined in Chapters 6 and 7 for its crop-side impacts, feed scheduling was among the systems affected. The Iowa cooperative managed feed distribution for livestock operations across the Midwest. Animals waited for deliveries that couldn't be coordinated through normal systems. Just-in-time inventory approaches work efficiently under normal conditions but create immediate pressure when systems fail.[40]

Ingredient supply chains extend globally. Feed mills depend on soybean meal from South America, fish meal from Peru and Chile, and premix ingredients from suppliers concentrated in specific regions. The United States depends on China for 78 percent of its vitamin imports, with individual vitamins, including biotin, sourced exclusively from Chinese suppliers.[41] The vulnerability isn't about any single country; it's about sole-source dependencies that create single points of failure across the entire industry.

What It Teaches Us: The Foundation Beneath Our Food

Before animals can feed us, they must eat. The systems that formulate, manufacture, and deliver their food have become digital, from the optimization algorithms that calculate least-cost rations to the PLCs that control feed mill equipment to the logistics software that coordinates deliveries. When those systems fail, animals suffer, including us humans. The biological clock doesn't pause while we troubleshoot. Poultry can't wait for feed deliveries to resume. Dairy cows can't postpone their nutritional requirements. And the working dogs and horses that make farms function need their meals on schedule.

The people who work in this industry understand these stakes intimately. They don't need to be convinced that cybersecurity matters. What they need are tools and resources proportionate to the risks they face: resources that have flowed to banking, healthcare, and energy but have never materialized for agriculture.

From the kibble in your dog's bowl to the integrated operations feeding millions, the feed industry connects every corner of animal agriculture. And attackers have noticed. South Africa, France, the American Midwest, British retailers, Canadian processors—the incidents span continents and business models. The pattern is unmistakable: an industry essential to food production, increasingly dependent on digital systems, and inadequately protected.

Moving Forward: Recognition Without Resources

Governments are beginning to acknowledge this vulnerability. The United Kingdom's Food Security Digest 2025 includes a dedicated section on cybersecurity for the first time. Following attacks on major British retailers in April 2025, the document explicitly positions food as "part of the UK Critical National Infrastructure."[42]

The UK's National Cyber Security Centre reported that nearly half of all incidents requiring its support during its 2024–2025 reporting year (September 2024 through August 2025) were classified as "nationally significant," more than double the proportion from the year before.[43] Separate surveys found that 43 percent of British businesses reported experiencing some form of cyber breach or attack in the previous 12 months.[44]

The European Union's NIS2 Directive extends cybersecurity requirements to food production and processing, treating these sectors as essential and requiring enhanced protection.[45] In the United States, proposed legislation would direct the USDA to assess vulnerabilities in the feed supply chain, including ingredient-sourcing dependencies.[46]

Academic researchers have begun mapping the threat landscape systematically. A 2025 study provides what may be the first cybersecurity roadmap specifically for dairy and poultry industries, warning that "compromised digital systems disrupt critical farm operations, including milking routines, feed formulations, and climate control, profoundly impacting animal health, productivity, and consumer trust." Recognition is growing, but resources have not followed.[47]

Animals don't exist in isolation; they interact with the environment in complex ways. Their waste becomes a resource or a pollution problem, depending on how it's managed. The crops that feed them require sustainable cultivation. Climate change affects both animal welfare and the feed ingredients they depend on. The next chapter examines agro-environmental management: how agricultural operations manage their environmental footprint through digital systems that are themselves becoming targets.

Notes

1. International Feed Industry Federation. (2026). *Annual Report 2024/2025.* **https://annualreport.ifif.org**
2. Ritchie, H., Rosado, P., & Roser, M. (2024). "Agricultural production." *Our World in Data.* **https://ourworldindata.org/agricultural-production**; using data from Poore, J. & Nemecek, T. (2018). "Reducing food's environmental impacts through producers and consumers." *Science*, 360(6392), 987–992.
3. Greenpeace European Unit. (2019). *Feeding the Problem: The Dangerous Intensification of Animal Farming in Europe.* **https://www.greenpeace.org/eu-unit/issues/nature-food/1803/feeding-problem-dangerous-intensification-animal-farming**
4. FAO. (2023). "Livestock and the environment." **https://www.fao.org/livestock-environment/en**
5. National Research Council. (2001). *Nutrient Requirements of Dairy Cattle: 7th Revised Edition* (National Academies Press). **https://doi.org/10.17226/9825**
6. Williams, R. B. (2005). "Quantification and estimation of the welfare losses to broiler chickens caused by coccidiosis." *Revue Scientifique et Technique (International Office of Epizootics)*, 24(1), 327–338.
7. Pereira, A. (2025). "The price of coccidiosis: understanding its economic impact on poultry farms." *Grand Valley Fortifiers.* Poultry Grist. April 24, 2025. **https://www.grandvalley.com/poultry-grist/the-price-of-coccidiosis-understanding-its-economic-impact-on-poultry-farms**
8. EW Nutrition. (2025). "Coccidiostats in the European Union: challenges and perspectives." **https://ew-nutrition.com/us/coccidiostats-european-union-challenges-perspectives**
9. National Research Council. (2016). *Nutrient Requirements of Beef Cattle:* 8th Revised Edition (National Academies Press). **https://www.nationalacademies.org/read/19014**
10. Nutrena Animal Nutrition. (2026). "Essential nutrition for working farm dogs." **https://nutrenaworld.com/blog/dogs-cats/essential-nutrition-for-working-farm-dogs**
11. Mad Barn. (2026). "Ionophore toxicosis in horses: causes, symptoms and treatment." **https://madbarn.com/ionophore-toxicosis-in-horses**
12. American Pet Products Association. "Industry trends and stats." **https://americanpetproducts.org/industry-trends-and-stats**
13. FDA. (2025). "Import alert 99-29." **https://www.accessdata.fda.gov/cms:ia/importalert_267.html**
14. FDA. (2021). "FDA Alert: certain lots of Sportmix pet food recalled for potentially fatal levels of aflatoxin." **https://www.fda.gov/animal-veterinary/outbreaks-and-advisories/fda-alert-certain-lots-sportmix-pet-food-recalled-potentially-fatal-levels-aflatoxin**
15. Oregon Department of Agriculture. (2024). "Morasch Meats voluntary recall of feline raw pet food due to HPAI." **https://apps.oregon.gov/oregon-newsroom/OR/ODA/Posts/Post/morasch-meats-voluntary-recall-feline-raw-pet-food-hpai**
16. American Feed Industry Association. (2025). "Feed manufacturing technology." **https://fmt.afia.org/event-data**
17. Pesti, G. M., & Miller, B. R. (1993). *Animal Feed Formulation: Economics and Computer Applications* (Van Nostrand Reinhold).
18. Greig, J. (2025). "Cyberattack causes delays for South Africa's largest chicken producer." *The Record.* **https://therecord.media/cyberattack-delays-south-african-chicken-producer**
19. Astral Foods Limited. (2025). "Voluntary trading update and initial trading statement." *Sharenet.* **https://www.sharenet.co.za/v3/sens_display.php?tdate=20250324091000&seq=34**
20. Feedinfo. (2023). "INSIGHT: cyber threats in animal agriculture industry still loom." **https://www.feedinfo.com/our-content/insight-cyber-threats-in-animal-agriculture-industry-still-loom/338497**

21. Gardizy, A. (2022). "Hackers hit Hood. Dairy shut down milk production this week after "cyber security event." *Boston Globe.* **https://www.bostonglobe.com/2022/03/18/business/school-milk-could-be-short-supply-after-hood-plants-hit-by-cyber-event**
22. Bohnert, K. (2025). "Multiple DFA manufacturing plants experienced ransomware attack." *Dairy Herd Management.* **https://www.dairyherd.com/news/business/multiple-dfa-manufacturing-plants-experienced-ransomware-attack**
23. Malone, T. (2025). "Why cybersecurity must be a top priority for agribusiness in 2025." *Purdue University Center for Food and Agricultural Business.* **https://agribusiness.purdue.edu/2025/02/20/why-cybersecurity-must-be-a-top-priority-for-agribusiness-in-2025**
24. FAO and IFIF. (2010). *Good Practices for the Feed Industry: Implementing the Codex Alimentarius Code of Practice on Good Animal Feeding.* **https://www.fao.org/4/i1379e/i1379e.pdf**
25. European Parliament and Council of the European Union. (2005). "Regulation (EC) No 183/2005 of 12 January 2005 laying down requirements for feed hygiene." **https://www.legislation.gov.uk/eur/2005/183/body/2020-01-31?view=plain**
26. European Federation of Feed Manufacturers (FEFAC). "Feed safety standards." **https://fefac.eu/priorities/feed-safety/feed-safety-standards**
27. European Commission. (2008). "Report from the Commission to the Council and the European Parliament on the use of coccidiostats and histomonostats as feed additives." *EUR-Lex.* **https://eur-lex.europa.eu/legal-content/EN/TXT/HTML/?uri=CELEX:52008DC0233**
28. FDA. (2015). "Veterinary feed directive." *Federal Register* 80(106): 31707–31735. **https://www.federalregister.gov/documents/2015/06/03/2015-13393/veterinary-feed-directive**
29. American Feed Industry Association. (2026). "Safe Feed/Safe Food certification program." **https://www.afia.org/resources/safe-feed-safe-food-certification**
30. FAO. (2021). Looking at Edible Insects from a Food Safety Perspective: Challenges and Opportunities for the Sector. **https://openknowledge.fao.org/server/api/core/bitstreams/33de5ff0-3b21-4108-98bc-d2f6e190e992/content#:~:text=Page%20v-,PREFACE,producers%2C%20as%20well%20as%20consumers%20**
31. Zulkifli, N. F. N. M., Seok-Kian, A. Y., Seng, L. L., Mustafa, S., Kim, Y. S., & Shapawi, R. (2022). Nutritional value of black soldier fly (Hermetia illucens) larvae processed by different methods. *PLoS One*, 17(2), e0263924. **https://doi.org/10.1371/journal.pone.0263924**
32. Newsroom Panama. (2025). "The United States releases millions of flies over Panama every week." **https://newsroompanama.com/2025/05/10/why-the-united-states-releases-millions-of-flies-over-panama-every-week**
33. USDA Animal and Plant Health Inspection Service. (n.d.). "Eradicating new world screwworm with sterile insect technique." **https://www.aphis.usda.gov/sites/default/files/factsheet-eradicating-nws-sit.pdf**
34. USDA Animal and Plant Health Inspection Service. (2017). "New world screwworm: Ready reference guide—sterile insect response." **https://www.aphis.usda.gov/sites/default/files/nws_rrg_sterileinsectresponse.pdf**
35. National Geographic. (2025). "Inside the program that breeds billions of screwworm flies each year." **https://www.nationalgeographic.com/animals/article/north-american-screwworm-barrier**; USDA Animal and Plant Health Inspection Service. (2026). "New world screwworm." **https://www.aphis.usda.gov/livestock-poultry-disease/cattle/ticks/screwworm**
36. Panama-United States Commission for the Eradication and Prevention of Screwworm (COPEG). (n.d.). "Situación Actual." **https://develop1.webstudiopanama.com/copeg/en/situacion-actual**
37. USDA Animal and Plant Health Inspection Service. (2026). "New world screwworm."

38. USDA Animal and Plant Health Inspection Service. (2025). "New world screwworm domestic readiness and response policy initiative." **https://www.usda.gov/sites/default/files/documents/nws-visit-policy-brief.pdf**; American Veterinary Medical Association. (2025). "USDA unveils Texas screwworm facility, eradication strategy." **https://www.avma.org/news/usda-unveils-texas-screwworm-facility-eradication-strategy**
39. Feed Strategy. (2024). "5 grain and feed logistics platforms." **https://www.feedstrategy.com/feed-mill-of-the-future/article/15667628/5-grain-and-feed-logistics-platforms**
40. Kass, D. H. (2021). "New cooperative ransomware attack timeline: status updates." *MSSP Alert*. **https://www.msspalert.com/news/new-cooperative-ransomware-attack-timeline-status-updates**
41. Institute for Feed Education and Research. (2025). Strategic Assessment on the Impact of Vitamin and Amino Acid Supply Chain Disruptions on U.S. Food Security. **https://www.ifeeder.org/research/analyzing-the-vitamin-and-amino-acid-supply-chain**
42. Department for Environment, Food & Rural Affairs. (2025). "United Kingdom food security digest 2025." **https://www.gov.uk/government/statistics/united-kingdom-food-security-digest-2025**
43. National Cyber Security Centre. (2025). *NCSC Annual Review 2025*. **https://www.ncsc.gov.uk/collection/ncsc-annual-review-2025**
44. Poireault, K. (2025). "UK: 130% Spike in 'nationally significant' cyber incidents." *Infosecurity Magazine*. **https://www.infosecurity-magazine.com/news/uk-ncsc-spike-national-cyber**
45. Finnish Food Authority. (2025). "Cybersecurity Directive (NIS2) in the food industry." **https://www.ruokavirasto.fi/en/foodstuffs/food-sector/setting-up-a-food-business/kyberturvallisuusdirektiivi-nis2-elintarviketoimialalla**
46. American Feed Industry Association. (2025). "AFIA applauds Senate introduction of Innovative FEED Act." **https://www.afia.org/news/news-release-items/afia-applauds-senate-introduction-of-innovative-feed-act**
47. Neethirajan, S. (2025). "Safeguarding digital livestock farming: a comprehensive cyber security roadmap for dairy and poultry industries." *Frontiers in Big Data*, 8, 1556157. **https://www.frontiersin.org/journals/big-data/articles/10.3389/fdata.2025.1556157/full**

CHAPTER 15

Agro-Environmental Management: Sustainable Farming, Climate-Resilient Practices, and Waste

The Wheelbarrow Isn't Enough Anymore

A regenerative farm in the American Midwest doesn't just practice cover cropping and reduced tillage. It monitors soil organic carbon with sensors, tracks microbial activity through laboratory analysis, and uses satellite imagery to measure soil health outcomes across seasons. A climate-smart greenhouse operation in the Netherlands manages temperature, humidity, carbon dioxide supplementation, and lighting through integrated control systems responding to real-time environmental data. A rancher in Montana checks an app instead of reaching for a rifle when wolves approach cattle. This is what sustainable farming looks like now. Not a return to preindustrial methods, but a forward leap into data-driven agriculture that happens to align with ecological principles. The wheelbarrow isn't going away, but it's no longer enough. You cannot farm sustainably at scale without technology, and that technology must be secure.

The World Business Council for Sustainable Development (WBCSD) now tracks 11 measurable outcomes for regenerative agriculture, from soil carbon sequestration to water pollution reduction to farmer well-being.[1] Measuring those outcomes requires data collection infrastructure: sensors, monitoring systems, cloud platforms, and analytics tools.

The World Economic Forum's 2023 Smarter Climate Farmers Challenge awarded 12 innovators transforming global agriculture, and nearly all depend on connected technology. AMINI builds environmental data infrastructure using satellites, artificial intelligence (AI), and machine learning for African smallholders. Dvara E-Registry uses remote sensing to assess crop health in India. Instacrops provides what they call "Precision Agriculture as a Service," virtual advisors translating processed data into real-time recommendations for farmers across Latin America.[2]

Regenerative agriculture currently covers only about 1.5 percent of the 5.3 billion hectares of agricultural land globally. Scaling it will require somewhere between $210 billion and $240 billion in cumulative investment by 2030, according to the Food and Land Use Coalition, much of it flowing into technology and monitoring infrastructure.[3] The question nobody seems to be asking is, who secures all of this?

Conservation Partnerships: When Technology Enables Coexistence

In the northern Rocky Mountains, a quiet transformation is reshaping how ranchers and predators share the landscape. GPS collars on wolf packs feed location data through platforms like EarthRanger, developed by the Allen Institute for AI. When a pack approaches cattle, the system sends SMS alerts to ranchers, who can then move livestock, deploy range riders, or activate deterrents. Virtual fencing systems from companies like Nofence (Norway) and Vence (now owned by Merck Animal Health) take this further. Solar-powered GPS collars on cattle deliver audio cues and mild electrical signals when animals approach virtual boundaries, keeping livestock away from wolf dens, sensitive riparian zones, and wildlife corridors without a single strand of barbed wire blocking wildlife movement. Conservation Northwest has piloted these systems in Washington State to protect both wolves and the ranching operations they share landscapes with.[4] These partnerships took years to build. Conservation organizations invested in relationships with ranching communities, earning trust one conversation at a time. Technology simply makes coexistence practical.

Across Africa, similar technology enables humans and elephants to share agricultural landscapes. EarthRanger integrates GPS collar data from elephant herds with virtual geofences around farms, sending SMS and WhatsApp alerts to farmers when herds approach their crops. Programs run by the International Fund for Animal Welfare, Save the Elephants, and African People & Wildlife have deployed these systems across Zimbabwe, Malawi, Zambia, Kenya, and Tanzania.[5]

In West Bengal, India, researchers tested AI-powered cameras that detect approaching elephants and send real-time alerts. During the 2023–2024 study period, the system achieved 266 detections with an average alert time of just 42 seconds. Response teams arrived in under 20 minutes on average. The results: zero human fatalities during the study period, compared to four deaths and five serious injuries in the prior two years.[6]

Kenya's Big Life Foundation has deployed drones with thermal cameras to guide elephant herds away from farms using sounds only they can hear, causing the elephants to turn back from the buzzing they associate with bees. "Beehive fences," developed through a nine-year Oxford University and Save the Elephants study, achieve 86 percent deterrent effectiveness during peak crop seasons while providing honey income for farmers.[7] In Upper Kitete, Tanzania, the approach has reduced human-elephant conflict by 91 percent over six years.

But here's the question these programs don't always address: what happens if the app gets hacked? If the rancher stops getting alerts and loses cattle, does the conservation partnership collapse? If false alerts send farmers chasing phantom elephants, will they abandon the program within a season? Conservation organizations asking communities to trust their platforms have a duty to secure them: two-factor authentication, strong passwords, and encrypted communications. One breach could unravel decades of relationship-building and endanger both humans and animals.

Adapting to Climate: When Technology Becomes Survival

Climate change isn't a future threat for agriculture; it's a present reality requiring daily technological response. The systems enabling adaptation are sophisticated, interconnected, and entirely dependent on reliable operation.

In California's wine country, frost protection has evolved far beyond smudge pots and prayer. Wind machines equipped with temperature sensors activate automatically when readings drop toward critical damage thresholds, mixing warmer air from above the inversion layer downward around vulnerable vines. Bouchaine Vineyards installed Cisco industrial Internet of Things (IoT) sensors throughout its operation in 2021, automatically collecting and visualizing data on temperature, soil humidity, and wind speed.[8] Automated overhead sprinklers create protective ice coatings on buds counterintuitively: the latent heat released as water freezes keeps plant tissue just warm enough to survive. These systems must work perfectly at 3:00 AM on the coldest night of spring. A single failure sensor misread, a network outage, or a compromised temperature threshold can mean losing an entire vintage.

Heat stress presents the mirror-image challenge. The U.S. dairy industry loses an estimated $1.5 billion annually to heat stress through decreased milk production, reduced reproductive efficiency, and, in extreme cases, fatalities.[9] Precision livestock farming (PLF) technologies now enable continuous real-time monitoring. On-animal sensors track respiration rate, body temperature, location, and activity. Barn environmental sensors monitor temperature, humidity, ammonia levels, and carbon dioxide concentrations. AI systems detect early signs of heat stress through changes in rumination patterns and lying behavior, triggering automated ventilation and cooling systems before animals suffer.

Research from the University of Reading modeled dairy housing conditions during the 2021 UK heatwave and projected forward to the 2080s. Current heatwave conditions already push milking parlors to "Level 3: Severe" heat stress. Projections for 2080s heatwaves indicate that dairy facilities will experience "Level 4: Emergency" conditions day and night for extended periods and that current mitigation strategies will be insufficient.[10] The systems protecting livestock today will need to be more sophisticated, more reliable, and better secured tomorrow.

When outdoor conditions become too unpredictable, agriculture moves inside. A commercial greenhouse is essentially a building-sized IoT installation: climate control systems managing temperature, humidity, and carbon dioxide; hydroponic systems monitoring water quality, pH levels, nutrient concentrations, and flow rates; LED lighting tuned to specific growth stages; and AI-driven analytics optimizing every parameter. The Netherlands achieves the highest tomato yields per hectare globally through exactly this kind of tech-intensive controlled environment agriculture (CEA).[11]

Singapore, which imports more than 90 percent of its food, has committed to producing 30 percent of its nutritional needs locally by 2030 through its "30 by 30" initiative, and vertical farming is central to that strategy. Companies like Sky Greens use stacked systems producing up to 10 times more food per square meter than traditional farms.[12] Vertical farms can achieve up to 98 percent water savings compared to traditional agriculture.[13]

These aren't farms with technology added; they are technological systems that happen to grow food. A research review published in the journal *Sensors* notes explicitly that "increasing reliance on IoT networks exposes greenhouse infrastructures to cybersecurity vulnerability, including unauthorized access, data manipulation, and service disruption."[14] The academic literature acknowledges the risk. But the industry has yet to address it systematically.

Water, Pollinators, and Waste: The Supporting Infrastructure

Agriculture consumes roughly 70 percent of global freshwater withdrawals.[15] In an era of drought and scarcity, technology enables precision. Soil moisture sensors provide real-time data on actual field conditions; weather integration aligns irrigation schedules with rainfall events, and IoT platforms enable remote monitoring and control from anywhere with cellular coverage. Research from Oklahoma State University found that soil moisture controllers achieved 72 percent irrigation savings compared to manual scheduling and still delivered 34 percent savings even during drought conditions.[16] Studies supported by the USDA Natural Resources Conservation Service demonstrated 30 percent water savings while maintaining the same marketable yields and crop quality.[17] Mediterranean and South Asian countries, the regions most vulnerable to drought, depend increasingly on smart irrigation for food security.

Bees pollinate approximately 70 of the 100 crop species that feed 90 percent of the world.[18] Colony Collapse Disorder (CCD) and other threats have devastated populations, making the surviving colonies precious resources. Connected beehive monitoring now tracks temperature, humidity, and acoustic signatures inside hives. Weight sensors detect nectar flow and food consumption. AI analyzes sounds to identify colony stress, queen loss, or swarming before beekeepers notice anything wrong. BeeHero, using Silicon Labs IoT sensors with Bluetooth connectivity, reports that its technology can reduce colony losses by up to 30 percent.[19] ApisProtect in Ireland uses machine learning to provide actionable insights and alerts. The bees themselves become environmental sentinels; their health reflects the health of the ecosystem around them. But that data flows through connected systems, cloud platforms, and mobile applications.[20]

Even waste management has gone digital. Anaerobic digesters converting manure and crop residue to energy and fertilizer require environmental monitoring to ensure optimal digestion temperatures and safe operation. Control systems managing temperature, pressure, and gas collection must operate reliably because methane accumulation presents genuine safety hazards beyond the agricultural context.[21]

Moving Forward: The Responsibility Gap

All this technology requires protection. But whose job is it?

Conservation organizations and platform providers bear responsibility when they ask communities to trust their systems. If you're asking ranchers to depend on your wolf-tracking app, you need to secure it. If farmers rely on your elephant alert system, you need to protect it. Security failures don't just compromise data; they compromise partnerships built over decades, putting lives and livelihoods at risk.

Technology vendors face their own questions. Are greenhouse management systems, irrigation controllers, and beehive monitors designed with security as a foundation, or is security bolted on as an afterthought by a third party? Who provides updates, patches, and ongoing maintenance? When AgTech startups fold or pivot (and many do), what happens to the systems farmers have come to depend on?

Farmers and ranchers become system administrators by default, not by training. The person managing a virtual fencing app is most likely managing livestock, equipment, crops, and finances. Nobody teaches sustainable farmers about phishing attacks that could compromise their grazing boundaries or credential security for the platforms monitoring their soil health. The gap is widening. Conservation programs install technology without security assessments. Environmental consultants focus on compliance, not cybersecurity. Agricultural extension services are only beginning to address digital security, and their resources are stretched thin already. The workforce managing these systems needs training that largely doesn't exist.

Climate-smart agriculture, regenerative farming, and conservation technology all represent what's possible when farmers work with their environment and community. The innovations are real, and the outcomes are measurable, but these systems exist only because of connected infrastructure. Someone needs to protect that infrastructure. Right now, the answer to "Who?" remains unclear, and the gap between innovation and security grows wider each season.

Chapter 16 moves into food science and innovation, where laboratory research developing tomorrow's crops and novel foods faces its own questions about data integrity, intellectual property protection, and research security.

Notes

1. World Business Council for Sustainable Development. (n.d). "A global framework for regenerative agriculture." **https://www.wbcsd.org/actions/a-global-framework-for-regenerative-agriculture**
2. la Luz Gonzalez, B. (2023). "12 climate-smart technologies that could transform the way we grow food." *World Economic Forum*. **https://www.weforum.org/stories/2023/10/12-climate-smart-technologies-that-could-transform-the-way-we-grow-food**
3. Zakomirnyi, A. (2025). "How regenerative agriculture is reshaping our food system and the path forward." *Roland Berger*. **https://www.rolandberger.com/en/Insights/Publications/Regenerative-agriculture.html**
4. Conservation Northwest. "Range Rider pilot project." **https://conservationnw.org/our-work/wildlife/range-rider-pilot-project**
5. EarthRanger. (2026). "Implementation details from International Fund for Animal Welfare and Save the Elephants program reports." **https:// www. earthranger. com**
6. Rastogi, S., Bhattacharyya, S., Vimal, K., Dinerstein, E., Shah, W. A., Nagano, Y., Hirari, S., Gulick, S., Lee, A., & Sekar, N. (2025). Promoting human–elephant coexistence through integration of AI, real-time alerts, and rapid response. *Conservation Science and Practice*, 7(12), e70186. **https://doi.org/10.1111/csp2.70186**
7. King, L. E., Arkajak, R., Arkajak, J., Ceccarelli, V., Sluiter, L., Doornwaard, S. M., Praet, V., Owen, D., & Matteson, K. (2020). Beehive fences as a sustainable local solution to human-elephant conflict in Thailand. *Conservation Science and Practice*, 2(10), e260. **https://doi.org/10.1111/csp2.260**
8. Kania, S. (2024). "Vineyards of the 21st century: IT solutions for winemaking." *Webmakers*. **https://webmakers.expert/en/blog/vineyards-of-the-21st-century-it-solutions-for-winemaking**
9. Kim, Y., Choi, C., Brounts, S., & Van Os, J. (2025). "Mitigating heat stress in dairy cattle using a physiological sensingbehavior analysis-microclimate control loop." Poster. *CPS Virtual Organization*. **https://cps-vo.org/sites/cps-vo.org/files/2025-04/Choi_2038452-CPS2025-Poster_0.pdf**
10. Liu, C. & Cao, Y. (2025). Heat stress monitoring, modelling, and mitigation in a dairy cattle building in reading, UK: impacts of current and projected heatwaves. *Building and Environment*, 260, 112527. **https://centaur.reading.ac.uk/122537/9/1-s2.0-S036013232500527X-main.pdf**
11. Eurostat. (2024). "Agri-environmental indicator - energy use." Statistics explained. **https://ec.europa.eu/eurostat/statistics-explained/index.php?title=Agri-environmental_indicator_-_energy_use#:~:text=A%20widely%20adopted%20denominator%20for,with%202022%E2%80%93%20see%20Figure%204**
12. Funk, B. (2025). "The future of vertical farming: transforming global food production." *Eden Green Technology*. **https://www.edengreen.com/blog-collection/future-of-vertical-farming**
13. World Economic Forum. (2023). "How vertical farming can save water and support food security." **https://www.weforum.org/stories/2023/06/how-vertical-farming-can-save-water-and-support-food-security**
14. Bicamumakuba, E., Reza, M. N., Jin, H., Samsuzzaman, Lee, K. H., & Chung, S. O. (2025). "Multi-sensor monitoring, intelligent control, and data processing for smart greenhouse environment management." *Sensors*, 25(19), 6134. **https://doi.org/10.3390/s25196134**
15. UNESCO. (2024). "Water for prosperity and peace: statistics." **https://www.unesco.org/reports/wwdr/en/2024/s**

16. Oklahoma State University, "Smart irrigation technology: controllers and sensors." HLA-6445, Oklahoma Cooperative Extension Service. **https://extension.okstate.edu/fact-sheets/print-publications/hla/smart-irrigation-technology-controllers-and-sensors-hla-6445.pdf**
17. Dong, Y., Werling, B., Cao, Z., & Li, G. (2024). "Implementation of an in-field IoT system for precision irrigation management." *Frontiers in Water*, 6, 1353597. **https://doi.org/10.3389/frwa.2024.1353597**
18. Katumo, D. M., Liang, H., Ochola, A. C., Lv, M., Wang, Q. F., & Yang, C. F. (2022). "Pollinator diversity benefits natural and agricultural ecosystems, environmental health, and human welfare." *Plant Diversity*, 44(5), 429–435. **https://doi.org/10.1016/j.pld.2022.01.005**.
19. Silicon Labs. (2022). "BeeHero uses IoT sensor technology to save bees." **https://www.silabs.com/applications/case-studies/beehero-uses-iot-sensor-technology-to-save-bees**
20. Irish Tech News. (2019). "ApisProtect to double bee monitoring in 2019." **https://irishtechnews.ie/apisprotect-to-double-bee-monitoring-in-2019**
21. EPA. (2020). *Anaerobic Digester/Biogas System Operator Guidebook: A Guidebook for Operating Anaerobic Digestion/Biogas Systems on Farms in the United States*. EPA 430-B-20-003 (Office of Water, AgSTAR Program). **https://www.epa.gov/agstar/anaerobic-digesterbiogas-system-operator-guidebook**

CHAPTER 16

Food Science and Innovation

Novel Foods, Sensory Research, and Future-Forward Packaging

The UPSIDE facility in Emeryville, California, looks nothing like a farm. No fields stretch to the horizon. No animals graze, cluck, or moo. Instead, stainless-steel bioreactors stand in climate-controlled rooms, tended by scientists in lab coats who monitor growth curves on screens displaying temperature, pH levels, and dissolved oxygen concentrations. What emerges from these vessels is chicken: real chicken, cellularly identical to what you'd find at any grocery store, grown without a single bird.[1]

UPSIDE Foods received clearance from the United States Food and Drug Administration (FDA) in November 2022, confirming that its cultivated chicken was safe for human consumption. Seven months later, the United States Department of Agriculture (USDA) granted inspection approval, making UPSIDE and its competitor, GOOD Meat, the first companies authorized to sell cultivated meat to American consumers.[2] Singapore had approved GOOD Meat's product two years earlier, in December 2020, making the city-state the first jurisdiction anywhere to permit the commercial sale of meat grown from cells.[3]

This chapter ventures into territory that may feel like science fiction: meat produced without animals, 3D printers extruding personalized nutrition one layer at a time, packaging that dissolves safely in your mouth, and artificial intelligence (AI) systems designing flavors before any human ever tastes them. It might feel like the future, but everything described here exists today. These technologies operate in laboratories, pilot facilities, and increasingly, commercial production. They represent where the food industry is heading.

The cybersecurity implications differ fundamentally from those of the processing plants and distribution networks examined in earlier chapters. Here, the assets requiring protection include intellectual property worth hundreds of millions of dollars, research data accumulated over years, consumer preference databases revealing intimate details about taste and behavior, and prototype technologies that haven't yet developed security maturity. This sector represents food's future, and the future needs protection before it arrives.

Defining the Terrain

Before we explore specific technologies, some definitions will help orient any readers unfamiliar with this domain. *Food science* applies chemistry, biology, physics, and engineering to understand how food behaves, why bread rises, how milk spoils, what makes chocolate melt on your tongue but not in your hand, etc. Food scientists work in laboratories, but their discoveries shape every product on grocery shelves. *Food innovation* takes that scientific understanding and develops new products, ingredients, processes, or packaging. Sometimes innovation means incremental improvements: a yogurt with better texture, a cracker that stays crisp longer. Sometimes it means breakthroughs that redefine categories entirely.

Sensory science studies how humans perceive food through sight, smell, taste, touch, and even sound: the crunch of a chip, the fizz of a carbonated drink. Trained panels evaluate products using standardized vocabulary. Consumer tests measure whether people like what scientists create. Increasingly, biometric measurements capture responses that participants themselves may not consciously recognize, such as pupil dilation, changes in heart rate, and facial microexpressions.

Novel foods, in regulatory terms, are foods or ingredients without a significant consumption history in each region. The European Union's Novel Food Regulation requires premarket authorization for foods that Europeans haven't traditionally eaten, which include cultivated meat, insects for human consumption, and products manufactured through processes so new that regulators must evaluate safety from scratch.[4]

Where Food Science Happens

Food innovation emerges from an interconnected ecosystem spanning universities, corporations, contract laboratories, and startups. Each node generates and stores sensitive data. Each connection between nodes creates potential exposure to this data.

Academic research centers anchor the ecosystem. University food science departments at institutions like Cornell, UC Davis, Wageningen University in the Netherlands, and the University of Reading in the United Kingdom conduct fundamental research that corporate partners eventually commercialize. Land-grant universities in the United States have explicit missions that connect agricultural research to public benefit.

Corporate research and development operations then scale academic discoveries into commercial products. Nestlé's research center in Lausanne, Switzerland, employs hundreds of scientists across disciplines from nutrition to packaging.[5] Ingredient suppliers' companies—Kerry, Givaudan, International Flavors & Fragrances (IFF), and DSM-Firmenich—develop the building blocks that food manufacturers assemble into finished products. Their formulation databases represent decades of accumulated knowledge.[6]

Contract research organizations (CROs) fill gaps that internal teams cannot. Companies such as Eurofins Scientific, SGS, and Intertek conduct testing, clinical trials for functional food claims, and regulatory studies on behalf of clients. A startup developing a novel protein might lack the infrastructure for consumer sensory testing; a CRO provides it. This means sensitive formulation data and consumer research results reside on third-party systems, governed by contracts rather than direct control.

The ecosystem's distributed nature creates security challenges that no single organization can solve. Research data moves between academic collaborators via shared drives. Formulation details travel to contract manufacturers that will produce at scale. Consumer testing results flow from CROs to clients to marketing agencies. At each handoff, the security posture of the weakest participant becomes everyone's vulnerability.

Cultivated Meat: Growing the Future

Cultivated meat, also called cell-based meat, cultured meat, or lab-grown meat, is animal tissue produced without raising and slaughtering whole animals. The science involves extracting cells from a living animal through a small biopsy, developing those cells into lines that can proliferate in controlled environments, feeding them nutrient-rich growth media in bioreactors, and harvesting the resulting tissue for processing into consumer products. The technology stack resembles pharmaceutical manufacturing more than agriculture. Bioreactors—stainless steel vessels ranging from benchtop scale to tens of thousands of liters—maintain precise conditions for cell growth. Temperature must be held within narrow ranges. pH levels require continuous monitoring and adjustment. Dissolved oxygen concentrations affect how cells metabolize nutrients. And contamination from bacteria or fungi can destroy an entire batch.[7] Process control systems, like those discussed in Chapter 5's examination of operational technology, manage these variables automatically.

The intellectual property stakes are staggering. A single cell line capable of efficient proliferation might represent more than $100 million in development investment. Growth media formulations—the proprietary recipes for nutrient broths that feed cells—determine production economics. Some early efforts in cultivated meat failed because growth media costs exceeded what any consumer would pay for the final product. Companies that have solved this problem guard their formulations jealously.[8]

UPSIDE Foods, GOOD Meat, Mosa Meat in the Netherlands, Aleph Farms in Israel, and Believer Meats represent the sector's leading edge. They have collectively raised billions in venture capital from investors, including Singapore's Temasek and SoftBank Vision Fund, and established food companies Tyson, Cargill, and JBS (the same meat processor whose ransomware incident we examined in Chapter 8).[9] Believer Meats is constructing what it calls the world's largest cultivated meat production facility in North Carolina, designed to produce thousands of metric tons annually.[10]

The third-party dependencies multiply as companies scale. Cell lines may originate from academic collaborators. Growth media components come from specialty chemical suppliers. Bioreactor equipment arrives from vendors who also serve pharmaceutical clients. Contract manufacturers may handle production while the cultivated meat company focuses on research and marketing. Each relationship extends the attack surface beyond what any single security team controls. Also, it is important to mention that the ties to these huge organizations as third parties or investors further increase the attack surface, opening them up to threat actors hoping to disrupt and infiltrate those larger brands.

Consider what a determined adversary might target: cell-line databases containing proprietary genetic information, process parameters that took years of optimization, quality control records required for regulatory compliance, market research data, or the operational technology systems that keep bioreactors functioning. The sector is young enough that documented cyber incidents remain rare, which may say more about reporting practices than actual security posture, sadly.

Precision Fermentation: Brewing Without Beer

Precision fermentation uses engineered microorganisms, typically yeast or bacteria, to produce specific proteins, fats, or other compounds. The concept isn't new; insulin for diabetics has been produced through fermentation since the 1980s. What's changed is applying these techniques to food ingredients at a commercial scale.

Perfect Day, founded in 2014, produces whey and casein proteins—the same proteins found in cow's milk—using engineered fungi. The proteins are molecularly identical to those from dairy cows, enabling ice cream, cream cheese, and protein powders that taste like dairy products because, at the molecular level, they *are* dairy products. No cows required.[11] The EVERY Company, formerly Clara Foods, makes egg white proteins the same way. Impossible Foods produces heme, the iron-containing molecule that makes their plant-based burgers "bleed," through fermentation in engineered yeast.[12]

Fermentation facilities operate industrial-scale vessels, sometimes exceeding 100,000 liters, containing engineered organisms that excrete target molecules into the surrounding liquid. Downstream processing then purifies the desired compounds from the fermentation broth. The entire operation depends on strain libraries' proprietary collections of engineered organisms, which represent years of genetic optimization.[13]

Those strain libraries are crown jewels, and unlike a physical product that competitors might reverse-engineer, an engineered organism's genetic modifications can be sequenced if samples are obtained. Protecting strain libraries means controlling physical access to facilities, securing digital records of genetic sequences, managing which employees can access what information, and vetting the security practices of partners that might handle samples.

The third-party exposure risk parallels that of cultivated meat but with additional complexity. Precision fermentation companies often license enabling technology from universities or other biotech firms. They purchase fermentation capacity from contract manufacturers when their own facilities cannot meet demand. They ship samples to analytical laboratories for testing. And they engage regulatory consultants to navigate novel food approvals in multiple jurisdictions. Each relationship creates dependencies on partners whose security practices vary widely by organization.

3D Printed Food: Layer by Layer

3D food printing is additive manufacturing applied to edible materials to build food products layer by layer from digital instructions. Technology enables shapes, textures, and nutritional profiles that conventional manufacturing cannot achieve.

Healthcare applications lead the current adoption. Patients with dysphagia (difficulty swallowing) often face puréed diets that look unappetizing, regardless of taste. 3D printing can create foods with textures safe for compromised swallowing that still resemble recognizable meals: for instance, a printed chicken breast that dissolves safely in the mouth rather than requiring chewing. Hospital nutrition programs in Europe have piloted these approaches to improve patient intake and restore dignity.[14,15]

NASA has funded research on 3D-printed food for long-duration space missions, where carrying premade meals becomes impractical. BeeHex, originally a NASA-backed project, developed technology capable of printing pizzas.[16] Fine-dining restaurants use 3D printers to build intricate chocolate sculptures and garnishes that are impossible to create by hand.[17]

The infant-nutrition sector represents another emerging frontier. Research demonstrates that early allergen introduction can dramatically reduce allergy development. A landmark Learning Early About Peanut Allergy (LEAP) study showed an approximately 80 percent reduction in peanut allergies among high-risk infants who regularly consumed peanut-containing foods beginning in early infancy, compared to those who avoided peanuts entirely. In response, companies have started developing products designed to safely expose babies to common allergens.[18] Precision dosing matters enormously when feeding allergens to infants, and the technology platforms enabling exact measurement and controlled delivery share characteristics with 3D printing approaches.

A food 3D printer contains multiple ingredient cartridges, precision extrusion systems, and software that translates digital designs into physical products. Recipe files hold the instructions governing what gets deposited where, determining everything about the final product. Altering a recipe file could change nutritional content, introduce allergens, or modify textures in ways that create safety hazards for vulnerable consumers like dysphagia patients, not to mention food safety concerns.[19] Connected printers compound the risk. Many commercial food printers feature cloud connectivity for recipe downloads, remote monitoring, and usage analytics. The same Internet of Things vulnerabilities discussed in previous chapters (and that we'll touch on again in later chapters) come into play. Devices designed for functionality in professional kitchens may lack security hardening appropriate for networked operation.

Sensory Science and the Data It Generates

Understanding how humans experience food has become a data-intensive discipline. Traditional sensory science relied on trained panel groups of experts who learned standardized vocabularies to describe texture, aroma, flavor, and appearance with precision that ordinary consumers lack. These methods remain foundational. But new technologies capture responses that panelists cannot articulate.

Not all sensory testing involves trained experts in laboratory settings. When I worked at a bakery ingredients company, employee sensory tastings were a regular occurrence. They weren't mandatory, but it definitely felt as though we were voluntold to be there. When I was first hired, I was excited about it—who doesn't love tasting frosting? Years later, with a bit more weight on my body, I loathed these tastings. We were given a cup of water, a few salted crackers, and several variations of the same icing, frosting, fruit filling, cake, or whatever they were testing that day, along with a rating card. Our opinions were fed into quality decisions on these products. We were data points, whether we thought of it that way or not.

I now avoid store-bought cake, frosting, and icing like the plague. I can still smell it as I write this, and I'm making a face. But the experience taught me something relevant to this chapter: sensory data collection doesn't just happen in formal research studies. It happens in break rooms and conference rooms, with employees who become part of databases that track preferences across products, formulations, and time. The line between "participant" and "colleague" blurs when your employer is also gathering data about your taste responses.

Those rating cards captured what I could put into words. They couldn't capture the involuntary grimace, the slight hesitation before a second bite, or the microexpressions that betrayed my real feelings before I'd consciously formed an opinion. That's where the technology has gone now. Biometric measurements during tastings track heart rate variability, galvanic skin response, and pupil dilation—the physiological reactions that occur before conscious evaluation. Facial expression analysis uses computer vision to read microexpressions as people eat, identifying emotional responses they might not report. Some research programs have employed neuroimaging, placing participants in fMRI machines while they taste and mapping which brain regions activate in response to different foods.[20]

AI now shapes flavor development at some major companies. Givaudan, the world's largest flavor and fragrance company, uses machine learning through platforms it calls "Crea" to analyze combinations that human flavorists might never consider.[21] McCormick has partnered with IBM to apply AI to product development, drawing on decades of consumer testing data and ingredient interaction records.[22] These systems learn what people prefer, what disgusts them, and what drives premium pricing from databases containing millions of data points.

The data underlying these systems spans multiple sources. Sensory testing generates detailed preference profiles linked to demographics. Retail transaction data reveals what consumers purchase. Social media monitoring captures sentiment about products and brands. Food-tracking apps provide insights into eating patterns. Wearables and connected kitchen appliances contribute information about meal timing, portions, and preparation methods.

Once aggregated, this information paints intimate portraits of consumer behavior. Individually, a single taste-test response reveals little. But when combined across platforms, products, and years, patterns emerge that companies consider proprietary competitive intelligence. The databases powering AI flavor development represent assets that took decades to accumulate: assets residing in systems that weren't designed with data protection as a primary objective.

Third parties permeate this ecosystem. Consumer research firms recruit and manage taste panels. Analytics providers process raw data into actionable insights. Marketing agencies access preference information to craft campaigns, and academic collaborators receive datasets for research projects. Each access point extends exposure beyond direct organizational control.

And here's what rarely gets asked: where's the opt-out checkbox? Your grocery loyalty card, your fitness tracker logging meals, your smart refrigerator reporting what you've run out of . . . Do you know which vendors touch that data, how long they keep it, or whether it gets bundled into datasets sold downstream? The terms of service you scrolled past to download a food-tracking or food-ordering app probably granted permissions you never consciously gave. These are questions cybersecurity professionals ask instinctively. Consumers should be asking them, too, but most don't know they should, and companies aren't rushing to tell them.

Packaging That Thinks and Talks

Packaging innovation has moved far beyond putting food in containers. Active packaging interacts with food to extend shelf life: oxygen absorbers slow spoilage, antimicrobial films inhibit bacterial growth, and ethylene absorbers delay fruit ripening. Intelligent packaging communicates information about product condition: time–temperature indicators reveal whether the cold chain was compromised, freshness sensors detect gases produced by spoilage, and color-changing labels signal when contents are past their prime.[23]

Connected packaging takes communication even further. Radio-frequency identification (RFID) tags and near-field communication (NFC) chips enable tracking through supply chains, authentication against counterfeiting, and consumer engagement through smartphone interactions. Scan a package with your phone and receive recipes, sourcing information, or loyalty rewards. The data flows bidirectionally: consumers get information, and companies learn about consumer behavior.

Sustainability has driven another innovation wave. Edible packaging eliminates waste: seaweed-based films from companies like Notpla dissolve in water or can be eaten along with their contents.[24] Mushroom-based materials from Ecovative replace petroleum-derived foams. Agricultural waste becomes feedstock for compostable containers. The European Union's Single-Use Plastics Directive and similar regulations worldwide have accelerated adoption.[25]

The cybersecurity implications of smart packaging remain underexplored. Freshness indicators that can be manipulated could allow expired products to reach consumers. RFID tags can be cloned, enabling counterfeit products to pass authentication checks. Connected packaging generates data about when and where consumers interact with products: information valuable for marketing but sensitive if breached. QR codes on packaging could redirect consumers to malicious sites rather than legitimate brand content.

Packaging supply chains involve their own third-party complexity. A food company might source containers from one supplier, labels from another, smart sensors from a third, and closure systems from a fourth. Each component arrives with its own provenance. The packaging vendor providing RFID tags may have sourced chips from manufacturers whose security practices remain opaque.

What It Teaches Us: The Vendor Web

Across every technology examined in this chapter, a pattern emerges: innovation depends on networks of partners, suppliers, collaborators, and contractors whose security postures vary enormously. This third-party risk represents perhaps the sector's greatest vulnerability.

A startup making cultivated meat might depend on an academic institution where foundational research occurred, equipment vendors supplying bioreactors and sensors, specialty chemical companies providing growth media components, contract manufacturers handling production, analytical laboratories conducting quality testing, regulatory consultants navigating approval processes, cloud service providers hosting data and applications, and software vendors whose platforms manage everything from formulation to finance.

Each relationship involves data sharing. Formulation details go to contract manufacturers. Testing samples go to analytical labs. Financial projections go to investors and their advisors. Regulatory submissions go to government agencies across multiple countries. And employment records go to payroll processors. At every handoff, sensitive information leaves direct organizational control.

The problem compounds because food innovation companies are typically young, capital-constrained, and focused on technical challenges rather than security infrastructure. A precision fermentation startup with 50 employees and $30 million in funding cannot afford a dedicated security team, yet it handles intellectual property that established competitors would pay dearly to access.

Vendor assessment in this environment often means checking boxes on questionnaires rather than conducting genuine due diligence. Does your partner have a security policy? Check. Do they encrypt data in transit? Probably. How do they vet their own vendors? Unknown. What happens to your data when the contract ends? Unclear.

Chapter 19 will examine third-party risk across the food sector in detail. Here, the point is simpler: the most innovative segment of the food industry may also be its most exposed, precisely because innovation requires collaboration and collaboration requires trust in partners whose security investments may not match the sensitivity of shared information.

Moving Forward: The Innovation-Security Gap

Food science and innovation face a structural security challenge rooted in the sector's origins and incentives. Speed matters more than security in startup culture. Venture capital rewards companies that reach milestones quickly: first prototype, first regulatory approval, first commercial sale. Security investments don't accelerate those timelines. They may even slow progress by introducing friction into research workflows. So, security gets deferred until later and often never arrives before an incident forces attention.

Research culture emphasizes openness. Academic food scientists build careers by publishing findings, presenting them at conferences, and collaborating across institutions. Information security conflicts with these norms. Restricting access to data or requiring cumbersome authentication interrupts research momentum. Scientists accustomed to academic environments may resist security measures that feel bureaucratic or distrustful.

Talent scarcity creates impossible trade-offs. Food scientists and biotechnologists command premium salaries in a tight labor market. Cybersecurity professionals are equally scarce and expensive. A company struggling to hire enough researchers to advance its core technology is unlikely to prioritize hiring security specialists even though its research data represents the company's primary value.

Regulatory frameworks haven't caught up. Food safety regulations address contamination, labeling, and manufacturing practices, but they don't address data protection for novel food development. The FDA evaluates whether cultivated meat is safe to eat, not whether the company's intellectual property is safe from theft. GDPR and similar privacy regulations apply to consumer data, but enforcement attention focuses elsewhere. The sector operates in a regulatory gap where traditional food safety rules don't contemplate cyber risks and cybersecurity regulations don't specifically address food innovation. The irony isn't lost: although these facilities perfect the science of cultivating protein, they may also be cultivating the ideal conditions for a security breach.

Looking Ahead

This chapter concludes Part II's examination of where risks exist across the food system. From Chapter 6's farm fields through precision technology, processing facilities, food safety laboratories, distribution networks, restaurants, seafood operations, beverage production, animal feed, environmental management, and now research innovation, we have traced how technology has transformed every stage of food production while security has struggled to keep pace.

Certain themes occur across sectors. Technology adoption follows efficiency and cost pressures rather than security requirements. Vendor consolidation creates single points of failure that propagate risk across entire industries. Regulatory frameworks designed for earlier eras leave gaps that threat actors exploit. Workforce training lags behind the cyber-physical realities of modern food systems. The people who grow, process, and distribute food understand their domains deeply but rarely receive preparation for the digital threats now targeting their operations.

We've spent Part II touring the food system from seed to science lab, watching technology weave itself into every stage. Now comes the part where things go wrong. Part III gets into the meat and potatoes of cybersecurity in food and agriculture: the control systems that can be hijacked, the physical consequences when digital defenses fail, and the vendor relationships that turn one company's weakness into everyone's problem. You've seen what we've built and what's at risk. Next, we'll see what happens when someone decides to break it.

Notes

1. Brill, J. (2021). "Upside Foods opens cultivated meat engineering, production and innovation center." *Quality Assurance & Food Safety Magazine.* **https://www.qualityassurancemag.com/news/upside-foods-opens-cultivated-meat-engineering-production-and-innovation-center**
2. FDA. (2022). "FDA completes first pre-market consultation for human food made using animal cell culture technology." **https://www.fda.gov/food/hfp-constituent-updates/fda-completes-first-pre-market-consultation-human-food-made-using-animal-cell-culture-technology**; Reiley, L. (2023). "USDA gives two companies the green light to sell lab-grown chicken." *Washington Post.* **https://www.washingtonpost.com/business/2023/06/21/usda-cultivated-meat-approval**
3. Good Food Institute Europe. (2020). "World's first regulatory approval for cultivated meat begins new space race for the future of food." **https://gfieurope.org/blog/worlds-first-regulatory-approval-for-cultivated-meat-begins-new-space-race-for-the-future-of-food**
4. European Union. (2015). "Commission Regulation (EU) 2015/2283 of 8 December 2015 Amending Regulation (EC) No 1881/2006 as Regards Maximum Levels of Nivalenol in Certain Foodstuffs." *EUR-Lex.* **https://eur-lex.europa.eu/legal-content/EN/TXT/?uri=celex%3A32015R2283**
5. Nestlé. (2018). "Nestlé strengthens its research capabilities in Switzerland." **https://www.nestle-caribbean.com/research-capabilities-switzerland**
6. GluCorp. (2023). "How do food ingredient suppliers and food manufacturers work together?" **https://www.glucorp.com/blog/how-do-food-ingredient-suppliers-and-food-manufacturers**
7. Good Food Institute. (2025). "The science of cultivated meat." **https://gfi.org/science/the-science-of-cultivated-meat**
8. Specht, L. (2021). "An analysis of culture medium costs and production volumes for cultivated meat." Good Food Institute. **https://gfi.org/wp-content/uploads/2021/01/clean-meat-production-volume-and-medium-cost.pdf**
9. Nestlé. (2018). "Nestlé strengthens its research capabilities in Switzerland"; Good Food Institute. (2024). 2023 State of the Industry Report: Cultivated Meat and Seafood. **https://gfi.org/wp-content/uploads/2024/04/2023-State-of-the-Industry-Report-Cultivated-meat-and-seafood.pdf**
10. Poinski, M. (2023). "Believer Meats breaks ground on commercial-scale cultivated meat facility." *Food Dive.* **https://www.fooddive.com/news/believer-meats-commercial-scale-cultivated-meat-groundbreaking-future-meat-technologies/638263**
11. Nay, K. (2023). "Beyond GHG: study finds animal-free dairy sustainable in multiple metrics." *Perfect Day.* **https://perfectday.com/blog/beyond-ghg-study-finds-animal-free-dairy-sustainable-in-multiple-metrics**
12. Simon, M. (2016). "The impossible burger." *Wired.* **https://www.wired.com/story/the-impossible-burger**; Tuomisto, H. L. & Teixeira, Joyce D. (2013). "Environmental impacts of cultured meat production." *Environmental Science & Policy*, 26, 87. **https://www.sciencedirect.com/science/article/abs/pii/S016378270300064X?via%3Dihub**
13. GEA. (2015). "Time to reinvent meat." **https://www.gea.com/en/new-food/cultivated-meat**
14. Sun, J., Peng, Z., Yan, L., Fuh, J. H., & Hong, G. S. (2015). "3D food printing—An innovative way of mass customization in food fabrication." *Trends in Food Science & Technology*, 48, 230–240. **https://www.sciencedirect.com/science/article/abs/pii/S092422441500045X**

15. Gómez, C. & Sanchís, R. (2023). "Could 3D food printing help to improve the food supply chain resilience?" *Journal of Future Foods*, 3(1), 1–5. **https://research.wur.nl/en/publications/could-3d-food-printing-help-to-improve-the-food-supply-chain-resi**
16. Terfansky, M. L., & Thangavelu, M. (2014). "3D printing of food for space missions." *ResearchGate*. **https://www.researchgate.net/publication/268459458_3D_Printing_of_Food_for_Space_Missions**
17. Huzaifa, Q. (2018). "Edible carving: The world's first 3D printing restaurant." *The Spoon Tech*. **https://thespoon.tech/edible-carving-the-worlds-first-3d-printing-restaurant**
18. Du Toit, G., Roberts, G., Sayre, P. H., Bahnson, H. T., Radulovic, S., Santos, A. F., Brough, H. A., Phippard, D., Basting, M., Feeney, M., Turcanu, V., Sever, M. L., Gomez Lorenzo, M., Plaut, M., & Lack, G. (2015). "Randomized trial of peanut consumption in infants at risk for peanut allergy." *New England Journal of Medicine*, 372(9), 803–813. **https://www.nejm.org/doi/full/10.1056/NEJMoa1414850**
19. Smart Food Safe. "3D printed food safety." **https://smartfoodsafe.com/3d-printed-food-safety**
20. Spence, C. (2015). "Multisensory flavor perception." *Cell*, 161(1), 24–35. **https://doi.org/10.1016/j.cell.2015.03.007**
21. Givaudan. (2021). "Givaudan launches ground-breaking AI tools for next generation product development." **https://www.givaudan.com/media/trade-media/2021/givaudan-launches-ground-breaking-ai-tools-next-generation-product**
22. McCormick & Company, Inc. (2021). "McCormick Company and IBM announce collaboration pioneering use of AI in the flavor industry." **https://ir.mccormick.com/news-releases/news-release-details/mccormick-company-and-ibm-announce-collaboration-pioneering-use**
23. Yam, K. L., Takhistov, P. T., & Miltz, J. (2005). "Intelligent packaging: concepts and applications." *Journal of Food Science*, 70(1), R1–R10. **https://doi.org/10.1111/j.1365-2621.2005.tb09052.x**
24. Notpla. **https://www.notpla.com**
25. European Union. (2019). "Directive (EU) 2019/904 of the European Parliament and of the Council of 5 June 2019 on the reduction of the impact of certain plastic products on the environment." *EUR-Lex*. **https://eur-lex.europa.eu/legal-content/EN/TXT/PDF/?uri=CELEX:32019L0904**

PART III

Core Cybersecurity Challenges in the Food Sector

CHAPTER 17

Operational Technology and Industrial Control Systems

The Invisible Backbone

When I worked at a bakery ingredients and finished goods company, I pulled on-call duty at least twice a month. The good old days. And I swear, every single time (at least twice during my week), my phone would blow up at 3:00 AM, which was always third shift rolling into the factory with production lines down. "Hang on, let me get online and see what's going on." I'd stumble into the living room so I wouldn't wake my partner. The cats would pop their heads up like "Whoa, Mom, is it time to eat?" I'd grab my laptop, remote into the facility, and try to make sense of what the production lines were doing.

By the time I got back on the phone, the production lead was screaming. At me, at the world, at technology, at his schedule, at the fact that he couldn't manually bypass whatever had locked him out. It's 3:00 AM. I hadn't slept a full night. No coffee. I was using every ounce of self-control I have not to scream right back. One night, the line was down because of a configuration error in Systems, Applications & Products in Data Processing (SAP). Somebody had shifted the line the day before and didn't bother telling anyone. Not his fault. Not mine. But I was the one fixing it at 3:00 AM in my pajamas, and the production lead was starting to sound like the lead singer from a metal band. A few configuration tweaks later, the line went green. He hung up without saying thank you. I fired off a couple of emails so the day shift would know what happened and then tried to get a few more hours of sleep. The next morning, I walked the floor with the incoming production lead, checking every line and making sure the configurations matched the schedule. I started doing that morning walkthrough religiously, and sometimes another one before I left for the day. And yeah, I logged in before bed to make sure everything looked good.

If you've worked in operational technology (OT), you know that feeling. The heart-drop when something shows offline. Or, worse, when it doesn't even show up. Here's the thing

I didn't understand back then: it wasn't really about technology. Those production leads weren't screaming at me because they hated computers. They were screaming because product was about to spoil. Because they had deadlines that didn't care about SAP configurations. Because if Quality Assurance (QA) couldn't test until morning and the batch sat too long, the whole thing was garbage. Because if the line stayed down, their crew got sent home without pay. There was more on the line than schedules. Crew safety. Food safety. A batch that represented somebody's paycheck and somebody else's celebration cake.

I didn't fully get that at first. I was just the person trying to make the screens turn green again. But those 3:00 AM calls taught me something that's shaped everything since: technology is only part of it. The people running these systems, the pressure they're under, the processes that must hold even when nothing else does, that's the real story. This is OT in food and agriculture. Not abstract systems. The invisible infrastructure that is standing between raw ingredients and safe food on tables. Managed by people whose knowledge runs deeper than any documentation. And increasingly connected to networks that were never part of the original design.

Building on What We Know

In Chapter 5, we walked through the foundations of OT and how it differs from information technology (IT), the systems most people encounter daily. We defined the key components: programmable logic controllers (PLCs), the rugged industrial computers that act like recipe cards for entire facilities; supervisory control and data acquisition systems (SCADA), which pull data from sensors and PLCs across operations and display it for human operators; and human–machine interfaces, the screens where operators interact with all of it. We introduced Ag-OT, a term for the OT tailored specifically to agriculture, from irrigation valves to grain dryers to milking parlors.

Now that we understand what these systems are, we need to recognize why they remain so vulnerable despite decades of warnings. The answer lies not in technical shortcomings alone but in the collision between how food systems operate and the assumptions embedded in most cybersecurity thinking. It lies in equipment built for reliability that predates cybersecurity as a discipline. In connections that accumulate organically, each sensible on its own, collectively erasing isolation that was never there. And in the economic and biological realities that make traditional security approaches impractical or impossible.

The Equipment That Keeps Running

When I toured a major food processing facility in the Midwest, I encountered equipment from the 1980s operating alongside systems from the 2010s. The operations manager expressed a sentiment I have heard repeatedly across the industry: they cannot justify scrapping equipment that still produces quality products just because it lacks connectivity features or modern security capabilities. This perspective is economically sound and, frankly, environmentally responsible. But it creates a cybersecurity reality that few outside the industrial industries appreciate.

According to research from the Manufacturing Leadership Council, over 70 percent of manufacturing equipment in North America is more than 20 years old.[1] The food processing sector is no exception. A study from Food Engineering magazine found that the lifespan of an automation industrial control system (ICS) can range from 10 to 30 years.[2] Many facilities push equipment well beyond those timelines. Walk into a dairy plant, a bakery, or a meat processing facility, and you will find PLCs that predate the smartphone era still running production lines.

The Allen-Bradley SLC 500, introduced in 1991,[3] remains operational in food facilities across North America. Siemens S5 controllers, with roots in the 1980s,[4] continue managing processes in plants worldwide. These are not outliers; they represent normal operating conditions in an industry where equipment replacement competes with other capital expenditures, including new product lines, regulatory compliance upgrades, and roof repairs.

The challenge is that IT devices typically have a lifespan of three to five years, whereas OT devices operate for decades.[5] Office laptops are replaced regularly, but the PLC controlling a pasteurizer does not. That pasteurizer still heats milk to exactly 161 °Ffor precisely 15 seconds, exactly as designed.[6] The PLC still sends commands and receives sensor readings. It works, but it runs software that hasn't received a security patch since before anyone worried about ransomware.

I want to be clear about something: this is not a problem to be condemned but a reality to be understood and managed. The people running these systems are not negligent. They are operating within genuine constraints that most IT security professionals never encounter. A cheese-aging facility cannot shut down its climate control during a maintenance window. A poultry processing line cannot pause mid-shift for a firmware upgrade. A dairy cannot tell cows to hold off on producing milk while systems are updated.

And here is the part that gets lost in most cybersecurity discussions: this older equipment often has advantages. It has a historical track record. Plant operators know its failure modes, its maintenance needs, and its quirks. New equipment introduces unknown failure patterns and learning curves. In food safety environments, where consistency matters and surprises can be dangerous, the proven reliability of older systems has genuine value.

The Air Gap That Was Never Really There

"Our control systems aren't connected to the Internet." I have heard this statement from plant managers, operations directors, and maintenance supervisors across multiple facilities. They believe it, and in a narrow technical sense, they may be right: there is no cable running directly from a PLC to a router with a public IP address. But the air gap they imagine, which would provide complete isolation from connected networks, rarely exists in practice. Connections accumulate through decisions that make operational sense, collectively erasing whatever isolation once existed.

Consider how this happens: the refrigeration vendor needs remote access to monitor compressor performance and schedule preventive maintenance. A cellular modem has been installed. Now the freezer system is reachable from outside the facility. A major customer requires real-time temperature data from cold storage to satisfy their QA protocols. An integration is built to automatically transmit readings. Now the monitoring system is connected. The agronomist recommends variable-rate application software that optimizes fertilizer use and saves thousands annually. It requires connectivity to field equipment, weather data, and cloud analytics. The planter, sprayer, and applicator are now connected. A major retailer mandates supply chain visibility. Product tracking gets implemented from receiving through shipping. Now, inventory management is connected. The maintenance supervisor downloads the equipment manufacturer's app to get alerts and check the status from home. Her phone is on the facility's Wi-Fi, giving the equipment one hop from the Internet.

None of these individual connections is unreasonable. Vendor remote access reduces response times and maintenance costs. Customer data requirements are nonnegotiable for major contracts. Precision agriculture tools deliver measurable returns on investment. Traceability satisfies regulatory and market demands. Mobile monitoring apps are genuinely useful. Each decision makes sense in isolation. But nobody maps the cumulative effect or, honestly, even notices.

Going back to the facility manager who says ICSs are air-gapped: they may be technically correct about direct connections, but there are indirect pathways that nobody has documented. The network boundary between "connected" and "isolated" has become porous. That boundary usually stays invisible: nobody maps it until an incident does it for them.

When IT Falls, OT Follows

Most ransomware attacks on food operations do not directly compromise ICSs. The attackers go after IT networks: email servers, file systems, business applications, the digital infrastructure that runs every modern organization. The PLCs controlling the processing lines often remain untouched, yet production still stops. Understanding why reveals everything about the IT/OT interdependence that defines modern food operations.

In Chapter 8, we discussed that in May 2021, JBS S.A., the world's largest meat processor, discovered ransomware on its network. The attackers, later found to be the Russian hacker group REvil, had not accessed the ICSs running JBS slaughter lines and processing equipment. Those systems could potentially still operate, but JBS shut down plants across the United States, Canada, and Australia anyway.[7] Why? Because modern meat processing cannot function as an isolated physical operation. Without IT systems, JBS could not access production schedules, document food safety compliance, process customer orders, coordinate shipping logistics, or verify inventory status. The plants went dark not because the equipment broke, but because operating blind violated food safety requirements and made business operations impossible.[8]

The scale was staggering. JBS controls roughly 20 percent of American beef and pork slaughter capacity.[9] When its facilities stopped, it represented a sudden, significant reduction in national meat processing. The company stood down approximately 7,000 Australian employees. The U.S. Department of Agriculture could not publish wholesale beef and pork prices because there was not enough market activity to calculate them. JBS ultimately paid $11 million in ransom to restore operations.[10]

As we discussed in Chapter 4, the JBS attack became a headline case that made cybersecurity in food impossible to ignore. But the pattern it revealed—IT compromise cascading into OT shutdown—has repeated across the sector. When Campbell's Soup's Napoleon, Ohio, plant experienced a cyber intrusion in 2023, production halted for three days, even though the intrusion started on the IT side. If there is even a possibility that kettles or sterilizers might be compromised, the lines stop. Nobody was willing to risk the soup's food safety.

In November 2024, German poultry producer VOSSKO fell victim to ransomware that encrypted internal systems and databases.[11] The Black Basta ransomware group claimed responsibility, asserting that it had exfiltrated 800 gigabytes of data, including financial records, employee information, and project documents.[12] The malware affected operational processes, although the company reported that affected systems and production were subsequently restored with help from internal IT staff, external specialists, and law enforcement, including IT forensics experts. The VOSSKO case illustrates that this pattern extends beyond North America; the food system is global. German food producers face the same IT/OT interdependence and the same vulnerability to attacks that target business systems but functionally disable operations.

Biology Sets the Timeline

I've already mentioned it many times throughout the book, but it's a point that warrants reiteration: food and agriculture operate on biological clocks that do not negotiate with attackers, system administrators, or anyone else. This fundamental reality shapes what security approaches can work. Perishability creates urgency that most industries never face. Fresh produce begins declining the moment it is harvested. Dairy has regulatory time limits from cow to cooler. Meat processing operates under USDA time-temperature requirements that are not suggestions. When systems go down, biology keeps moving. The strawberries do not pause their ripening. The milk does not stop its journey toward spoilage. The chicken on the processing line does not wait for IT to restore backups.

Think about making bread at home. Once you have mixed the dough and it starts rising, you are committed. You cannot decide to finish tomorrow or pause the yeast. The bread happens on the bread's schedule. Food and agriculture operations manage this reality at an industrial scale, for thousands of products simultaneously. Fermentation continues regardless of whether monitoring is working. Cheese cultures do not pause for IT maintenance. Yogurt incubation will over-acidify if not stopped at the right moment.

Living inventory adds another dimension. Livestock require daily care regardless of system status. As we discussed in Chapter 6, poultry houses require continuous climate control. Birds will die within hours if ventilation fails in summer.[13] Aquaculture systems require constant water quality management. The inventory is literally alive, but only with continuous operational support.

Margins compound the pressure. A grain elevator might handle millions of bushels during harvest, earning somewhere between 7 and 27 cents per bushel depending on volume.[14] The ICSs managing grain drying, storage, and movement must operate flawlessly during a six-week window that accounts for the year's revenue. There is no budget for redundant systems when margins are measured in fractions of pennies. A poultry processor runs birds through the line at 140 per minute every minute the plant operates.[15] Downtime is not an inconvenience. It is birds backing up in holding, worker hours wasted, customer orders missed, and live inventory requiring care and feeding while processing waits—or, even worse, safety issues.

Regulations Shape What Is Possible

Food safety regulations create constraints that cybersecurity professionals often fail to appreciate. When a canning operation validates its retort process with the Food and Drug Administration (FDA), the validation covers the entire system: equipment, software, and procedures. Changing the software requires more than just IT approval. It may require revalidating the entire food safety process, including microbial studies that can take months and cost tens of thousands of dollars.[16]

Continuous compliance adds another layer. Food safety monitoring must be continuous. A meat processor cannot turn off Hazard Analysis and Critical Control Points (HACCP) monitoring to apply security patches. Temperature logs must be complete. Critical control point verification must be documented. Gaps in the record raise regulatory questions regardless of the reason for those gaps.

Audit trails matter in ways that create tension with common IT practices. Food safety regulations require documentation of who did what, when.[17] Shared credentials that work fine for production-floor access become compliance problems when regulators ask who verified the cooking temperature at 3:47 AM on a specific date. The intersection of cybersecurity best practices and food safety documentation requirements is not always smooth.

In Europe, the NIS2 Directive has expanded cybersecurity obligations to include food production and processing as essential sectors. Companies must assess risks, implement security measures, and report significant incidents quickly.[18] This represents a significant shift toward treating food system cybersecurity as a regulated necessity rather than a voluntary aspiration. But regulations alone do not change the fundamental operational constraints that make security difficult in food environments.

Why Food Systems Are Targets

In April 2022, the Federal Bureau of Investigation (FBI) issued a private industry notification that should have been a wake-up call for the entire food and agriculture sector. The Bureau warned that ransomware actors may be more likely to attack agricultural cooperatives during critical planting and harvest seasons, deliberately timing attacks to create maximum leverage.[19] The FBI documented ransomware attacks against six grain cooperatives during the fall 2021 harvest and two additional attacks in early 2022 that could have impacted the spring planting season by disrupting seed and fertilizer supply. The attackers understood agricultural timing. They knew that cooperatives facing harvest deadlines and with perishable products at risk would face intense pressure to pay the ransom. Cyber threat actors, the FBI noted, may perceive cooperatives as lucrative targets, given their willingness to pay due to the time-sensitive role they play in agricultural production.[20] Between September 15 and October 6, 2021, alone, six grain cooperatives experienced ransomware attacks using variants including Conti, BlackMatter, Suncrypt, Sodinokibi, and BlackByte.[21] Some targets had to halt production entirely, while others lost administrative functions essential to operations. The pattern was clear: attackers had studied the agricultural calendar and understood where pressure would be greatest.

Criminal ransomware groups are not the only concern. Nation-states know what food infrastructure represents: a country's people must eat. Disrupting food supply creates social instability more effectively than military attacks on military targets. Agricultural exports are a key economic pillar for many nations. Undermining confidence in a country's food supply affects trade, commodity prices, and economic stability. Food infrastructure connects to energy, water, transportation, and telecommunications. Attacking food creates cascading impacts across critical sectors.

Nation-state actors do not always attack immediately. They map infrastructure for future targeting. They establish persistent access for potential activation. They gather intelligence on food supply chain dependencies. They test defenses to understand capabilities. A grain elevator or dairy processor may not seem strategically significant on its own, but multiplying it by hundreds of similar operations and disrupting them simultaneously becomes strategically meaningful.

What It Teaches Us: Bridging Two Worlds

If you come to this chapter from a food and agriculture background, your operations are extraordinary. The complexity you manage daily, including biological processes, tight margins, regulatory requirements, and seasonal pressure, represents expertise that cybersecurity professionals rarely understand. But that expertise exists in an increasingly connected environment. The equipment and systems you rely on have become attack surfaces, not just production assets. The vendor access that keeps your equipment running also creates pathways for attackers. The data systems that coordinate your operations have become single points of failure.

Understanding these risks does not require becoming a cybersecurity expert. It requires bringing your operational expertise to conversations about risk and security. Security professionals need your knowledge to design controls that work in your environment. They need to understand why their recommendations might not fit, why certain downtime windows are impossible, why some changes require months of regulatory validation, and why those decisions carry consequences for both employee and food safety.

If you come to this chapter from a cybersecurity background, food and agriculture are not waiting for you to bring security enlightenment. These operations have been managing complex risks—biological, mechanical, and regulatory—for decades. The people running them are experts in their domain. Your job is not to impose security frameworks designed for other industries. It is your responsibility to understand what makes food and agriculture unique, respect the existing constraints, and design security that works within operational reality. That means learning enough about food operations to understand why your recommendations may not fit: listening when operators tell you a control will not work in their environment, designing for 24/7 operations where downtime has biological and safety consequences, and accepting that legacy equipment is a reality and building security around it rather than pretending it can be wished away or ignored.

Moving Forward: Approaches That Actually Work

Instead of condemning legacy equipment, protect it appropriately. Network segmentation ensures that legacy systems can only communicate with what they need to communicate with. That 2008 climate control system does not need internet access, so make sure it cannot reach the Internet and the Internet cannot reach it. Compensating controls work around equipment limitations. When you cannot install security software on the equipment itself, install monitoring around it. Watch the network traffic to and from that system for anomalies. You can do that using any technology you see fit, but again, your security controls cannot affect production.

Operational awareness treats human expertise as a security asset rather than an obstacle to overcome. The operators who know these systems best notice when something is not right. They detect anomalies that have no sensor captures. Create channels for them to report concerns without navigating IT bureaucracy. Build relationships among security, operations, production, and food safety teams before incidents occur.

Document baseline behavior for legacy equipment so you know what normal looks like. Understand typical network connections, traffic patterns, and standard operational characteristics to identify deviations. This documentation becomes invaluable during incident response when teams need to distinguish between malicious activity and normal operations.

Accept what does not work. You cannot install modern endpoint protection on a PLC running 15-year-old firmware. Do not pretend otherwise. Demanding equipment replacement when capital budgets are allocated to production needs is wishful thinking, not a security strategy. A 20-year-old system that is properly isolated and monitored may be more secure than a new system with default passwords and Internet exposure.

Looking Ahead

The OT running food and agriculture represents one of the most sophisticated and least appreciated infrastructure systems in the world. It converts raw ingredients into safe food for ingestion. It coordinates supply chains spanning continents. It operates every hour of every day because people need to eat.

That infrastructure was built for operational excellence, not cybersecurity. Connections accumulated for good operational reasons, and legacy equipment persists because it works. Skilled operators manage complexity that would overwhelm any purely automated system. These are not weaknesses but characteristics of an industry that has successfully fed growing populations for generations.

But these characteristics create vulnerabilities when adversaries view food systems as targets. The connections that enable efficiency create attack pathways. The legacy equipment that works reliably cannot defend itself against threats its designers never imagined. Sadly, the skilled operators who keep everything running can be deceived, manipulated, or undermined.

Chapter 18 examines what happens when these vulnerabilities are exploited. When digital attacks create physical consequences across food systems, spoilage clocks start ticking, and safety decisions must be made without complete information. The ripple effects of a single incident spread through communities that depend on food systems continuously working. Understanding OT in food and agriculture is the foundation. Understanding what is at risk when it fails comes next.

Notes

1. Pal, N. (2025). "Tackling Industry 4.0 with legacy equipment: a practical approach." *IndustryWeek*. **https://www.industryweek.com/technology-and-iiot/systems-integration/article/55284102/tackling-industry-40-with-legacy-equipment-a-practical-approach**
2. Labs, W. (2017). "Legacy automation systems will soon require replacement." *Food Engineering*. **https://www.foodengineeringmag.com/articles/97083-legacy-automation-systems-will-soon-require-replacement**
3. Automation Networks. (2025). "The history of the SLC-500." **https://automation-networks.com/dh-info-base/the-history-of-the-slc-500**
4. SZCT Automation. (2025). "A brief history of Siemens PLC technology: how it evolved." **https://www.szct-automation.com/info/a-brief-history-of-siemens-plc-technology-how-102869775.html**
5. Cisco. "What is OT vs. IT?" **https://www.cisco.com/site/us/en/learn/topics/industrial-iot/what-is-ot-vs-it.html**
6. International Dairy Foods Association. (n.d.). "Pasteurization." **https://www.idfa.org/pasteurization**
7. Polansek, T., & Mason, J. (2021). "U.S. says ransomware attack on meatpacker JBS likely from Russia." *Reuters*. **https://www.reuters.com/world/us/some-us-meat-plants-stop-operating-after-jbs-cyber-attack-2021-06-01**
8. Kapko, M. (2022). "Food supplier cyber risk spreads 1 year after JBS attack." *Cybersecurity Dive*. **https://www.cybersecuritydive.com/news/food-supplier-cyber-risk-spreads-jbs/624800**

9. Collier, K. (2021). "Meat supplier JBS paid ransomware hackers $11 million." *CNBC*. **https://www.cnbc.com/2021/06/09/jbs-paid-11-million-in-response-to-ransomware-attack-.html**
10. Harwell, D., & Lerman, R. (2021). "JBS paid $11 million in ransom after hackers shut down meat plants." *Washington Post*. **https://www.washingtonpost.com/technology/2021/06/09/jbs-11-million-ransom**
11. Kaspersky ICS-CERT. (2025). "Q4 2024: a brief overview of the main incidents in industrial cybersecurity." **https://ics-cert.kaspersky.com/publications/reports/2025/04/08/q4-2024-a-brief-overview-of-the-main-incidents-in-industrial-cybersecurity**
12. ForeNova. (2025). "Recap of the largest ransomware attacks in 2024." **https://www.forenova.com/blog/recap-of-the-largest-ransomware-attacks-in-2024**
13. Oluwagbenga, E., & Fraley, G. (2023). "Heat stress and poultry production: a comprehensive review." *Poultry Science*, 102(12), 103141. **https://doi.org/10.1016/j.psj.2023.103141**
14. Kenkel, P. (2008). *Grain Handling and Storage Costs in Country Elevators* (Oklahoma State University). **https://extension.okstate.edu/programs/agribusinessand-cooperative-management/site-files/docs/software-and-templates/grain-handling-and-storage-costs-in-country-elevators.pdf**
15. USDA-FSI. (2026). "Maximum line speed rates for young chicken and turkey establishments operating under the new poultry inspection system." *Federal Register*, 91(34). **https://www.federalregister.gov/documents/2026/02/19/2026-03227/maximum-line-speed-rates-for-young-chicken-and-turkey-establishments-operating-under-the-new-poultry**
16. FDA. (2017). *Deciding When to Submit a 510(k) for a Software Change to an Existing Device*. **https://www.fda.gov/media/99785/download**
17. FDA. (2025). "FSMA final rule for preventive controls for human food." **https://www.fda.gov/food/food-safety-modernization-act-fsma/fsma-final-rule-preventive-controls-human-food**
18. European Commission. (2026). "NIS2 Directive: securing network and information systems." **https://digital-strategy.ec.europa.eu/en/policies/nis2-directive**
19. FBI. (2022). Ransomware attacks on agricultural cooperatives potentially timed to critical seasons. **https://www.ic3.gov/CSA/2022/220420-2.pdf**
20. Gatlan, S. (2022). "FBI warns of ransomware attacks targeting US agriculture sector." *Bleeping Computer*. **https://www.bleepingcomputer.com/news/security/fbi-warns-of-ransomware-attacks-targeting-us-agriculture-sector**
21. Homeland Security Today. (2022). "Ransomware attacks on agricultural cooperatives potentially timed to critical seasons." **https://www.hstoday.us/subject-matter-areas/cybersecurity/ransomware-attacks-on-agricultural-cooperatives-potentially-timed-to-critical-seasons**

CHAPTER 18

Cyber-Physical Threats and Systemic Risk

The Company You've Never Heard Of

Most British consumers have never heard of Peter Green Chilled. They know Tesco or Sainsbury's, and they recognize the logos, trust the brands, and fill their trolleys without thinking much about how the food got there. Peter Green Chilled is one of the invisible companies that make the visible ones work: a cold-chain logistics provider that moves refrigerated goods from warehouses to store shelves across the UK. The company began 60 years ago as a haulage operation for Somerset dairy suppliers and has grown into a critical part of the nation's food infrastructure.[1]

On the evening of May 14, 2025, Peter Green Chilled was hit with ransomware.[2] Within hours, deliveries to major supermarket chains slowed to a crawl, chilled products sat in trucks that couldn't be dispatched, temperatures drifted, and retailers scrambled for alternatives. Shoppers who had never given a thought to Peter Green Chilled walked into grocery stores and found empty spaces where the milk and meat should have been.

Wilfred Emmanuel-Jones, founder of The Black Farmer food company, spoke publicly about the chaos. He had 10 pallets of meat products sitting in Peter Green's warehouse, unable to move. "There's no information," he told the BBC. "Everything along the chain has to be stopped, and then there are thousands of pounds worth of product that are just wasting away."[3] For a small producer, watching perfectly good food spoil because a computer system was locked is devastating.

In the previous chapter, we examined OT and ICS in depth and explained why they matter to keep our food system running. We have defined the problem; now let's see what happens when things go wrong. And I need to be honest: things have been going wrong for a while now, and we still haven't learned from it. Shelves emptied not because the food didn't exist, but because it couldn't get there: this is what cyber-physical risk looks like.

When Digital Problems Become Physical Ones

A cyber-physical attack doesn't stop at computers. That's the whole point. It crosses from the digital world into the physical one, affecting equipment, processes, products, and people. In food systems, we're dealing with perishable goods, living organisms, and processes that can't pause while we troubleshoot. You've heard me say this before, and I'm going to keep saying it because it's the constraint that makes everything else more difficult. A cold storage facility loses refrigeration for 12 hours during a ransomware lockout, and you're not looking at a business inconvenience: you're looking at entire shipments headed for disposal. In the UK, Peter Green's attack meant exactly that.

The Cold Chain Federation, which represents the UK's food storage and distribution sectors, has been tracking this problem. Their chief executive noted a "huge increase" in cyberattacks and ransomware over the past year, with at least 10 member companies targeted. About half of the food consumed in the UK passes through the cold chain. When hackers figured that out, they started treating these companies as high-value targets because the ticking clock of spoilage creates pressure to pay.[4]

In March 2025, Astral Foods, South Africa's largest poultry producer, suffered a cyberattack that disrupted its processing division for days. The company reported losses of approximately R20 million, or around $1.1 million, due to the revenue hit and costs associated with clearing the production backlog. But the attack also came while the broader poultry industry was already under pressure from drought-driven feed costs and reduced consumer spending. One incident compounded everything else because chickens can't wait. Astral warned investors that the attack would contribute to a profit decline of up to 60 percent for the half-year period.[5]

The timing matters: attackers now have a playbook. These aren't isolated stories; they're a pattern. And the pattern keeps repeating because we keep hoping the next attack won't happen to us.

People Get Hurt

Here's something we don't talk about enough: cybersecurity in food systems reaches into physical safety: the safety of people working around heavy equipment, extreme temperatures, animals, and industrial processes that can injure or kill if something goes wrong.

In 2014, a German steel mill became only the second known cyberattack in history to cause confirmed physical destruction, after Stuxnet, a 2010 cyberweapon that physically destroyed Iranian nuclear centrifuges by silently manipulating their spin speeds until they tore themselves apart.[6] Attackers used spear-phishing emails—targeted messages designed to trick specific employees into handing over their credentials—to gain access to the plant's office network. From there, they worked their way into the production systems controlling the blast furnace. The malware disrupted multiple components and, critically, prevented the furnace from being shut down through its normal, carefully sequenced process. The result was massive structural damage to the facility. No fatalities were publicly reported, but the physical destruction was real.[7]

Now consider what didn't happen but could have. A blast furnace holds molten metal heated to temperatures exceeding 1500 °C. Shutting one down safely requires a precise, multistage sequence that can take weeks.[8] Bypass that sequence, whether through human error or malicious interference, and you are looking at a potential structural fire in an industrial facility surrounded by stored materials, compressed gases, and heavy equipment. And a fire of that scale can spread to neighboring buildings in the industrial zone. Secondary explosions become possible depending on what's stored nearby. Toxic smoke from burning industrial materials doesn't check property boundaries before it moves into surrounding neighborhoods. Soil contamination, air quality alerts, and community evacuations are not hypothetical overreactions. They are the documented consequences of large industrial fires, cyber-caused or otherwise. The digital intrusion is the first domino, but the last domino to fall has nothing to do with computers.

I was working for a bakery company when I first read about the attack. Industrial ovens don't generate molten metal, but they run continuously, they connect to networked control systems, and when something goes wrong at scale, the consequences move fast. Standing in those facilities, considering risk, I wasn't just thinking about the oven. I was thinking about the people working next to it, the building around it, and the neighborhood outside it. A facility exists within a community, and the community has no say over what happens when the systems within that fence line fail.

This is what cyber-physical risk means in food and agriculture. It isn't just the technology that's affected. It's everything that the technology is near and connected to. It's product, people, animals, infrastructure, environment, and community trust. The risk profile of a facility extends well past its fence line, and the systems we're protecting aren't just digital. They're woven into the physical world in ways that don't announce themselves until something goes wrong.

When Attackers Target the Controls Directly

In May 2024, the RansomHub ransomware group claimed to have compromised the supervisory control and data acquisition (SCADA) system at Matadero de Gijón, a meat processing and bioenergy plant in Spain. Screenshots published by the group appeared to show access to the facility's digestor controls and to critical components of the heating system for both the meat processing operations and the biogas plant that manages waste.[9] The attack forced operations to shut down, and workers were sent home. In addition, more than 40 cattle were stopped from being slaughtered while the facility dealt with the intrusion.[10]

This wasn't just ransomware encrypting files on a server somewhere. The attackers claimed direct access to the industrial controls that manage physical processes. Whether they could have manipulated those systems to cause harm isn't entirely clear from public reporting. But the fact that they got that far, that they had visibility into SCADA controls at a meat plant, should concern everyone in this industry. RansomHub appeared to be purchasing compromised access from initial access brokers on Russian-language forums.[11] They weren't necessarily sophisticated hackers who understood industrial systems from the ground up. They bought their way in, and then they used what they found. Ransomware groups are increasingly targeting OT environments, particularly those with Internet-exposed controls or those connected via remote access tools.

When Food Stops Flowing

On Easter weekend 2025, Marks & Spencer (M&S), one of Britain's most trusted retail names, started experiencing glitches. Contactless payments failed, click-and-collect orders couldn't be processed, and customers noticed something was wrong before the company publicly said a word.[12] Within days, the situation had escalated into one of the most damaging cyberattacks on a UK retailer in years. M&S stopped taking online orders entirely, a shutdown that lasted 46 days. The company's food halls ran low on stock, and shelves emptied. Automated inventory systems were taken offline to contain the damage, forcing staff to track stock with pen and paper, as if it were 1985. The estimated hit to operating profit: £300 million.[13]

The attack originated with a group called Scattered Spider, which used DragonForce ransomware. What made the breach particularly painful was how it started: the attackers didn't exploit a sophisticated zero-day vulnerability. They simply called the IT help desk. M&S's IT support was handled by a third-party contractor, and the attackers impersonated employees to trick the help desk staff into resetting credentials and disabling multifactor authentication. Once inside, they extracted the company's entire Active Directory database, cracked the password hashes offline, and gained access to systems across the organization. By the time the ransomware was deployed, they had been inside the network for months.[14]

This is cyber-physical impact at scale. M&S has more than 1,000 stores and 64,000 employees.[15] When the company's digital systems went dark, physical food stopped flowing to physical customers. Products spoiled because logistics couldn't coordinate. Waste increased because manual processes couldn't match the efficiency of automated systems. Customers who depended on M&S for their weekly groceries had to go elsewhere, assuming they could find what they needed, because the attack was part of a wave hitting multiple UK retailers simultaneously.[16]

M&S chose not to pay the ransom. It rebuilt from backups, brought in CrowdStrike and Microsoft for forensics, and gradually restored operations. But the attack showed how a single phone call to a help desk can bring a £10 billion retailer to its knees and leave food sitting where customers can't buy it.[17]

Concentration Amplifies Everything

The structure of the food system makes every disruption worse. In the United States, four companies control more than 80 percent of beef processing.[18] A handful of grain traders dominate global commodity flows. When one major player gets hit, there isn't much slack in the system.

We saw this with the JBS attack in 2021, when the world's largest meat processor was taken offline by ransomware. All U.S. beef plants shut down for days. Nearly a quarter of American beef processing capacity went dark overnight. The company paid $11 million to restore operations, but the damage didn't stay inside JBS's facilities.[19]

Independent ranchers who had animals ready for market suddenly had nowhere to send them. Livestock that had been raised and fed for months, representing real financial investment and real animal welfare obligations, had no destination. Some producers faced the impossible choice of continuing to feed animals at a loss or selling into a market that had no capacity to receive them. Smaller operators who couldn't absorb the disruption felt it for weeks after JBS's systems came back online.

That's what concentration without resilience looks like. It isn't just a corporate problem or a cybersecurity problem. It's a structural problem that runs straight through to the people and animals at the beginning of the supply chain, the ones with the least ability to absorb a shock they had no part in creating.

Concentration isn't inherently bad. Large operations can invest in equipment and efficiency that smaller ones can't. But concentration without resilience means a single point of failure can affect the entire market. And we haven't built the resilience to match the concentration.

What It Teaches Us: Why We Keep Getting Surprised by Attacks

Ransomware actors have been timing attacks on agricultural cooperatives to coincide with critical planting and harvest seasons.[20] Attacks have completely halted production at some facilities while crippling administrative functions at others. The timing isn't coincidental. Attackers have learned that food and agriculture companies face unique pressure to pay quickly because their products spoil and their operations can't simply pause. Seems like a movie script, where the threat actor lies in wait to spring the trap; sadly, it's kind of like that in food and agriculture. More than one-third of food and beverage companies report that a single hour of operational downtime costs at least $1 million.[21] That kind of pressure creates exactly the leverage ransomware operators are looking for.

Agriculture isn't exempt from this conversation; any farm equipment connected to the Internet is vulnerable to malware and data theft.[22] Four major cyber threats face the agriculture sector: ransomware attacks, foreign malware, data and intellectual property theft, and bioterrorism. Ransomware leads the list—no surprise to anyone.

Groups like Akira have collected more than $244 million in ransom payments since March 2023, specifically targeting sectors such as food and agriculture, as well as manufacturing and healthcare.[23] These aren't opportunistic attacks hitting random targets. The attackers have done their homework, and they know which industries pay.

I've spent enough time in this industry to recognize what happens next. After each major incident, there's a burst of attention. Companies review their security postures, industry groups publish guidance, regulators issue concerned statements, lawyers' statements go out as press releases, and people post online; then, gradually, attention drifts back to production targets and cost pressures. The food industry still treats cybersecurity as a cost center rather than a production enabler or an opportunity center. When budgets tighten, security investments get deferred. Hackers have learned to time attacks for maximum leverage, hitting during planting seasons, holidays, and peak demand periods. And we keep secretly hoping the next attack will hit another industry.

Every year brings more attacks, and every year the industry says it needs to do better. But the attacks keep working because the defenses don't change and the mindset remains the same. We're not keeping up; we're only responding, not being proactive enough. That's not a technology problem. It's a priority problem.

Moving Forward: Hope to Action

Cyber-physical threats in food systems are real, present, and worsening. The attacks aren't theoretical scenarios for tabletop exercises. They're happening right now to companies large and small, across every sector of food and agriculture. The physical consequences are real, too. Food spoils when refrigeration fails. Products get contaminated when processing controls malfunction. Workers face danger when safety systems can't be trusted. Food security is threatened when communities go hungry due to supply chain disruptions.

Addressing these risks requires moving beyond the hope that it won't happen here. It requires treating cybersecurity as integral to operations rather than separate from them. It requires investing in resilience, even when budgets are tight, or at least getting more creative with budgets. It requires accepting that the systems we've built are vulnerable and doing the work to protect them.

That work includes understanding where vulnerabilities enter through relationships with vendors, partners, and service providers. Many of the incidents in this chapter didn't start inside the food company itself. They started with a trusted third party whose systems connect to the food company's operations: a logistics provider, an equipment dealer who remotely connects to your production line, or a software vendor that handles your scheduling. When any of these partners are compromised, you feel the effects.

The food industry has spent decades building efficiency through interconnection: just-in-time delivery, integrated supply chains, and shared platforms. All of it works beautifully until one link breaks, and then the whole system shudders. We built these connections because they made business sense. Now we must secure them or accept that every partner's weakness is our weakness, too.

In the next chapter, we turn to third-party and vendor risk. How do you assess whether a supplier's security practices are adequate before you sign the contract? What happens when their breach becomes your crisis? And how do you maintain the operational flexibility that makes modern food systems work without leaving the door open for someone else's attacker to walk through?

Here's the uncomfortable truth: you can do everything right inside your own walls and still get taken down by someone you trusted. That's the world we're operating in now. And pretending otherwise is just another way of hoping the next attack hits someone else.

Notes

1. Peter Green Chilled. (2026). **https://petergreenchilled.co.uk/company**
2. Scroxton, A. (2025). "Retail cyber attacks hit food distributor Peter Green Chilled." *Computer Weekly*. **https://www.computerweekly.com/news/366624212/Retail-cyber-attacks-hit-food-distributor-Peter-Green-Chilled**
3. Espiner, T. (2025). "Supplier to major supermarkets hit by cyber attack." *BBC*. **https://www.bbc.com/news/articles/czr88myp570o**
4. Ibid.
5. Reuters. (2025). "South Africa's Astral Foods hit by cybersecurity incident." **https://www.reuters.com/world/africa/south-africas-astral-foods-hit-by-cybersecurity-incident-2025-03-24**
6. Fruhlinger, J. (2022). "Stuxnet explained: the first known cyberweapon." *CSO*. **https://www.csoonline.com/article/562691/stuxnet-explained-the-first-known-cyberweapon.html**; Zetter, K. (2014). *Countdown to Zero Day: Stuxnet and the Launch of the World's First Digital Weapon* (Crown).
7. Zetter, kim. (2015). "A cyberattack on a German steel mill has caused confirmed physical damage for the second time ever." *Wired*. **https://www.wired.com/2015/01/german-steel-mill-hack-destruction**
8. IspatGuru. (2015). "Methods of shutting down a blast furnace." **https://www.ispatguru.com/methods-of-shutting-down-a-blast-furnace**
9. ICS STRIVE. (2024). "Ransomhub targets SCADA system of Spanish meat processing plant." **https://icsstrive.com/incident/ransomhub-targets-scada-system-of-spanish-meat-processing-plant**
10. La Nueva España. (2023). "Sociedad que gestiona el matadero de Gijón declara la insolvencia." **https://www.lne.es/gijon/2023/09/14/sociedad-gestiona-matadero-gijon-declara-92088263.html**
11. Cyble. (2024). "Ransomware menace amplifies for vulnerable industrial control systems: heightened threats to critical infrastructure." **https://cyble.com/blog/ransomware-menace-amplifies-for-vulnerable-industrial-control-systems-heightened-threats-to-critical-infr**
12. Krishnan, H. (2025). "M&S ransomware attack: cybersecurity lessons for the retail sector." *ManageEngine*. **https://www.manageengine.com/eu/log-management/cyber-security/marks-and-spencer-ransomware-retail-sector-cybersecurity.html**
13. Kovacs, E. (2025). "Marks & Spencer expects ransomware attack to cost $400 million." *SecurityWeek*. **https://www.securityweek.com/marks-spencer-expects-ransomware-attack-to-cost-400-million**
14. White, M. (2025). "M&S ransomware hack: Service Desk & Active Directory security lessons." *Specops* Software. **https://specopssoft.com/blog/marks-spencer-ransomware-active-directory**
15. Statista. (2024). "Marks and Spencer - statistics & facts." **https://www.statista.com/topics/3707/marks-and-spencer**
16. Flack, A. (2025). "Marks & Spencer cyberattack: what happened and what it means." *Conosco*. **https://conosco.com/in-the-news/marks-and-spencer-what-it-means**
17. Davey, J., & Young, S. (2025). "M&S' slow recovery from cyberattack puts it at risk of lasting damage." *Reuters*. **https://www.reuters.com/business/retail-consumer/ms-slow-recovery-cyberattack-puts-it-risk-lasting-damage-2025-05-19**

18. Mordor Intelligence. (2026). "United States red meat market size & growth to 2031." **https://www.mordorintelligence.com/industry-reports/united-states-red-meat-market**
19. Kobialka, D. (2023). "JBS meat supplier cyberattack timeline, payment and recovery." *MSSP Alert*. **https://www.msspalert.com/news/jbs-status-updates**
20. FBI. (2022). "Ransomware attacks on agricultural cooperatives potentially timed to critical seasons." **https://www.ic3.gov/CSA/2022/220420-2.pdf**
21. Claroty. (2021). *The Global State of Industrial Cybersecurity 2021: Resilience Amid Disruption*. **https://web-assets.claroty.com/resource-downloads/claroty_the_state_of_industrial_cybersecurity_2021.pdf**
22. Markovic, S. (2025). "The food supply chain has a cybersecurity problem." *Help Net Security*. **https://www.helpnetsecurity.com/2025/07/30/agri-food-sector-cybersecurity**
23. Cybersecurity and Infrastructure Security Agency. (2025). "#StopRansomware: Akira ransomware." Joint cybersecurity advisory AA24-109A. **https://www.cisa.gov/news-events/cybersecurity-advisories/aa24-109a**

CHAPTER 19

Third-Party and Vendor Risk

How Partners Can Open the Floodgates

The week before Thanksgiving 2024, Starbucks store managers across North America pulled out pens and paper. The scheduling software was down. The system that tracked barista hours—who worked when and who was owed what—had gone dark. Eleven thousand stores were now calculating payroll by hand.[1] Three thousand miles away, Morrisons warehouse staff in the United Kingdom were watching fresh produce sit. The systems that coordinated shipments to 500 stores had frozen. Backup processes kicked in, but the company acknowledged the obvious: the "smooth flow of goods" wasn't flowing.[2] The connection between American coffee shops and British grocery stores wasn't immediately obvious. They shared no suppliers, no products, no customers. What they shared was a vendor: Blue Yonder, a supply chain management platform most consumers have never encountered. When the Termite ransomware group hit Blue Yonder's managed services environment, the blast radius crossed industries and oceans within hours.[3] None of these companies had chosen poorly. Blue Yonder was a market leader, a Panasonic subsidiary with more than 3,000 customers, trusted by Microsoft, Tesco, and DHL. The problem wasn't due diligence. The problem was structural: your security depends on decisions made by companies you may never evaluate.

This chapter examines what happens when the relationships that make modern food systems function also become the pathways through which disruption travels.

When Your Vendor's Vendor Becomes Your Problem

The Kaseya VSA attack of July 2021 demonstrated how third-party compromise can cascade through layers of business relationships. Kaseya provided remote monitoring and management software to managed service provider companies that handle IT functions for businesses lacking internal technical staff. When attackers exploited a zero-day vulnerability (a secret security bug that the software maker doesn't know about yet, so there is no fix for it[4]) in Kaseya's platform, they didn't just compromise Kaseya's systems: they gained the ability to push ransomware through legitimate update mechanisms to approximately 1,500 downstream organizations simultaneously.[5]

The victims hadn't had any direct relationship with Kaseya. Many probably didn't know their managed service provider used that platform. The attack reached a Swedish grocery chain, schools and kindergartens in New Zealand, and small towns in Maryland whose IT systems went dark on a holiday weekend.[6] The attackers, affiliated with the REvil ransomware group, initially demanded $70 million for a universal decrypter, a price tag reflecting the scale of disruption they'd achieved through a single point of compromise.[7]

The Cyber Peace Institute's analysis of the incident noted its impact on nurseries, schools, pharmacies, and supermarkets across 17 countries.[8] These weren't organizations that had made poor technology choices, but small businesses and public institutions that had sensibly outsourced IT management to local providers that, in turn, had sensibly adopted an industry-standard remote monitoring platform. Each layer made reasonable decisions. The aggregate created catastrophic exposure.

Food and agriculture weren't the primary target, but the sector felt the effects. The attack's timing over the Fourth of July weekend in the United States meant many organizations discovered the compromise when they had a skeleton staff and limited ability to respond. More importantly, it established a pattern: compromise the technology providers that serve multiple clients, and you can affect entire industries without ever directly attacking the companies you ultimately disrupt.

This pattern has become increasingly attractive to threat actors. Gartner projects that the financial impact of supply chain attacks will grow from $40 billion in 2023 to $138 billion by 2031.[9] The economics are compelling from an attacker's perspective: breach one platform and harvest hundreds of victims. The Cl0p ransomware group has built an entire operation model around this approach, accounting for 64 percent of attributable third-party breaches in recent years through mass exploitation of vulnerabilities in file-transfer solutions and other shared infrastructure.[10]

The Remote Access Accumulation

Food production facilities have quietly accumulated dozens of permanent network connections without anyone fully mapping their extent. Consider the equipment in a typical dairy processing plant: pasteurizers with remote monitoring capabilities, automated filling lines with vendor diagnostics, refrigeration systems with predictive maintenance sensors, and packaging equipment with firmware update requirements. Each piece of equipment may come from a different manufacturer, and each manufacturer maintains remote access for support, troubleshooting, and software updates.

The logic behind each individual connection is sound. Equipment downtime is expensive and annoying. Having vendor technicians able to diagnose problems remotely before dispatching a service call saves money and accelerates repairs. The alternative, air-gapping equipment from all external connections, would mean slower maintenance response times and potentially longer production interruptions.

But the aggregate effect creates something no one specifically designed for: a facility with 20, 30, or more permanent pathways from the outside world into operational technology (OT) systems. The pasteurizer vendor can see the pasteurizer. The filling line vendor can see the filling line. The refrigeration vendor can see the refrigeration controls. Collectively, these vendors have visibility into critical processes that determine food safety outcomes, and the security practices governing that access vary widely.

A useful exercise: attempt to inventory every remote-access connection to your facility. Include equipment vendors, software providers, managed service companies, and any contractor with VPN credentials. Then consider: when was each account last audited? What happens when a technician at one of those vendors leaves their job? Who has the authority to change credentials, and how often is that authority exercised? What controls are in place to monitor them?

Most organizations discover gaps as a result of this exercise. The account created for a vendor three years ago that is still active, although no one remembers why. The shared credentials used by multiple technicians at a service company. The VPN connection that was supposed to be temporary but somehow became permanent. Each of these represents not malicious intent but accumulated convenience that becomes vulnerability.

Earlier chapters examined how researchers discovered vulnerabilities at John Deere and CNH Industrial systems that could have exposed equipment owners' identities and locations. What's relevant here isn't the specific technical findings but what they revealed about vendor-ecosystem security more broadly. Security researcher Sick Codes found that after creating a developer account with Deere, vulnerabilities surfaced within hours of casual exploration. Default credentials remained unchanged. Monitoring pages were publicly accessible. APIs exposed more data than intended. These weren't sophisticated zero-day exploits; they were the kind of configuration oversights that proliferate when security isn't a primary design consideration.[11]

Multiply this across every equipment vendor, software provider, and managed service company with access to food production systems. Each vendor faces its own resource constraints, security priorities, and pressure to prioritize functionality over protection. The security of your operation becomes an aggregate of decisions made by dozens of organizations, most of which you've never evaluated and some of which you've never heard of.

Shared Platforms, Shared Fate

Blue Yonder revealed a structural problem that traditional vendor risk management can't address: too many companies depend on the same platforms. When a single software provider serves thousands of customers across multiple industries and geographies, a compromise at that provider becomes an industry-wide event.

Blue Yonder wasn't obscure or specialized. The companies affected—Starbucks, Morrisons, Sainsbury's, Procter & Gamble, Walgreens, and BIC—chose Blue Yonder because it represented best-in-class supply chain management capabilities.[12] They weren't cutting corners on technology investment. They were concentrating operational dependency on a platform that, when it faltered, took down pieces of multiple supply chains simultaneously.

This concentration appears throughout food and agriculture. A handful of companies dominate agricultural equipment manufacturing. A limited number of platforms handle commodity trading. A few major providers offer cold-chain monitoring. When consolidation improves efficiency and capability, it also creates single points of failure that affect everyone when things go wrong.

The standard response is to diversify vendors and maintain alternatives, but this collides with operational reality. Supply chain software isn't interchangeable; switching platforms requires significant migration effort. Equipment vendors can't be swapped mid-season. Cold-chain monitoring systems integrate with logistics operations in ways that make rapid substitution impractical. Food's biological constraints compound the challenge. When Morrisons' warehouse management systems went dark, fresh produce didn't politely wait for restoration. The company acknowledged that the "smooth flow of goods to our stores" had been impacted. In an industry where temperature abuse and delivery delays translate directly to spoilage, system outages carry costs that go beyond inconvenience.

The Data That Flows Without Asking

Vendor relationships involve information flows that may not be obvious to the organizations sharing them. Consider the data generated by a food safety laboratory providing testing services. Every sample submitted carries information about the submitting company's products, suppliers, and production practices. A laboratory serving multiple food companies accumulates competitive intelligence across the industry, not because anyone intended to create an intelligence repository, but because the testing relationship inherently requires sample information.

Quality management systems hold similar accumulations. Cloud-based platforms for food safety documentation contain specifications, supplier evaluations, corrective action records, and complaint histories. A breach at such a provider exposes operational details that could damage reputation, inform competitive strategy, or enable targeted attacks on identified weaknesses.

The Mondelez case from 2023 also demonstrated how indirect relationships can create exposure pathways that vendor risk programs rarely anticipate. Attackers didn't breach Mondelez directly; they compromised Bryan Cave, a law firm handling sensitive matters for the company. The breach affected more than 50,000 current and former employees, and the full extent of the damage took months to determine.[13] Although the legal relationship existed to manage risk, it became a vector for the very harm it was supposed to address.

Supply chain visibility platforms designed to provide transparency across complex networks create their own concentration risks. These platforms aggregate data from multiple participants: who supplies whom, at what volumes, on what schedules, with what dependencies. The intelligence value of such aggregation is precisely why the platforms exist. It's also why they represent attractive targets for actors interested in mapping and potentially disrupting food supply networks.

The Trust Problem

Food and agriculture operate on relationships that often span generations. The farmer who buys equipment from the same dealer that their parents used. The processor that sources ingredients from suppliers its company has worked with for decades. The cooperative membership passed from grandparents to grandchildren. These relationships aren't just commercial arrangements; they're interwoven with community, identity, and mutual obligation.

This trust-based culture creates blind spots that technical security measures can't easily address. Asking a long-term partner about their cybersecurity practices feels like questioning their integrity. The farmer who's worked with their equipment dealer for 30 years isn't going to demand penetration test results before the next service call. The processor whose quality manager has a personal relationship with the ingredient supplier's sales representative isn't going to insert security questionnaires into that dynamic.

The discomfort is understandable. Trust enables the collaboration that keeps food systems functioning. The informal information shared among farmers about weather patterns, pest pressures, and market conditions depends on relationships built over time. The responsive service that distinguishes good vendors from commodity suppliers depends on relationships that can't be reduced to contractual terms.

But trust in one domain doesn't transfer automatically to another. A supplier's product quality says nothing about its password policies. An equipment dealer's service responsiveness reveals nothing about its network architecture. An ingredient vendor's consistent specifications don't indicate anything about its data-handling practices. The qualities that make someone a trusted partner in traditional senses are entirely separate from the qualities that make them a secure partner in cybersecurity terms.

One approach that preserves relationship quality while addressing security needs is to frame the conversation around shared risk rather than evaluation. "We're both vulnerable if your systems are compromised" positions cybersecurity as a mutual interest rather than a judgment. Sharing your own security challenges and how you're addressing them invites reciprocal transparency without creating adversarial dynamics.

Contracts, Questionnaires, and Their Limits

The standard toolkit for managing third-party risk—vendor questionnaires, contractual security requirements, audit rights—documents due diligence without necessarily creating protection.

Questionnaires capture point-in-time snapshots. A vendor's practices when they complete the questionnaire may differ from its practices six months later. Staff changes. Budget pressures shift priorities. New systems introduce new vulnerabilities. The completed questionnaire sits in a file while conditions evolve.

Contractual requirements establish expectations without ensuring implementation. Security provisions typically include representations about practices, notification requirements for incidents, and indemnification for damages. But indemnification works only if the vendor can pay, incident notification works only if the vendor detects breaches, and representations about practices have meaning only if the implementation is in place.

Audit rights theoretically allow verification. In practice, they are rarely exercised. Audits are expensive and require technical expertise that many food companies don't have in-house. They strain vendor relationships, and they cover only what auditors think to examine, which may not include the specific vulnerabilities that attackers ultimately exploit.

The deeper problem is scope. Vendor risk programs evaluate direct relationships. They don't evaluate vendors' vendors, the infrastructure providers that underlie multiple vendor relationships, or the shared platforms that create correlated exposure across industries. The questionnaire sent to your ingredient supplier says nothing about the precision agriculture platform used by the farms that grow those ingredients, or the cloud provider hosting that platform, or the telecommunications carrier connecting farm equipment to cloud services.

Each participant in food supply chains faces identical limitations. The farmer can't meaningfully evaluate the cybersecurity of their equipment manufacturer's software development practices because they lack expertise and access. The processor can't evaluate security at every farm supplying ingredients through its suppliers because it doesn't have relationships that far back. The retailer can't evaluate every step in every supply chain for every product on its shelves—the scale is impossible.

Everyone has a limited ability to see upstream. Everyone pushes responsibility to contracts and questionnaires. The result: no one truly knows how secure the complete supply chain is. Responsibility diffuses until no one holds accountability for systemic risk.

When Regulation Follows the Supply Chain

The limitations of voluntary vendor risk management haven't gone unnoticed by regulators. Europe's NIS2 Directive, which took effect in October 2024, represents the most significant shift in how governments approach supply chain cybersecurity, and its reach extends far beyond European borders.

NIS2 explicitly designates food production, processing, and distribution as essential sectors subject to mandatory cybersecurity requirements. But the directive's most consequential provision may be its focus on supply chains: covered entities must ensure that their suppliers and service providers meet appropriate security standards. This isn't a suggestion or a best practice recommendation. It's a legal obligation with penalties of up to €10 million or 2 percent of global annual turnover, whichever is higher.[14] For food companies based outside the European Union, the instinct might be to consider NIS2 someone else's problem, and that instinct would be wrong.

Consider a U.S.-based spice processor selling to food manufacturers in Germany and France. Those European customers are now legally required to assess their suppliers' cybersecurity practices and ensure that adequate protection is in place. The American company isn't directly regulated by NIS2, but its European customers will start asking questions—detailed questions—that they've never asked before, with contractual requirements attached. Fail to satisfy those requirements, and market access disappears.

The dynamic works in the other direction as well. A U.S. food manufacturer sourcing packaging equipment from Italy, ingredients from the Netherlands, or cold-chain monitoring technology from Denmark will find that those European suppliers now operate under mandatory security obligations. When those suppliers experience incidents, they face reporting requirements that may surface problems previously kept quiet. The transparency that regulation imposes on European vendors benefits their global customers, assuming those customers are paying attention.

NIS2 also addresses the concentration risks we discussed earlier in this chapter. The directive requires covered entities to consider the overall resilience of supply chains and account for vulnerabilities specific to individual suppliers. Regulators have recognized that depending on a single platform that serves an entire industry creates systemic risk that individual vendor assessments miss.

The United States, on the other hand, has no equivalent comprehensive framework for food sector cybersecurity. The Food Safety Modernization Act (FSMA) focuses on preventing contamination, not cyber intrusion. Voluntary guidance is offered by the Cybersecurity and Infrastructure Security Agency (CISA) and the Food and Drug Administration (FDA), but voluntary frameworks struggle against the economic pressures that have historically deprioritized security investment in food systems.

The result is an emerging two-track reality. Companies with European market exposure or European supply chain dependencies will increasingly operate under de facto security requirements, whether they're directly regulated or not. Companies without those connections may continue to treat cybersecurity as optional until an incident or eventual U.S. regulation forces the issue. Later chapters will examine this regulatory landscape in detail, but the immediate implication for vendor risk is clear: the questionnaires and contracts that feel optional today may become mandatory tomorrow.

What It Teaches Us: Collective Action for Systemic Risk

Some third-party risks exceed what individual organizations can address. When an entire industry depends on the same platforms, individual companies can't solve the concentration risk alone. When vulnerabilities affect shared infrastructure, individual patches don't address systemic exposure.

The Cyber Tractor Challenge, now operating as a nonprofit with participation from John Deere, CNH Industrial, and AGCO, represents one model for collective action. Competitors that might resist sharing security information directly can collaborate on workforce development, vulnerability research, and best-practice development through neutral structures.[15] The agricultural equipment manufacturers recognize that they face common adversaries and share customers who mix equipment across brands. Collective security investment benefits everyone; security weaknesses at one manufacturer affect the entire sector's reputation.

Information sharing and communicating about threats and vulnerabilities without revealing competitive information helps industries learn and adapt collectively. When one organization identifies a vulnerability in shared infrastructure, alerting others prevents the same mistake from propagating. When one company experiences a novel attack technique, sharing indicators enables others to detect similar attempts.

Standards development creates baseline expectations that all vendors must meet. Individual food companies have limited leverage to demand security improvements from major technology providers. Industry associations that coordinate requirements among members carry more weight, creating accountability that individual contracts struggle to establish.

Moving Forward: Designing for the Inevitable

Third-party risk in food and agriculture isn't a category on a risk register. It's the fundamental condition of modern food systems. Every farm, processor, distributor, and retailer exists within webs of relationships extending across the globe, touching technology platforms, service providers, infrastructure operators, and regulatory systems. These webs enable the food system to feed billions of people daily. They also create exposure that traditional vendor management can't capture.

Traditional vendor risk management assumes that threats can be prevented through proper evaluation and contractual protection. Systems thinking suggests a different approach: design for failure rather than assuming prevention will succeed.

The organizations that weathered Blue Yonder's disruption best had prepared for exactly the scenario they faced. Sainsbury's had contingency processes in place and activated them successfully.[16] Procter & Gamble built and deployed a manual shipping system within 12 hours, an effort described as "resource-intensive," which is corporate shorthand for a holiday-period all-hands emergency.[17] That turnaround reflects something both organizations had done long before Blue Yonder: mapping supplier failure scenarios and preserving operational flexibility substantial enough to absorb a sudden failure.

Designing for vendor failure means asking different questions during risk assessment. Instead of "Is this vendor secure?" the question becomes "What happens when this vendor fails?" Instead of evaluating vendors in isolation, the question becomes "Where do multiple critical functions depend on the same vendor or platform?"

Redundancy—having alternative capabilities for critical functions—requires investment that tight margins make difficult. But redundancy doesn't necessarily mean maintaining duplicate systems. It might mean preserving manual processes alongside automated ones, maintaining relationships with alternative suppliers even when primary suppliers serve adequately, or ensuring that data is available in formats that can operate outside proprietary platforms.

Segmentation that limits what any single vendor can access reduces the blast radius when a compromise occurs. Not every vendor needs access to everything. The equipment maintenance provider doesn't need access to financial systems. The logistics platform doesn't need access to product formulations. Compartmentalizing access ensures that a single compromised relationship doesn't expose the entire operation.

Detection capabilities matter more when prevention can't be guaranteed. Monitoring for anomalies in vendor-connected systems, unusual data flows, unexpected access patterns, and authentication from unfamiliar locations can identify compromises that prevention measures missed. The investment isn't in keeping attackers out, but in knowing when they've gotten in.

Response planning acknowledges that vendor failures will occur and prepares specific protocols for managing them. How would production continue if the enterprise resource planning system went down? How would shipments be managed if the transportation management platform failed? How would food safety documentation be maintained if the quality management system were compromised? Working through these scenarios before they occur transforms chaotic emergency response into structured business continuity.

The organizations that navigate this reality successfully share certain characteristics. They understand their actual vendor ecosystem: not just the names on their approved vendor list, but the dependencies that extend through multiple tiers. They identify concentration points where multiple critical functions depend on single platforms or providers, design for vendor failure, maintain contingency capabilities that don't depend on the same infrastructure as primary systems, and participate in collective action on risks that exceed the scope of individual organizations.

What they don't achieve—what no one can achieve—is certainty that their extended supply chain is secure. The honest assessment is that complex food systems involve too many relationships, too many technologies, and too many decision-makers for any single participant to control. Accepting that limitation isn't defeatism; it's a realistic assessment that enables better decisions about where to focus limited security resources.

Third-party risk assumes the threat comes from outside relationships. But what about the threats that emerge from inside: from people who already have access, who already have trust, and who already know how your operations work? The next chapter examines insider threats and the often-overlooked vulnerabilities that emerge when trusted relationships end badly or trust proves misplaced from the start.

Notes

1. Ikeda, S. (2024). "Blue Yonder ransomware attack ripples through supply chain to impact big names in retail." *CPO Magazine*. **https://www.cpomagazine.com/cyber-security/blue-yonder-ransomware-attack-ripples-through-supply-chain-to-impact-big-names-in-retail/**
2. Schwartz, M. J. (2024). "Ransomware attack on supply chain provider causes disruption." *BankInfoSecurity*. **https://www.bankinfosecurity.com/ransomware-attack-on-supply-chain-provider-causes-disruption-a-26907**
3. SOCRadar. (2024). "Termite ransomware attack on Blue Yonder." **https://socradar.io/termite-ransomware-attack-on-blue-yonder/**
4. Kaspersky. (n.d.). "Zero-day exploits & zero-day attacks." **https://www.kaspersky.com/resource-center/definitions/zero-day-exploit**

5. National Counterintelligence and Security Center. (2021). "Kaseya VSA supply chain ransomware attack." **https://www.dni.gov/files/NCSC/documents/SafeguardingOurFuture/Kaseya%20VSA%20Supply%20Chain%20Ransomware%20Attack.pdf**
6. Olenick, D. (2021). "List of victims of Kaseya ransomware attack grows." *BankInfoSecurity*. **https://www.bankinfosecurity.com/list-victims-kaseya-ransomware-attack-grows-a-17013**
7. Sason, D. (2021). "REvil ransomware attack on Kaseya VSA: what you need to know." *Varonis*. **https://www.varonis.com/blog/revil-msp-supply-chain-attack**
8. Radu, R. (2021). "What we learned from the Kaseya attack: recommendations for a human-centric approach to curb ransomware." *CyberPeace Institute*. **https://cyberpeaceinstitute.org/publications/what-we-learned-from-the-kaseya-attack-recommendations-for-a-human-centric-approach-to-curb-ransomware/**
9. Patnaik, A. (2024). "Rising tide of software supply chain attacks." *Dark Reading*. **https://www.darkreading.com/vulnerabilities-threats/rising-tide-of-software-supply-chain-attacks**
10. Poireault, K. (2025). "SecurityScorecard observes surge in third-party breaches." *Infosecurity Magazine*. **https://www.infosecurity-magazine.com/news/securityscorecard-surge-third/**
11. Roberts, P. (2021). "DEF CON: security holes in Deere, Case IH shine spotlight on agriculture cyber risk." *The Security Ledger*. **https://securityledger.com/2021/08/def-con-security-holes-in-deere-case-ih-shine-spotlight-on-agriculture-cyber-risk/**
12. Klappholz, S. (2024). "Blue Yonder ransomware attack disrupts grocery, retail and hospitality firms." *ITPro*. **https://www.itpro.com/security/cyber-attacks/blue-yonder-ransomware-attack-disrupts-grocery-retail-and-hospitality-firms**
13. TXOne Networks. (2024). "Understanding cyber threats in the food manufacturing industry." **https://www.txone.com/blog/understanding-cyber-threats-in-food-manufacturing-industry/**
14. European Union. (2022). "Directive (EU) 2022/2555 on measures for a high common level of cybersecurity across the Union (NIS2)."*EUR-Lex*. **https://eur-lex.europa.eu/eli/dir/2022/2555/oj**
15. Kell, J. (2025). "Why Deere paid hackers $1.5 million to detect vulnerabilities and keep the agriculture equipment maker safe." *Fortune*. **https://fortune.com/2025/06/04/why-deere-paid-hackers-1-5-million-to-detect-vulnerabilities-and-keep-the-agriculture-equipment-maker-safe/**
16. Robb, B. (2024). "Ransomware meets retail: Blue Yonder attack." *BlackFog*. **https://www.blackfog.com/ransomware-meets-retail-blue-yonder-attack/**
17. Stroh, K. (2025). "Procter & Gamble operations unhindered by Blue Yonder disruption." *Supply Chain Dive*. **https://www.supplychaindive.com/news/procter-gamble-operations-tms-outage-blue-yonder/738293/**

CHAPTER 20

Insider Threats

When a Needle Changes Everything

In September 2018, a six-year-old boy in Queensland, Australia, bit into a strawberry and felt something sharp pierce his mouth. His mother found a sewing needle embedded in the fruit, one of several that had been deliberately inserted into strawberries from multiple Australian farms.[1] What followed was one of the most damaging food sabotage incidents in Australian history. Within days, needles were found in strawberries across Queensland, New South Wales, Victoria, and Western Australia. Consumers posted photographs on social media, and panic spread faster than investigation could keep pace. Supermarkets pulled strawberries from shelves nationwide. Some destroyed entire shipments as a precaution, and export markets closed. The Australian strawberry industry, which had been approaching peak season, watched $160 million in value evaporate in weeks.[2]

The investigation eventually focused on a former farm supervisor who had worked at one of the affected strawberry farms and been dismissed in circumstances she reportedly considered unfair. According to prosecutors, her motive was revenge against the employers she blamed for her termination. She knew the operations intimately: the packing processes, the quality control gaps, the moments when strawberries were vulnerable. That knowledge, combined with her grievance, allegedly became the weapons she used against an industry that had employed her.[3]

She was charged with contaminating strawberries with needles, but all charges were dropped before trial. The damage extended far beyond the initial cases, as more than 200 reports followed nationwide, believed to be hoaxes or copycat events.[4] The entire Australian strawberry industry suffered for one person's alleged act of revenge. Farmers who had never met the accused lost their livelihoods, and workers across the supply chain lost jobs. Ultimately, consumers lost trust in a beloved fruit.

One person, feeling wronged and knowing how the system worked, caused an entire national industry to crater. She didn't need sophisticated hacking or organized terrorism, just needles, strawberries, and the insider access that comes from having worked the job. This is what insider threat looks like in food and agriculture: a human being with human grievances, using the access and knowledge their employment provided to cause harm that ripples across an entire supply chain.

What Is an Insider Threat?

An *insider threat* is any risk to an organization that originates from people who have or had legitimate access: current employees, former employees whose access wasn't fully removed, contractors, vendors, partners, or anyone else whose position grants them knowledge or entry that outsiders don't have.[5] The term sounds clinical, but we're talking about people.

In food and agriculture, insider threats create consequences across three interconnected domains. First, they threaten food safety and integrity—the contamination cases that make headlines, the tampering that can sicken or kill. Second, they threaten operational continuity: sabotaged equipment, disrupted production, damaged capability. Third, they threaten information and systems: stolen formulations, manipulated data, and compromised monitoring systems that enable other harms.

Food Defense programs, now required under the FDA's Food Safety Modernization Act (FSMA) Intentional Adulteration Rule for certain facilities, specifically address insider threats through vulnerability assessments and mitigation strategies.[6] These programs recognize that people with legitimate access pose different risks than external actors; they know the operations, they understand the gaps, and they have the opportunity that outsiders lack.

I saw this firsthand during my years at a bakery factory. An employee, angry about not receiving a pay increase, walked from the peanut production area into the non-peanut area and back again. It was intentional. The production line made cookies for a major restaurant chain, and that single act of contamination triggered a full recall of the company's chocolate chip cookies, warning signs posted in stores, and, eventually, the loss of that account. Thankfully, no one was harmed.

Food Defense was involved in the investigation, but what stuck with me was a question I couldn't shake: why was it so easy for the employee to walk between those areas? Years later, at a Food Defense meetup, I told this story and asked their opinions. The answer was sobering: food safety culture often depends on people simply doing the right thing. I pushed back: what about physical access controls, keycards, biometrics? That's where cyber-physical security enters the picture, and where cybersecurity can help prevent these incidents before they happen. A badge reader doesn't care if someone is having a bad day.

These domains blur together in practice. Someone who manipulates temperature records in a food safety database creates a cyber-physical threat: digital action, physical consequence. Someone who shares credentials with an external attacker may not intend harm but may still enable it. Someone who disables a safety monitoring system removes the safeguard that would catch contamination.

Insider threats aren't limited to malicious actors with deliberate intent to harm. The spectrum includes three distinct categories that organizations must address differently.

Malicious insiders are people who deliberately cause harm. They may be motivated by revenge, like the strawberry saboteur allegedly seeking payback for her termination. They may be motivated by money, accepting payment to provide access or look the other way. They may be motivated by ideology, believing that harming food systems serves some larger cause. They may be coerced, acting under threat to themselves or their families. Whatever the motivation, they intend to cause damage.[7]

Negligent insiders cause harm through carelessness rather than intention. The worker who shares their password because it's convenient. The supervisor who skips a food safety check because they're running behind schedule. The employee who plugs an infected USB drive into a production system without thinking. They don't want to cause problems, but their actions create vulnerabilities that others can exploit or that lead directly to harm.[8]

Compromised insiders are people whose legitimate access is exploited by external actors, often without their full awareness. Their credentials get stolen through phishing. They are manipulated through social engineering. They provide information to someone who seems trustworthy but isn't. They may not even know they've become a threat vector until damage is done.[9]

You have keys, passwords, and access codes to various places and systems. You know things about your workplace that outsiders don't, such as schedules, vulnerabilities, workarounds, the door that doesn't latch properly, and the system that hasn't been updated. If you wanted to cause harm, you'd know where to start. If someone manipulated you skillfully, they could extract that knowledge without you realizing its significance. That's what every organization faces with every person who has access. Not that everyone will become a threat; the vast majority never will. But the potential exists wherever access exists.

Why People Become Threats

Understanding why people become insider threats isn't about excusing their actions. It's about recognizing patterns that enable prevention. When we understand what drives people from being trusted employees to becoming threat actors, we can intervene earlier, build systems that reduce opportunity, and create cultures where harmful impulses find less fertile ground.

Grievance is the most common pathway. An employee feels wronged: they were passed over for promotion, disciplined unfairly, denied a raise they expected, or terminated in circumstances they consider unjust. The grievance may be legitimate. Maybe they were treated badly. Maybe management made poor decisions. Maybe the organization broke implicit promises.

But a legitimate grievance doesn't justify a harmful response. The gap between "I was treated unfairly" and "therefore I will contaminate product" or "therefore I will sabotage equipment" is the space where insider threats develop. Most people with grievances don't cross that line. They complain to colleagues, update their resumes, and eventually move on. What distinguishes those who cause harm is usually some combination of perceived powerlessness, the feeling that legitimate channels won't work, awareness of opportunity from their position, and an emotional tipping point where resentment becomes action.

Consider the strawberry case. If the accounts are accurate, a woman who felt wronged by her termination used her knowledge of packing operations to insert needles into fruit.

Her grievance may have felt entirely justified to her. The unfairness she experienced may have been real. None of that justifies endangering children or destroying an industry. But understanding the pathway from grievance to action helps organizations recognize warning signs and address problems before they escalate.

Financial pressure drives different patterns. Many food production and agriculture workers earn modest wages. An unexpected medical bill, a car breakdown, or a family emergency can create genuine desperation. When legitimate options seem exhausted, illegitimate options become thinkable. This might mean stealing products that can be sold. It might mean accepting payment in exchange for information or access. It might mean participating in schemes orchestrated by others, such as when someone approaches you, offering money for "small favors" that gradually escalate into serious compromise.

If you've seen *Willy Wonka and the Chocolate Factory*, you've watched this pattern play out. Charlie's family survives on cabbage soup. He wins a chance to visit the Wonka chocolate factory, and a competing candy maker named Slugworth (who knows Charlie is desperate) offers him money for one small thing: an Everlasting Gobstopper, smuggled out of the factory. Charlie doesn't do it; that's the point of the story. But the recruitment attempt is textbook: identify someone with access, someone under financial pressure, and make an offer that feels safe. That's how it works in real life, too, except real people don't always have Charlie's integrity, and the recruiters aren't testing them.

Most organizations don't like discussing this. When a worker earning $15 an hour (or less) is offered $5,000 to leave a door unlocked or share an access code, they face a choice that looks different to them than it does for someone with financial security. That doesn't excuse the choice of people at all income levels not to maintain integrity under pressure. But pretending that financial desperation doesn't affect decision-making ignores reality. Organizations that want to reduce insider threats must grapple with what they pay people and how they support employees in their factories or fields who are facing hardship.

Ideology occasionally motivates insider threats in food systems. Someone comes to believe that industrial food production is immoral, that exposing conditions justifies any means, and that harming the system serves a righteous cause. They may have been radicalized through online communities, by a nation-state, through association with activist groups, or as a result of personal experiences that shifted their moral framework.

The radicalization pattern often follows a recognizable path. Existing moral discomfort with industry practices meets an external narrative that validates and intensifies that discomfort. Over time, the employer transforms in the employee's mind from "imperfect workplace" to "irredeemable evil." Actions that would once have seemed unthinkable become justifiable, even obligatory, within the new moral framework.

Coercion creates insider threats from employees who may desperately wish they weren't causing harm. Threats to family members. Blackmail based on personal information. Exploitation of immigration status or legal vulnerability. Financial debts being called in. The person being coerced may appear normal because they're not acting on their own anger or beliefs: they're acting because someone has leverage over them.

One pattern deserves particular attention: it's often easier to radicalize or recruit someone with existing access than to develop that access yourself. An external actor who wants to harm food systems doesn't need to get a job at a processing plant. They need to identify and influence someone who already works there. Social engineering, which we'll examine more closely later in this chapter, is often the bridge that turns an ordinary employee into an insider threat.

Across the Supply Chain: Where Insider Threats Live

Insider threats aren't limited to food processing facilities. They span the entire global food and agriculture supply chain, with distinct characteristics and vulnerabilities at each stage.

On the farm

The people with access include farm owners and family members, seasonal and migrant workers, equipment operators, veterinarians and consultants, delivery drivers, and service providers. The vulnerabilities include contamination of crops during growing or harvest, adulteration of livestock feed, tampering with agricultural inputs, manipulation of precision agriculture systems, and theft of genetic materials or proprietary data.

Farms present unique challenges. They operate across vast areas with minimal supervision. A worker alone in a field has opportunities that don't exist in a factory where activity is visible. Seasonal workers may have limited background verification and minimal ties to the operation. The trust that enables agricultural work—handing someone keys to a tractor and trusting them to operate across hundreds of acres unsupervised—is the same trust that creates vulnerability.

After a talk I gave, a dairy farmer approached me. He'd been nodding along during the discussion of insider threats, and when we spoke afterward, I understood why. A former employee, who'd worked on his operation for barely a week before being let go, had called health inspectors to report that the dairy's milk was contaminated. The milk wasn't contaminated. The inspectors came, tested, found nothing, and had a feeling it was most likely a disgruntled employee from the beginning. But the farmer still had to stop everything, accommodate the inspection, and wonder what else this former employee might do.

The farmer suspected the brief-tenured employee might have been an activist who took the job specifically to cause problems, but the employee requested his job back after termination, which struck the farmer as inconsistent with a genuine grievance. Whether that suspicion was accurate or not, it revealed something important: this dairy farmer understood social engineering. He understood that someone might seek employment not to work, but to gain insider knowledge and access. He understood that the threat might be not physical contamination but weaponized regulatory systems and reputation damage.

He told me his worries about perception, about misinformation and disinformation targeting his dairy, and about becoming a target both digitally and physically. These aren't abstract concerns for him. They're operational realities he thinks about while also managing a herd, maintaining equipment, running a business, and hosting farm tours. Farmers know these threats exist. What they often lack isn't awareness; it's resources and support to address risks that extend far beyond their traditional expertise.

In food-processing facilities

Access extends to production line workers, maintenance and sanitation staff, quality control personnel, supervisors and managers, and temporary workers brought in for peak seasons. The vulnerabilities include direct contamination of products (as in the strawberry case, but at the factory level), sabotage of food safety controls, manipulation of processing parameters, tampering with documentation, and theft of formulations and trade secrets.

Processing facilities handle enormous volumes under time pressure. Workers have physical contact with products, and supervision can't be continuous across every station on a line moving at speed. The same access that enables efficient production enables potential harm. This is where most people picture insider threats occurring, but it's only one link in a much longer chain.

Within distribution and logistics

The people with access include warehouse workers, truck drivers, loading dock personnel, cold storage staff, and third-party logistics providers. Products pass through many hands between production and consumption. A driver alone with a refrigerated trailer for hours has access that is impossible to monitor continuously. A warehouse worker handling thousands of cases of product has an opportunity that aggregates across every touch. The vulnerabilities here include tampering with product during storage or transit, temperature abuse (intentional or negligent), theft and diversion, manipulation of shipment documentation, and introduction of contamination at points far from where product was manufactured.

At retail and foodservice locations

Access extends to store employees, kitchen staff, servers, receiving personnel, and cleaning crews. The Disney case in which a terminated employee manipulated allergen data that could have endangered consumers demonstrates that insider threats in this segment aren't limited to people who physically touch food.[10] Someone with access to information systems can alter data that consumers trust: allergen warnings, ingredient lists, and expiration dates.

In supporting functions

People with access include IT and systems personnel, food safety and quality staff, regulatory compliance personnel, procurement teams, and administrative staff. These roles may seem removed from physical product, but they often have access to systems that directly affect safety: food safety databases, supplier approval processes, testing and certification records, and security systems themselves.

The cyber-physical reality that we have been examining throughout the book becomes particularly relevant here. You don't have to touch food to harm food systems. Altering allergen information, falsifying test results, manipulating supplier qualifications, and disabling safety monitoring—these digital actions create physical consequences.

An insider with access to the right systems can enable contamination to go undetected or can create conditions for harm without ever setting foot on a production floor.

Agricultural input and equipment supply

Insiders have access to seeds, fertilizers, pesticides, and animal pharmaceuticals and the equipment that applies them. Contamination or adulteration at this level affects everything downstream. Manipulation of precision agriculture software can alter how inputs are applied across thousands of acres. Compromises of equipment control systems can affect operations at every farm using that equipment.

Financial and regulatory systems

People with access include commodity traders, crop insurance adjusters, food safety inspectors, and certification auditors. These roles have visibility into operations across many organizations. A corrupted inspector can enable unsafe operations to continue. A compromised trader can manipulate markets based on inside information. A fraudulent certification can send unsafe product into channels that trust the certification.

Research on insider threats in food and agriculture identifies consistent patterns across incidents: grievances over perceived unfair treatment, financial pressure that creates vulnerability to recruitment, and ideological motivation that reframes harm as justified action.[11] These patterns appear whether the threat manifests at the farm level, in processing, or in distribution. Every point in this chain has insiders. Every insider has knowledge and access that could be misused. The chain is only as strong as its weakest human link.

When Digital Meets Physical

Food and agriculture cybersecurity centers on cyber-physical risk: digital systems that control or monitor physical processes, where compromise creates tangible harm in the real world. Insider threats operate at this intersection in ways that traditional Food Defense and traditional cybersecurity often fail to recognize.

Consider what an insider with the right access can do digitally to cause physical harm:

Someone with access to pasteurization controls could alter temperature setpoints, reducing the kill step that makes dairy products safe. Someone with access to food safety monitoring systems could disable alerts that would catch equipment malfunctions or temperature excursions. Someone with access to formulation databases could modify recipes in ways that introduce allergen cross-contact or unsafe ingredient ratios. And someone with access to quality control systems could approve products that should have been rejected or flag safe products as unsafe, causing economic damage.

The physical access required by food operations also enables digital compromise. USB drives introduced to production systems can carry malware. Credentials shared for operational convenience become attack vectors. Physical access to network infrastructure enables digital attacks. Equipment installed without security vetting creates persistent vulnerabilities.

This is where Food Defense and cybersecurity must converge.

Food Defense, as traditionally conceived, focuses on intentional acts to contaminate or disrupt the food supply, primarily through physical threats. The framework imagines someone introducing a contaminant, tampering with product, or sabotaging equipment. It developed when digital systems were less central to food operations.

Cybersecurity, as traditionally practiced, focuses on protecting information systems, data confidentiality, system integrity, and service availability. It developed in contexts where the primary concerns were data theft, financial fraud, and system outages.

In modern food operations, these domains overlap extensively. Digital manipulation can enable physical contamination to go undetected. Cyberattacks can disable food safety monitoring, creating windows during which physical contamination would not be detected. Information theft can provide intelligence that enables targeted physical attacks. Social engineering that deceives people into revealing information or taking actions gathers the knowledge that makes physical intrusion possible.

Security professionals have increasingly recognized that Food Defense programs must address both physical and cyber-physical attack vectors.[12] Yet in many food organizations, Food Defense reports to Food Safety or Operations, whereas Cybersecurity reports to IT. They rarely coordinate systematically. Training happens in separate silos. Incident response plans don't span both domains. The organizational structure reflects a division that no longer matches operational reality.

Insiders don't recognize these organizational boundaries. Someone planning harm will exploit whatever access they have: physical, digital, or both. The dismissed employee who still has remote access credentials might manipulate safety documentation rather than physically tamper with product. The financially desperate worker might sell information that enables a cyberattack rather than steal physical product. The ideologically motivated insider might document operations for external release while also considering digital sabotage. Effective programs require joint threat assessments, integrated training, and coordinated incident response. The artificial division between physical and digital security creates gaps that insiders, or those who recruit insiders, can exploit.

Social Engineering: Creating Insiders from Outside

Social engineering deserves particular attention because it represents how external threat actors create insider threats without ever getting hired themselves. *Social engineering* is the practice of manipulating people into revealing information, providing access, or taking actions that compromise security. It exploits human psychology—trust, helpfulness, authority, urgency, or fear—rather than technical vulnerabilities. It's often the first step in attacks that culminate in insider compromise.

This is what social engineering looks like in food and agriculture:

Pretexting involves creating a false scenario to extract information. Someone calls a food facility claiming to be from a supplier, a regulatory agency, or a customer's quality team. They ask questions about operations, systems, and procedures, gathering intelligence that enables later attacks. The employee being questioned wants to be helpful, wants to provide good customer service, and doesn't recognize that the questions are building a picture of vulnerabilities.

Phishing targets employees with deceptive messages, typically via email, designed to steal credentials or install malware. The message might appear to be from a manager, a vendor, or an IT department. It creates urgency—"Your account will be locked if you don't verify immediately"—that overrides careful evaluation. An employee who clicks the wrong link or enters credentials on a fake login page has just handed attackers the keys to whatever systems they can access.

Relationship building takes longer but achieves a deeper compromise. Someone cultivates a connection with a food industry worker through social media or industry events and the gradual development of an apparent friendship. Over time, they extract information about operations, systems, and vulnerabilities. Eventually, they may recruit active assistance: "I just need you to share this one thing. It's not a big deal. I'll make it worth your while."

Authority exploitation uses the appearance of power to compel compliance. Someone impersonates an executive, a regulator, or a law enforcement officer. They demand information or action, with an implied threat regarding noncompliance. Employees trained to defer to authority follow instructions that compromise security.

The escalation pattern is predictable. Social engineering often follows a gradual path. Initial requests seem harmless: information that might be publicly available, small favors that don't seem significant. Each successful request tests reliability and builds relationships. Requests gradually increase in sensitivity. By the time someone realizes they're compromised, they're already in too deep, either because they've provided enough to be implicated or because the relationship feels too valuable to jeopardize.

Employees targeted by social engineering may never have intended to become insider threats. They may be satisfied workers, loyal to their organization, horrified to discover how they've been used. But skilled social engineering exploits normal human tendencies: wanting to be helpful, trusting apparent authority, reciprocating kindness, and avoiding conflict. Programs to prevent insider threats must address the external manipulation that creates insiders, not just the internal conditions that enable them.

What It Teaches Us: Reducing Risk Without Destroying Trust

Every food and agriculture organization faces the same tension: the trust that makes operations functional is the same trust that insider threats exploit.

Food production depends on people working together, often under pressure, with limited individual supervision. You can't watch everyone constantly. You can't verify every action. You must trust that the worker on the night shift will follow procedures even when no one is watching. You must trust that the driver will maintain temperature logs honestly. You must trust that the supervisor will conduct the checks they're supposed to conduct.

Creating a surveillance state that trusts no one destroys the collaborative culture that enables food operations. It also doesn't work because people who feel distrusted become disengaged at best, hostile at worst. The adversarial environment that constant surveillance creates is the same environment that breeds insider threats.

The goal isn't eliminating trust. It's building systems that don't depend on perfection. Federal guidance on insider-threat indicators emphasizes behavioral changes that supervisors and colleagues may observe: expressions of grievance, unusual interest in areas outside job responsibilities, attempts to access information beyond role requirements, and changes in attitude toward the organization or its mission.[13] These indicators aren't proofs of threats, but they warrant attention and, often, supportive intervention before situations escalate. The practical steps that reduce risk without destroying that trust belong to specific roles, addressed next.

Moving Forward

What You Can Do

Whether you're a food industry professional, a cybersecurity specialist, or someone who simply eats food and cares about its safety, insider threats are a reality that demands attention:

For everyone working in food and agriculture: The people most likely to notice warning signs are those who know their colleagues well. You'll notice when someone's attitude changes, when they express unusual bitterness, when they ask questions outside their normal scope. You're not being asked to spy on coworkers. You're being asked to recognize that early intervention helps everyone, including the person showing warning signs.

If you see something concerning, say something through appropriate channels. That might be a supervisor, a food safety team, an HR department, or a confidential reporting line. The goal isn't to get colleagues in trouble. It's to surface issues before they escalate to harm.

Be aware of social engineering attempts. The helpful stranger asking questions about your work might be gathering intelligence. The urgent email demanding immediate action might be a phishing attempt. The new friend showing unusual interest in your workplace might have motives beyond friendship. This isn't paranoia; it's awareness that external actors target food industry workers specifically because of the access they have.

Don't fall for disinformation or misinformation that might be designed to radicalize you or manipulate your actions. Later, the book examines how narrative warfare targets food systems. For now, be aware that the content you consume about food systems, particularly inflammatory content about industry practices, may be designed to shift your beliefs toward action that serves someone else's agenda.

For cybersecurity professionals: Insider threats in food and agriculture have physical dimensions that pure IT security doesn't address. The access controls you implement affect food safety, not just data confidentiality. The monitoring systems you design should look for food-safety-relevant anomalies, not just traditional indicators of compromise. The incident response plans you develop should account for the physical consequences of digital incidents.

Partner with Food Defense teams. They understand physical security and threat scenarios that your training may not have covered. They can help you understand which systems have food safety implications. You can help them understand the digital attack vectors that modern insider threats exploit.

Recognize that a trust-based culture isn't a weakness to be eliminated. It's the context you work within. Security controls must work with food operations to be adopted. Controls that fight operational culture will be bypassed, making everyone less secure.

For food safety and operations leaders: Your culture is your first line of defense. Organizations where people feel valued, where grievances are heard and addressed, and where financial hardship can be discussed without stigma have fewer insider threats. This isn't just nice HR talk; it's a security strategy.

Before someone is hired, proportionate screening matters. Background checks should be calibrated to the level of access and risk involved: not blanket maximum scrutiny for every role, but more thorough evaluation for positions involving unsupervised access to product, sensitive systems, or authority over safety-critical processes. Verify credentials and employment history. Reference checks work best when they ask specific questions about reliability and conduct, not just "Would you rehire this person?" For agricultural roles, reviewing publicly visible social media for expressions of radicalized or extreme viewpoints toward the industry, food systems, or specific companies is a reasonable precaution.

Onboarding that builds genuine investment does more long-term work than onboarding focused on control. Clear communication of expectations, introduction to food safety culture and why it matters, and treating new employees as valued team members rather than risks to be managed—these aren't soft extras. They're conditions that determine whether someone develops loyalty or builds a grievance.

During employment, role-based access should restrict exposure to what's needed for the job. Access levels need regular review; the person promoted from line worker to supervisor may continue to hold both sets of permissions if no one updates the system. For sensitive functions, separation of duties means that causing significant harm requires collusion rather than individual action, which is a meaningful deterrent.

Offboarding deserves more attention than it usually gets. The terminated employee who becomes a threat is often someone who walked out with knowledge that should have been secured and access that should have been removed. Review your separation procedures. Close the gaps you find. If Food Defense and cybersecurity are siloed in your organization, you have a gap that insiders can exploit.

Human Reality

Insider threats are fundamentally about people with grievances that feel valid to them, people under financial pressure, people who've been manipulated or radicalized, and people who've made choices they'll regret.

The woman who allegedly put needles in Australian strawberries may have felt genuinely wronged by her termination. The terminated Disney employee who manipulated allergen data may have experienced real injustice in how he was treated. Their grievances may have been legitimate. Their responses were not.

Understanding the human element, psychology, pressures, and pathways enables prevention. It doesn't excuse harm. Nothing excuses putting needles where children will bite into them or changing allergen data that families depend on.

But prevention requires understanding. Organizations that want to reduce insider threat must attend to the conditions that create it: unfair treatment, unaddressed grievances, financial desperation without legitimate relief, ideological radicalization without a counter-narrative, and coercion that finds vulnerable targets.

Most people never become threats. Building systems that prevent exceptions doesn't require treating everyone as suspicious. It requires building healthy organizations where people feel valued, where problems surface before they explode, and where the trust that makes food operations work is paired with the verification that keeps that trust from becoming vulnerable.

The next chapter examines a different human challenge: what happens when well-intentioned people make mistakes: the employee who clicks the wrong link at the end of a 12-hour shift, or the operator who takes a shortcut under production pressure. Because in food operations, human error is inevitable.

Notes

1. Food Standards Australia New Zealand. (2019). "Strawberry tampering incident." **https://www.foodstandards.gov.au/publications/Strawberry-tampering-incident**
2. Bavas, J. (2018). "Strawberry needle tampering crisis: Queensland growers call for calm and common sense." *ABC News.* **https://www.abc.net.au/news/2018-09-18/strawberry-fund-set-up-for-farmers-qld-premier-says/10262804**
3. Ibid; BBC News. (2018). Australia strawberry scare: accused saboteur 'motivated by spite'." **https://www.bbc.com/news/world-australia-46175243**
4. Food Safety News. (2021). "Prosecutors drop charges in strawberry tampering incident involving needles." **https://www.foodsafetynews.com/2021/07/prosecutors-drop-charges-in-strawberry-tampering-incident-involving-needles**
5. Proofpoint. (n.d.). "Insider threats." **https://www.proofpoint.com/us/threat-reference/insider-threat**
6. Code of Federal Regulations. (2016). "Mitigation strategies to protect food against intentional adulteration." 21 C.F.R. pt. 121. **https://www.ecfr.gov/current/title-21/chapter-I/subchapter-B/part-121**

7. Idensohn, C., Flowerday, S., Van der Schyff, K., & Chua, Y. T. (2026). Malicious insider threats in cybersecurity: A fraud triangle and Machiavellian perspective. *Computers in Human Behavior*, 174, 108809. **https://doi.org/10.1016/j.chb.2025.108809**
8. Khan, N., Houghton, R. J., & Sharples, S. (2021). "Understanding factors that influence unintentional insider threat." *Computers & Security*, 110, 102419. **https://pmc.ncbi.nlm.nih.gov/articles/PMC8550909**
9. Kong, I., & Bashir, M. (2021). "A closer look at insider threat research." In Ahram, T., & Karwowski, W. (eds) *Human Factors in Cybersecurity* (AHFE Open Access). **https://openaccess.cms-conferences.org/publications/book/978-1-958651-29-2/article/978-1-958651-29-2_3**
10. Hogan, B. (2025). "Winter Garden man sentenced to 3 years in federal prison for conducting series of cyber intrusions against former employer." *News 6*. **https://www.clickorlando.com/news/local/2025/04/24/former-disney-employee-sentenced-to-federal-prison-after-pleading-guilty-to-altering-allergen-info-on-menu-items**
11. Center for Development of Security Excellence. (2023). "Insider threat job aid: food and agriculture facilities and organizations." **https://www.cdse.edu/Training/Job-Aids/Insider-Threat-Job-Aids**
12. Summers, J. T., & Pisciotta, F. (2020). "Food defense and the insider threat." *Security Management Magazine*. **https://www.asisonline.org/security-management-magazine/articles/2020/01/food-defense-and-the-insider-threat**
13. Center for Development of Security Excellence. (n.d.). "Insider threat toolkit." **https://www.cdse.edu/Training/Toolkits/Insider-Threat-Toolkit**

CHAPTER 21

The Human Layer: Training, Fatigue, and Errors

The Error Nobody Saw Coming

In October 2024, a 15-year-old girl from Colorado ate a Quarter Pounder at McDonald's. Within days, she developed severe stomach cramps and bloody diarrhea, and then her kidneys began to fail. She was rushed to the hospital, where doctors diagnosed hemolytic uremic syndrome, a life-threatening complication of *E. coli* O157:H7 infection. She required intensive care. Her family filed a lawsuit. Somewhere along a supply chain that stretched from Washington state onion fields to a Colorado processing facility to a fast-food restaurant, the systems designed to protect her had failed.

By the time federal investigators declared the outbreak over, more than 100 people had been sickened across 13 states. Twenty-seven required hospitalization. One elderly man in Colorado died. The source was traced to slivered onions processed at a Taylor Farms facility in Colorado Springs and served on hamburgers at McDonald's locations throughout the Midwest and Mountain West.[1]

When Food and Drug Administration (FDA) inspectors examined the Taylor Farms facility, they found conditions that told a story of system failures long before any pathogen reached a single onion. Production room temperatures that could support bacterial growth. Biofilm and food debris on equipment after sanitation shifts. Preoperation inspections gave passing grades despite food-contact surfaces not being fully cleaned. Multiple areas throughout the facility had tested positive for Listeria at various points in 2024, and corrective actions either hadn't been taken or hadn't worked.[2]

The instinct is to ask, who did this? Who cut corners? Who failed to clean properly? Who let contaminated product ship? But that question, although understandable, misses the point. The real questions are harder: What conditions made those failures possible?

What pressures shaped those decisions? What systems were supposed to catch problems before they reached consumers, and why didn't they?

Every food safety incident has a human at its heart. Not as the villain, but as the final link in a long chain of organizational, technological, and environmental factors that converge in a single moment. The same is true for cybersecurity incidents, and increasingly, the two are intertwined. Cyber failure can trigger a food safety crisis; a stressed, undertrained workforce is vulnerable to both contamination and attack. This chapter is about that human layer: how we fail, why we fail, and what we need to succeed. Because securing what feeds us ultimately means supporting the people who make it happen.

The Psychology of Error: Why Humans Fail

In 1990, psychologist James Reason published a book called *Human Error* that fundamentally changed how industries think about accidents. Reason, working from decades of research into cognitive psychology and industrial disasters, proposed what became known as the Swiss Cheese Model of accident causation.[3]

The concept is simple. Picture an organization's safety defenses as slices of Swiss cheese stacked in a row. Each slice represents a protective barrier: a policy, a procedure, a training program, a piece of equipment, a supervisor's oversight. But every slice has holes—weaknesses that exist because no system is perfect. An accident occurs only when the holes in multiple slices align, allowing a hazard to pass through all defenses.

Reason differentiated between two types of failures. *Active failures* are the unsafe acts committed by people at the sharp end of operations: the worker who skips a handwash, the operator who overrides a sensor alarm, the farmer who ignores a system warning. These are the visible errors that show up in incident reports. But behind every active failure lie what Reason called *latent conditions*: the resident pathogens embedded in the system long before any incident, poor training programs, inadequate staffing, equipment designed without safety in mind, and management decisions that prioritize speed over thoroughness. Rather than being the main instigators of an accident, operators tend to be the inheritors of system defects created by poor design, incorrect installation, faulty maintenance, and bad management decisions.[4] The person who makes the final error is often the one who had the least control over the conditions that made it possible.

Apply this framework to a food production environment, and the implications become clear. I saw this play out at a facility that produced finished products like garlic butter. I'd been brought in for a technology refresh and had just installed a new terminal in the butter room for the company's new enterprise resource planning (ERP) system. Everything was running smoothly when I left for the evening. The next morning, I got a call saying that the terminal wasn't working. People were panicking.

I suited up, walked into the butter room, and noticed a large industrial fan that hadn't been there the night before. The terminal was completely unresponsive—no power at all. I opened the protective case, checked the wires, and traced the power cord to the wall outlet. It was unplugged; that fan was plugged in instead.

When I called the supervisor, he explained that the room needed to dry after cleaning, and there was no other outlet for the fan. The overnight crew made a choice. Production and safety require air circulation, and extension cords are a no-go. The terminal could wait, was their thought.

I almost had to laugh, because this is the reality of food production: people and processes will override technology every time, and workers will bypass systems that get in the way of them doing their jobs. The overnight crew hadn't reported the issue because they figured the terminal wasn't important if production wasn't running in that room. Reasonable logic. Wrong assumption.

The fix wasn't disciplinary. We added a second power drop to that room so the fan and the terminal could coexist. The system adapted to the operation instead of expecting the operation to adapt to the system. That's the difference between designing for humans as they are and designing for humans as we wish they were.

Cybersecurity defenses work the same way. Firewalls, access controls, monitoring systems, and training programs are all slices of Swiss cheese. A ransomware attack succeeds when an employee clicks a phishing link (*active failure*) in an organization with outdated email filters, minimal threat awareness training, and an unsegmented network (*latent conditions*).

James Reason also categorized the types of human error that occur at the sharp end. *Slips* are unintentional actions: the hand that reaches for the wrong valve, the eyes that glance past a warning light. *Lapses* are memory failures: forgetting a step in a familiar process. *Mistakes* are conscious decisions based on incorrect information or faulty reasoning. And *violations* are deliberate deviations from rules, sometimes well-intentioned ("I'm saving time"), sometimes reckless ("Nothing bad has happened before").

Research suggests that roughly 70 percent of workplace errors are slips and lapses: unintentional actions by people who know better. Another 20 percent are mistakes: wrong decisions made with inadequate information. Only about 10 percent involve deliberate violations.[5] Yet punishment systems often treat all errors as if they were violations, as if the worker who forgot a step and the worker who deliberately bypassed a safety system deserve the same response. This misalignment between how errors happen and how organizations respond to them creates its own hazards, a topic we'll return to later in this chapter.

The Food Industry's Unique Human-Factor Challenges

Food production and agriculture present human-factor challenges that differ from those in other industries in important ways. The physical demands are substantial. According to the U.S. Bureau of Labor Statistics, food processing workers experience injury and illness rates above the manufacturing sector average.[6] The Occupational Safety and Health Administration (OSHA) found that food production workers in Ohio had a 57 percent higher rate of amputations and a 16 percent higher rate of fractures compared to overall manufacturing rates.[7] From 2015 to 2022, food and beverage processing machinery and butchering machinery were the primary contributors to severe injury reports in the food processing industry.[8]

But statistics only hint at the daily reality. Food workers often stand for entire shifts on wet floors. They work in extreme temperatures, from blast-freezers to steam-filled processing rooms. They operate equipment with moving blades, grinders, and conveyors that demand constant attention. They handle chemicals for sanitation and pest control. And they do all of this under time pressure, because biology sets nonnegotiable deadlines.

These conditions increase the likelihood of human error. Fatigue, physical discomfort, and cognitive overload all degrade performance. Add to this the seasonal labor surges common in agriculture and food processing, when operations bring in temporary workers who may have less training and less familiarity with specific equipment, and the potential for errors multiplies.

Agriculture faces its own distinct challenges. The average age of farmers in the United States is 58 years; in Japan, it exceeds 68.[9] These are operators who learned their craft on analog equipment and now manage systems with embedded processors, cellular modems, and cloud connectivity. Technology arrived faster than training programs could adapt, and individual aptitude for learning new systems varies enormously. A farmer who can rebuild a tractor engine by feel may struggle with software interfaces designed by engineers who have never set foot on a farm.

Agricultural economics compounds the problem. Net farm income in the United States recently averaged just 8.6 percent of gross revenue, leaving little budget for security investment or dedicated IT staff.[10] Farmers serve simultaneously as mechanics, agronomists, accountants, and now de facto IT administrators. When something goes wrong with a precision agriculture system during planting season, there is no help desk to call. The farmer figures it out, or the crop waits, and crops don't wait well.

And then there's the culture. Food and agriculture operate on trust: handshake deals, longtime vendor relationships, and community mutual aid. These social norms are assets for collaboration but vulnerabilities for social engineering attacks. A phishing email that impersonates a local feed supplier exploits trust patterns established long before email existed. The same openness that allows farmers to share equipment and knowledge with neighbors creates pathways for attackers who understand how to exploit that openness.

Where Errors Happen: The Catalog of Failure

Understanding where errors occur helps target prevention efforts. In food production, the most common human errors cluster around several categories:

Hygiene failures

These remain persistent: inadequate handwashing, jewelry not removed, hair not restrained or put into a hair net, failure to change gloves between tasks. They aren't exotic failures; they're the mundane lapses that happen when people are tired, rushed, or simply not paying attention. Cross-contamination follows similar patterns: shared utensils between raw and cooked products, inadequate surface sanitation, improper storage. Temperature control errors can include product left in the danger zone between 40 and 140 °F, refrigeration not verified, and cooking temperatures not achieved; these happen when monitoring systems fail or when workers override them.

Labeling mistakes

Undeclared allergens were the leading cause of food recalls in 2023, with more than 30 instances tracked by the FDA.[11] These aren't cases of deliberate mislabeling; they're usually the result of incorrect labels applied during packaging changeovers, cross-contact during production, or communication failures between production and packaging teams. For a consumer with a severe allergy, a mislabeled product isn't just an inconvenience: it's a potential death sentence.

Agricultural errors

In agriculture, *chemical application* errors can carry serious consequences: wrong concentration, wrong timing, or drifting onto unintended crops. *Equipment operation mistakes* include overriding safety systems, improper maintenance, and operating while fatigued. *Animal welfare failures* arise from missed feeding schedules, failure to recognize signs of illness, or improper handling under stress. *Data entry errors*—incorrect field records, misidentified lots, wrong coordinates for precision equipment—create downstream problems that may not surface until harvest or processing.

Cyber-physical errors

The cyber-physical dimension adds new categories of error. *Automation overrides and errors* occur when workers disable safety interlocks to maintain production, ignore sensor warnings because "it's always doing that," or manually adjust automated processes without understanding downstream effects. *Data misinterpretation* leads to acting on incorrect sensor readings without verification or misunderstanding precision agriculture recommendations. *Access and authentication failures*—sharing passwords to avoid login delays, leaving systems logged in across shift changes, using default credentials on operational technology equipment—create security gaps that attackers exploit.

When Errors Occur: The Human Aftermath

We tend to discuss errors in clinical terms: root causes, contributing factors, corrective actions. What gets lost is the human experience of making a mistake that hurts someone.

Consider what happens psychologically when someone realizes they've made an error with real consequences. Shame attacks the self: "I am defective." Guilt focuses on the action: "I did something wrong." Fear of punishment, judgment, and job loss creates immediate survival pressure. Anger, both at oneself and at the system that allowed failure, often follows. Embarrassment at public exposure compounds the damage, and when people feel cornered, defensive rage can emerge, creating conflict that makes learning impossible.

These emotional responses don't just hurt the individual. They create secondary risks for the entire food system. When people fear punishment, they hide problems. They don't report near-misses. They cover up small errors before anyone notices. They fail to speak up about unsafe conditions. They work around problems rather than flagging them. The hiding instinct is understandable, even rational, from an individual perspective. But it means the organization never learns about the hazards that need to be addressed.

Shame and anger can also drive compensatory risky behavior. People take shortcuts to "make up" for lost time. They work through injuries or illness to prove their commitment. They skip breaks. They resist asking for help. The very emotions triggered by one error can set up conditions for the next.

In cybersecurity, the hiding instinct is especially dangerous. The employee who clicked a suspicious link and closed the browser, hoping no one noticed. The operator who saw a strange login attempt but wasn't sure it mattered. The farmer who noticed odd recommendations or a random blinking light from the precision ag system but assumed they'd misunderstood. Every unreported anomaly is a missed early warning, and the damage extends far beyond one person or one facility. When a food company suffers a ransomware attack and stays quiet, every other company in the sector loses the chance to learn from it. The equipment manufacturer never patches the vulnerabilities or even knows about the incident. So, the next target never sees it coming. Attackers count on shame keeping people silent, and that silence keeps the entire food community in the dark about how attacks happen.

This must change. Reporting a cyber incident doesn't mean you failed. It means you were targeted, which in today's environment is less a sign of weakness than a sign that you had something worth stealing. Attackers aren't going after random victims; they're selecting targets with valuable data, operational leverage, or limited defenses. Being attacked is the reality we live in now. What matters is how quickly you detect it, how effectively you respond, and whether you share what you learned so others can defend themselves.

This is where the culture of food and agriculture can become an asset. The same community trust that sends neighbors to help during a storm or cover chores after an injury can extend to sharing threat information. Reporting what happened to you helps the person down the road. That's resilience built the way rural communities have always built it: together.

Training: You Can Only Work with What You Know

There's a fundamental principle that sounds obvious but matters more than it seems: you can only work with what you know, not what you don't know. Workers cannot defend against threats they've never heard of. They cannot follow procedures they were never taught. They cannot recognize warning signs no one told them to watch for.

The training gaps between food and agriculture and cybersecurity are substantial. Generic cybersecurity training designed for office workers doesn't translate to farming contexts. A course that teaches password hygiene says nothing about GPS spoofing or precision agriculture platform compromise. Food safety training rarely addresses digital vulnerabilities. Cyber training rarely addresses food-specific constraints.

One-size-fits-all approaches fail because the risks are sector-specific. A dairy farmer needs to understand milking parlor ransomware risks and herd management system vulnerabilities. A food production operator needs to know about supervisory control and data acquisition system compromise and refrigeration control attacks. A warehouse worker needs to recognize inventory system manipulation and cold-chain sensor tampering. A restaurant manager needs to understand point-of-sale breaches and allergen database tampering.

The opportunity lies in integration rather than addition. Food and agriculture workers already receive regular safety training. Cyber-physical risk education should be woven into existing programs, not stacked on top as another burden. During food safety training on temperature monitoring, include: "What if someone hacked the temperature sensor to show safe readings when food is actually in the danger zone?" During equipment safety training, include: "What if someone gained remote access and started this equipment unexpectedly?" During chemical handling training, include: "What if the precision agriculture system was compromised to recommend wrong application rates?"

The goal is to make cyber-physical awareness as automatic as food safety awareness—part of how workers think about their environment rather than a separate checklist. This requires training that is specific to the actual systems workers encounter, delivered in languages workers understand, and reinforced through practice rather than just presentation.

The Mental Health Crisis We Cannot Ignore

We've talked about shame, training gaps, and the hiding instinct. But there's a deeper layer that rarely gets discussed in cybersecurity conversations: the mental health of the people we're asking to stay vigilant. Farmers are in crisis. The suicide rate among male farmers is 52.1 per 100,000, compared to 32.0 per 100,000 for male working-age adults across all occupations.[12] More than 50,000 Americans took their own lives in 2023, the nation's highest yearly total on record, and rural communities bear a disproportionate share of that loss.[13]

The contributing factors overlap with everything else in this chapter. Financial stress from volatile commodity prices, debt, and weather. Geographic and social isolation. Generational pressure to maintain family operations. Physical health problems that accumulate over years of demanding work. Climate uncertainty adds new weight to an already heavy load. And a culture that traditionally views mental health struggles as weakness persists—a farmer will readily share a story about a broken arm but not about feeling hopeless or unable to get out of bed.[14]

Food production workers face their own challenges, less studied but equally concerning. Shift work disrupts circadian rhythms and increases the risk of depression. Physical demands create chronic pain. Many workers are from immigrant or migrant populations and carry additional stressors: language barriers, documentation concerns, family separation. Job insecurity and seasonal employment can create constant financial anxiety. Workplace injuries can leave trauma that doesn't clock out at the end of a shift.

Why does this belong in a cybersecurity book? Because stressed, exhausted, desperate people make mistakes. They're more vulnerable to social engineering; a fraudulent text or spoofed vendor call hits differently when someone is already overwhelmed. They're more likely to take shortcuts. They're less able to maintain the steady vigilance that security requires. Mental health isn't separate from cybersecurity; it's foundational to it.

The wrong approach adds more requirements, more training, and more things to worry about without providing support. The right approach integrates cyber awareness into existing workflows, builds systems that are secure by design, so they don't depend on human perfection, and recognizes that supporting mental health is supporting food security.

If you or someone you know needs support, the AgriStress Helpline is available 24/7 at +1 833-897-2474 and is staffed by professionals who understand the pressures of farming and food production. The 988 Suicide and Crisis Lifeline provides free, confidential support anytime. And if you notice a colleague, neighbor, or family member showing signs of distress, reach out. Ask how they're doing. Really listen. You could save a life.

Fatigue

Mental health and fatigue feed each other. Stress disrupts sleep. Poor sleep worsens stress. And in food and agriculture, the conditions that create both are baked into the work itself. More than 43 percent of workers are sleep-deprived, according to the National Safety Council.[15] Fatigue is estimated to cost employers approximately $136 billion annually in health-related lost productivity.[16] And the costs aren't just financial. Fatigue reduces reaction time, impairs decision-making, decreases alertness, increases risk-taking, causes memory lapses, and triggers mood changes, including irritability and anxiety.

Food production and agriculture are especially vulnerable to fatigue effects. Extended shifts of 10–12 hours or more are common during processing peaks. Mandatory overtime when orders must ship keeps workers on their feet well past when they should have stopped. Seasonal intensity during planting and harvest demands marathon effort. Twenty-four-hour operations mean someone is always working the graveyard shift when human performance naturally declines. Rotating schedules disrupt sleep cycles and prevent adaptation. And the physical demands of standing, lifting, and working in temperature extremes compound the exhaustion.

I've seen this pressure firsthand. During peak season at a food distribution facility, the ERP system went down, and warehouse workers couldn't use their tablets or scanners.

They went back to pen and paper while trucks waited at the dock. An 8-hour shift turned into 10 or more. The grumbling turned to cursing, which turned to shouting. I was part of the technology team, which made me a target. People screamed at me even though I hadn't caused the failure. I watched my colleagues get screamed at too—watched them fight the instinct to yell back, absorb the hit, and keep working under impossible pressure. The technology eventually got fixed. It always does. But it's the hours and days in between that test people's mettle, and those are the moments when errors happen.

Consider dairy farming as another example. Cows must be milked roughly every 12 hours, regardless of holidays, weekends, or how tired the farmer is.[17] Delays cause animal suffering and health complications. Robotic milking systems help, but they require technical knowledge to manage; and when they fail, manual backup under time pressure creates ideal conditions for errors and additional safety concerns.

The intersection of fatigue and cyber-physical risk is dangerous. Fatigued workers are more likely to click on phishing emails. Exhausted operators are more likely to override safety systems "just this once." Sleep-deprived technicians are more likely to misconfigure equipment. Overtired farmers are more likely to dismiss system warnings as false alarms. Being awake for 17 hours produces impairment equivalent to a blood-alcohol content of 0.05, and being awake for 21 hours is equivalent to 0.08, the legal limit for driving in most jurisdictions.[18]

The solutions aren't complicated, but they require organizational commitment. Fatigue-risk management programs. Scheduling that prioritizes adequate rest between shifts. Limits on consecutive workdays. Adequate break time during shifts. Education on sleep hygiene. Monitoring for signs of fatigue. But in industries where production pressure is constant and staff and margins are thin, these solutions often lose out to the immediate demand to keep product moving.

Building a Just Culture Beyond Blame

The concept of Just Culture emerged from Reason's work on human error and was fully developed in his 1997 book *Managing the Risks of Organizational Accidents*.[19] It offers a framework for responding to errors that balances accountability with learning.

The key insight is that organizations face a choice about how they respond to errors. A blame culture asks, "Who caused this?" and punishes those who answer. A no-blame culture says no one is ever accountable, which isn't workable either. A Just Culture asks "What went wrong? Why did it happen? Was this an honest error, an at-risk behavior, or reckless conduct?" and responds proportionally.

Just Culture distinguishes three categories of behavior. *Human error* is unintentional: the slip, the lapse, the honest mistake. The appropriate response is to console the individual and redesign the system to prevent recurrence. *At-risk behavior* involves conscious choices that seem reasonable in context: taking shortcuts because the rules seem unnecessary or because everyone else does it. The appropriate response is coaching and removing the incentives or traps that make at-risk behavior attractive. *Reckless behavior* involves conscious disregard for risk despite knowing the potential consequences. This is where disciplinary action is proportionate.

The value of Just Culture for cybersecurity in food and agriculture is substantial. You cannot protect against threats you don't know about. If workers fear punishment for reporting, near-misses go unreported, strange system behaviors are ignored, security concerns stay quiet, and the organization learns nothing. The Disney allergen issue we've discussed involved a former employee who retained access after termination. But how many current employees might have observed something concerning and not reported it? A Just Culture creates the psychological safety needed for people to speak up. This matters for cybersecurity because the first sign of an attack is often something small: an unusual login, a system acting strangely, a strange request, an email that doesn't look quite right. If workers fear being blamed for clicking the wrong link, they won't tell anyone they clicked it, and the attack will spread.

Implementing Just Culture requires sustained effort. Leadership must visibly commit to the principles and apply them consistently. Managers need training in how to assess error types and respond appropriately. Reporting systems must be genuinely nonpunitive for honest errors. Learnings from incidents must be shared organization-wide. And perhaps most importantly, there must be a willingness to look at system factors rather than stopping at individual blame.

The challenge is that Just Culture cannot be imposed by policy alone. It requires genuine cultural change. And that change must survive the pressure that emerges when a serious incident occurs and stakeholders demand that someone be held responsible. Organizations that practice Just Culture in good times but revert to blame during crises don't have a Just Culture at all.

What It Teaches Us: The Human Responsibility

Every food safety statistic represents a human being who was probably doing their best under difficult circumstances. The worker who made the error. The consumer who was harmed. The investigators who had to piece together what went wrong. The families on all sides who lived with the consequences.

The 15-year-old girl from Colorado who developed kidney failure after eating a Quarter Pounder didn't care whether her illness resulted from a sanitation failure, a training gap, a staffing shortage, or inadequate food safety systems. She just wanted to be healthy. Her suffering was real regardless of what category we assign to the errors that caused it.

This is why the human layer matters. Humans are both the source of errors and the source of solutions. The creativity and adaptability that allow people to find workarounds when systems fail also allow them to spot emerging hazards before they cause harm. Social connections that make people vulnerable to social engineering also allow teams to support each other through difficult situations. Judgment that sometimes leads to at-risk behavior also allows experienced workers to recognize when something doesn't look right.

If you lead a team, create cultures where people can succeed, not just systems that catch failure. If you work on a farm or a production floor, take care of yourself, because your mental and physical health is food security. If you're making investment decisions, put money into people with the same commitment you put into technology. And wherever you sit in the food system, remember that behind every error is a human being who deserves understanding along with accountability.

Moving Forward: Supporting the Human Layer

Securing the human layer in food and agriculture requires approaches that acknowledge reality rather than fighting it. Humans get tired, distracted, and stressed. Designing systems that account for these limitations is more effective than hoping people will perform perfectly under imperfect conditions.

Several principles should guide this work. First, design for humans as they are, not as we wish they were. Build in verification steps that don't rely on human memory. Create defaults that are safe. Make the secure choice the easy choice. If logging into a system requires a 12-step process, people will find workarounds. If reporting a suspicious email takes 30 seconds and doesn't trigger an interrogation, people will report. Reduce cognitive load by simplifying procedures where possible, using checklists and visual reminders, and automating routine verification tasks.

STOP, BLOCK, TELL: A Response Protocol Anyone Can Remember

Ask any American what to do if their clothes catch fire, and they'll immediately say "stop, drop, and roll." This response is drilled into children from kindergarten onward. It's simple, memorable, and requires no decision-making under stress. It works because it's been practiced until it's automatic.

Ask a farmer what to do if their tractor starts behaving strangely at mid-harvest, or a food production worker what to do if the control system shows impossible readings, and you'll likely get uncertain looks. The response gap is dangerous. In the moment when something digital goes wrong, people need equally simple protocols that don't require them to think through unfamiliar decisions under pressure.

Here's what we need: STOP, BLOCK, TELL.

Stop. Don't touch anything else. Don't click around trying to figure out what's happening. Don't restart the computer. Don't power it off. This is critical: the instinct to restart is strong, but restarting can destroy the evidence investigators need to understand what happened and prevent it from happening again. Just stop. Hands off.

Block. Block the Internet connection. Turn off the Wi-Fi. Unplug the network cable (the one that goes to your router or wall jack, not the power cord). This isolates the affected system so that if something malicious is happening, it can't spread to other systems or communicate with attackers. The computer stays on; the Internet goes off.

Tell. Tell someone immediately. Your supervisor. IT support, if you have it. The emergency number posted near the equipment or the vendor. Don't wait to see if the problem resolves itself. Don't worry about looking foolish if it turns out to be nothing. Your job is to raise the alarm, not diagnose the problem.

That's it. Three words. Three actions. STOP, BLOCK, TELL.

This protocol needs to be posted near digital equipment, just like fire evacuation routes are posted. It needs to be part of new-employee orientation. It needs to be practiced during safety drills until it becomes automatic. Just as no one is embarrassed to stop, drop, and roll if their clothes catch fire, no one should feel embarrassed to Stop, Block, and Tell when something digital seems wrong. Better a false alarm than a contamination event or data breach.

Second, support rather than just train. Knowledge doesn't automatically become behavior. Skills degrade without practice. Stress undermines performance. Beyond training programs, build support systems: mentorship, peer networks, easy access to help when questions arise, and recognition for safety-positive behaviors.

Third, address the whole person. Mental health affects cybersecurity. Physical health affects decision-making. Financial stress affects risk tolerance. Personal problems affect work performance. Providing employee assistance programs, mental health support, financial wellness resources, and work-life balance policies is an effective security practice.

Fourth, create psychological safety. Workers need to feel safe speaking up about concerns, reporting mistakes, and asking questions without fear of punishment or humiliation. This is the foundation of error reporting, safety improvement, security awareness, and organizational learning. Leaders build it by modeling vulnerability, responding supportively to reports of mistakes, thanking people who raise concerns, and never punishing honest reporters.

Finally, recognize that technology helps but cannot replace human judgment and care. The most advanced precision agriculture system, the most sophisticated HACCP program, and the most robust cybersecurity controls all depend on people to implement, monitor, and maintain them. Investing in technology without investing in the people who operate it is a recipe for failure.

When incidents occur despite our best efforts, and they will, how we respond determines whether we recover or spiral. The next chapter examines incident response, crisis communication, and the organizational resilience that separates organizations that learn from those that repeat.

Notes

1. FDA. (2024). "Outbreak investigation of E. coli O157:H7: onions (October 2024)." **https://www.fda.gov/food/outbreaks-foodborne-illness/outbreak-investigation-e-coli-o157h7-onions-october-2024**
2. Hanacek, A. (2025). "Taylor Farms responds to FDA observations in wake of McDonald's 2024 slivered onions foodborne illness outbreak." *Food Processing.* **https://www.foodprocessing.com/food-safety/cleaning-sanitation-hygiene/news/55260673/taylor-farms-responds-to-fda-observations-in-wake-of-mcdonalds-2024-slivered-onions-foodborne-illness-outbreak**
3. Reason, J. (1990). *Human Error* (Cambridge University Press). **https://www.cambridge.org/highereducation/books/human-error/281486994DE4704203A514F7B7D826C0#overview**
4. Reason J. (2000). "Human error: models and management." *BMJ,* 320(7237), 768–770. **https://doi.org/10.1136/bmj.320.7237.768**
5. Boysen, P. G. (2013). "Just culture: a foundation for balanced accountability and patient safety." *Ochsner Journal,* 13(3), 400–406.
6. U.S. Bureau of Labor Statistics. (2025). "Employer-reported workplace injuries and illnesses—2023–2024." **https://www.bls.gov/news.release/pdf/osh.pdf**
7. U.S. Department of Labor, OSHA. (2022). "US Department of Labor initiative seeks to protect food processing workers in Illinois, Ohio amid significantly higher injury rates." **https://www.dol.gov/newsroom/releases/osha/osha20221109-0**
8. OSHA. (2024). *Severe Injuries in the Food Processing Industry.* Hazard alert HA-4407. **https://www.osha.gov/sites/default/files/publications/OSHA4407.pdf**
9. USDA, National Agricultural Statistics Service. (2024). "Census of agriculture: 2022." **https://www.nass.usda.gov/Publications/AgCensus/2022**; Japanese Government Statistics. (2022). "2020 census of agriculture and forestry in Japan." **https://www.e-stat.go.jp/en/stat-search/files?page=1&layout=datalist&toukei=00500209&tstat=000001032920&cycle=7&year=20200&month=0&tclass1=000001147146&tclass2=000001163067**
10. USDA, Economic Research Service. (2026). "Farm income and wealth statistics." **https://ers.usda.gov/data-products/farm-income-and-wealth-statistics**
11. AIB International. (2024). "Most common causes of 2023 food safety recalls (and how to avoid them in 2024)." **https://blog.aibinternational.com/most-common-causes-of-2023-food-safety-recalls-and-how-to-avoid-them-in-2024**
12. Peterson C., Sussell A., Li J., Schumacher P.K., Yeoman K., & Stone D.M. (2023). "Suicide rates by industry and occupation — national vital statistics system, United States, 2021." *Morbidity and Mortality Weekly Report,* 72, 1346–1350. **https://www.cdc.gov/mmwr/volumes/72/wr/mm7250a2.htm**
13. CDC. (2023). *Suicide Mortality in the United States, 2001–2021.* NCHS Data Brief 464. **https://www.cdc.gov/nchs/data/databriefs/db464.pdf**
14. Thompson, M., & Saltzman, M. (2024). "How rural communities are tackling a suicide and depression crisis among farmers." *PBS NewsHour.* **https://www.pbs.org/newshour/show/how-rural-communities-are-tackling-a-suicide-and-depression-crisis-among-farmers**
15. National Safety Council. (2017). *Fatigue in the Workplace: Risky Employer Practices.* **https://www.nsc.org/faforms/fatigue-in-the-workplace-risky-employer-practices**

16. National Safety Council. (n.d.). "Cost of fatigue at work." **https://www.nsc.org/workplace/safety-topics/fatigue/cost-of-fatigue-at-work**
17. Smiling Tree Farm. (n.d.). "Once or twice a day?" **https://www.smilingtreefarm.com/resources/once-or-twice-a-day**
18. Canadian Centre for Occupational Health and Safety. (2024). "Fatigue." **https://www.ccohs.ca/oshanswers/psychosocial/fatigue.html**
19. Reason, J. (1997). *Managing the Risks of Organizational Accidents* (Ashgate Publishing).

CHAPTER 22

Incident Response, Crisis Communication, and Defense

What to Do When the Production Line Grinds to a Halt

On January 18, 2023, Yum! Brands discovered ransomware in its systems. The company owns KFC, Pizza Hut, Taco Bell, and The Habit Burger Grill, roughly 55,000 restaurants across 155 countries generating $1.3 billion in annual profit. When the malware hit, nearly 300 restaurants in the UK went dark.[1] What happened next offers a lesson in how incident response should work.

Within hours of detecting the intrusion, Yum! Brands activated containment protocols. Systems went offline, and enhanced monitoring kicked in. The company engaged external cybersecurity forensics teams and notified federal law enforcement. By the end of that same day, Yum! Brands had filed a public disclosure with the Securities and Exchange Commission (SEC), telling investors and the public exactly what it knew: ransomware had impacted certain IT systems, data had been taken from the network, and an investigation was underway.[2] The 300 UK restaurants affected reopened the next day. One day of closure, no prolonged outage, no weeks of manual workarounds, and no cascading supply chain failures.

"The Company is actively engaged in fully restoring affected systems, which is expected to be largely complete in the coming days," the disclosure read.[3] And that's what happened. Within days, core systems were back online. The investigation continued for months, eventually revealing that employees' personal information—names, driver's license numbers, and identification card numbers—had been stolen. But no customer databases were compromised. Also, Yum! Brands did not pay the ransom. The company offered affected employees two years of credit monitoring and moved on.[4] Compare that outcome to incidents that stretched for weeks, cost hundreds of millions of dollars, and left shelves empty across entire regions. The difference wasn't luck; it was proactive preparation.

In Chapter 21, we examined the human factors that create vulnerability—fatigue, training gaps, error patterns—and the psychology of mistakes. This chapter addresses what happens when those vulnerabilities are exploited, when prevention has failed, and the clock starts ticking. The decisions made in the first hours and days determine whether an incident becomes a manageable disruption or an existential threat.

Not every operation has Yum! Brands' resources; truthfully, most don't. A family dairy doesn't have a crisis team on standby or forensics experts on retainer. A mid-size grain operation doesn't file SEC disclosures. But the principles translate: know what you'll do before you must do it, move quickly when it matters, and keep the things that can't wait, such as animal welfare, food safety, and perishable product, at the center of every decision.

In food and agriculture, the stakes are different from those in other industries. Food safety implications accompany every system failure. Regulatory obligations trigger the moment certain thresholds are crossed. The incident response frameworks developed for traditional IT environments don't account for these realities. Cybersecurity professionals coming from other sectors need to understand this: the playbooks that work in finance or healthcare will break here, and breaking them at the wrong moment means spoiled food, suffering animals, or worse.

When Standard Playbooks Meet Biological Clocks

The National Institute of Standards and Technology (NIST) publishes respected incident response guidance. The SANS Institute trains thousands of cybersecurity professionals in its six-step framework: preparation, identification, containment, eradication, recovery, lessons learned. These approaches work well in environments where systems can be isolated, taken offline for remediation, and restored according to IT schedules.

Food systems don't operate that way.

Consider a dairy processing facility hit by ransomware at 2:00 AM. Raw milk is arriving from farms within hours. That milk must be pasteurized within regulatory time limits—the clock starts when the milk leaves the cow. The pasteurization system runs on networked controls that are now showing error messages. The standard IT response would be to isolate the affected systems, image the drives for forensics, and begin methodical remediation.

But the milk keeps coming. Tanker trucks are already en route. The dairy can refuse delivery, which means farmers dump milk they've already produced, but dumping isn't simple either. Milk has a biological oxygen demand roughly 50 times higher than raw sewage; improper disposal depletes oxygen in waterways, kills fish, and violates Clean Water Act regulations.[5,6] So the dairy faces a genuine dilemma: dump the milk and risk environmental fines and ecological damage, attempt manual pasteurization if manual controls exist and operators know how to use them, or gamble that the networked pasteurization controls are unaffected and continue running, hoping forensics later confirm nothing was compromised. None of these options appear in standard incident response frameworks because standard frameworks assume the business can pause while IT figures things out.

For cybersecurity professionals new to this sector: your instinct to "isolate everything" will collide with operational realities you haven't encountered before. The production manager pushing back isn't being difficult or ignorant of security. They're weighing consequences you may not see: animal welfare obligations, regulatory timelines, perishability windows, and the livelihoods of farmers upstream who have nowhere else to send their product.

Food safety further complicates containment decisions. The standard playbook says to take compromised systems offline to prevent lateral movement. In food production, that might mean halting a clean-in-place (CIP) sanitation cycle partway through, leaving equipment in an indeterminate state that requires complete resanitation. It might mean stopping a thermal process at a point when product is neither fully cooked nor safely raw. It might mean breaking cold-chain integrity that took hours to establish.

Each of these creates food safety hazards that may be worse than continuing under compromised digital conditions. The incident responder in a food facility must think simultaneously about cybersecurity containment, food safety implications, and regulatory consequences, a combination that generic frameworks never address.

When Campbell's Soup's facility in Napoleon, Ohio, experienced a cyber intrusion in 2023, the company halted production for three days even though the attack appeared to target IT systems rather than operational technology controlling the soup-making equipment.[7] That decision prioritized food safety certainty over production continuity. When in doubt about whether food safety controls have been compromised, stopping is the right call. The IT instinct to maintain availability must yield to the food safety imperative to maintain safety.

No single framework covers what food and agriculture face. NIST 800-61 addresses cyber-incident response but assumes systems can pause for remediation. NIST 800-34 covers contingency planning but wasn't written for environments where inventory spoils or animals suffer. Food safety incident frameworks handle Hazard Analysis and Critical Control Points (HACCP) deviations and recall but don't account for compromised digital controls. Business continuity planning (BCP) addresses operational disruption but rarely contemplates cyber-physical convergence.[8]

Food and agriculture incidents live at the intersection of all of these, and the absence of a unified framework isn't just an academic gap; it's a coordination problem. When the cyber team speaks NIST 800-61, the food safety team speaks HACCP, operations speak BCP, and the farm speaks plain English, everyone's working from different playbooks. A common language grounded in a shared framework doesn't exist yet. Until it does, organizations must build their own bridges between disciplines, and this chapter attempts to sketch what those bridges look like.

Effective incident response in food and agriculture requires planning that accounts for biological constraints, regulatory requirements, and the reality that your inventory might be alive, spoiling, or both.

Start with authority. When an incident hits, production wants to keep the line moving, quality wants to halt until they're certain nothing has been compromised, and IT wants to contain the threat. At 3:00 AM, that's an argument you don't have time for; you need the person's name. One person, designated before anything goes wrong, who makes the call and owns it. For smaller operations, that's probably you. For larger ones, document who it is, and make sure everyone knows.

The first four hours set the trajectory. Life safety comes first, which in agricultural settings includes animal welfare as well as employee safety. When poultry house climate control fails in summer, you're not managing a business disruption; you're watching birds die by the thousands within hours, and no amount of system recovery undoes that. These decisions can't wait for IT's assessment to finish.

Food safety assessment follows immediately. But here's what trips up cyber professionals coming from other industries: you can't trust what the screen tells you. If the system might be compromised, so might its readings. Verify critical controls with independent instruments: handheld thermometers, physical inspection, anything not connected to the network you're worried about. Confirm pasteurization, refrigeration, and allergen segregation with your own eyes and equipment before deciding whether production continues.

Stopping production, when it comes to that, requires more care than speed. A pasteurizer mid-cycle, a smoker mid-cook, a CIP system halfway through sanitation—interrupt these at the wrong point, and you've created new hazards on top of the ones you're already managing. Shutdown sequences exist for reasons. They take time, and skipping steps to "contain faster" can leave products in uncertain states and equipment in conditions that complicate recovery.

Legacy equipment adds another layer. Some of these systems are decades old, still running because they work, and they weren't designed for hard restarts. Controllers lose calibration. Mechanical components stress in ways that don't show up until you try to run the next batch. Your incident plan needs to identify which systems require graceful shutdown, how long that takes, and make sure whoever's managing the cyber response knows it before they start pulling plugs.

Between 4 and 72 hours, the focus shifts to operational continuity and the stakeholders who need to know what's happening. Product-hold decisions become pressing: if you can't verify that controls held during the incident window, that inventory sits until you can. Work out in advance who has the authority to release held product and what verification they'll require, because you don't want to figure that out while Sales is pushing to ship and customers are asking where their orders are.

Supply chain conversations also start during this window. Upstream suppliers, downstream customers, cooperatives, grain elevators, and anyone expecting delivery need to hear from you. These calls are uncomfortable, but the alternative is worse: customers who find out from empty shelves or from a news story don't tend to stay customers.

Regulatory timelines run on their own clocks and vary depending on what happened. The FDA's Reportable Food Registry requires notification within 24 hours when there's a reasonable probability of serious health consequences.[9] The USDA maintains separate requirements for meat and poultry.[10] State data-breach laws covering employee or customer information typically mandate disclosure within 30 to 90 days, depending on jurisdiction.[11] Europe's NIS2 Directive gives essential food producers just 24 hours. An incident that compromises traceability may trigger recall procedures even when no contaminated product was shipped. Know your thresholds before you're working against them.[12]

Beyond 72 hours, recovery sequencing becomes critical, and this is where food and agriculture diverge sharply from other industries.

Recovery Prioritization: Food Safety First, Then Everything Else

Traditional IT disaster recovery prioritizes business value, so revenue-generating applications first, customer-facing systems next, and back-office functions when there's time. That logic inverts for food operations. The systems verifying safety must come back before the systems that generate revenue, because without safety verification, you can't legally or ethically ship anything. Revenue from product you can't prove is safe isn't revenue.

Refrigeration and cold-chain monitoring sit at the top of any recovery sequence. Temperature integrity is product safety; lose the ability to prove one, and you've lost the ability to prove the other. Operations with livestock face equivalent urgency around climate control and automated feeding; animals experience the outage in real time, regardless of whether your recovery plan accounts for them.

Batch tracking and lot coding follow close behind. Traceability makes products both sellable and, when something goes wrong, recallable. Products you can't trace become products you can't move, regardless of whether anything is wrong with them. HACCP records, sanitation logs, and allergen controls document that what's moving through your facility meets safety requirements. Shipping without them means shipping products you can't vouch for, which most customers and regulators won't accept.

Business systems come after all of this. ERP, order management, and invoicing matter, but a functioning invoicing system doesn't help when you have nothing safe to invoice.

This hierarchy scales down. For a smaller operation, the thermometer proving your cooler stayed cold matters more than the spreadsheet tracking invoices. The notebook documenting what you did during manual operations matters more than getting email back. Prove safety first and maintain animal welfare, and the rest will follow when it follows.

Cyber professionals from other sectors sometimes push for different sequencing; the recovery frameworks they've used elsewhere optimize around revenue. In food, that meets resistance, and the resistance is justified. This isn't a business unit advocating for its priorities. It's the foundation everything else depends on.

Evidence Preservation When Production Can't Stop

Earlier, we established what a worker does when something looks wrong: Stop, Block, Tell. Don't touch anything else, disconnect from the network or Wi-Fi, and escalate immediately. That guidance stands. The person who discovers a problem shouldn't be weighing forensic preservation against production needs; their job is to freeze the situation and get to someone who can make those calls. So, what happens next, after the escalation, when someone with authority is assessing the situation and realizes that keeping systems frozen isn't an option?

Forensic best practice says to preserve everything exactly as the attacker left it. Don't reboot, don't modify; image the drivers; maintain the chain of custody. That's reasonable when you're dealing with a compromised workstation in an accounting department. It collides with reality when the system that holds forensic evidence also controls refrigeration for 10,000 pounds of product, or when the server you're not supposed to touch holds the batch records you need to verify food safety.

Before any authorized changes, photograph everything. Screen states, error messages, indicator lights, display panels. This takes seconds and preserves information that forensics teams arriving hours later would otherwise never see. Whoever stopped and blocked the system may already have captured some of this; if not, do it now before anything else changes.

When systems must be modified to keep operations running, document what's happening and why. If refrigeration won't restore without rebooting the controller, note the exact time, what the screen showed before, what it showed after, and who authorized the decision. The original state may be lost, but a clear record of observations and decisions turns a gap in evidence into an explainable choice.

Network architecture matters here. Understanding how systems connect allows surgical responses, and sometimes you can isolate compromised components without bringing down everything that depends on them. Disconnect the workstation from the network while leaving production controllers running. This requires knowing your environment before the incident, which is worth building toward even if you're not there yet.

When preservation and safety conflict, safety comes first; regulators, insurers, and law enforcement all understand that products aren't going to spoil, and animals aren't going to suffer, to maintain forensic purity. What they'll want is documentation showing that you understood the trade-off and made a reasonable call. Evidence you couldn't preserve becomes defensible when you can explain why.

For smaller operations where the person who discovered the problem and the person making decisions might be the same, the principle still holds: stop first, assess second. Take those photographs before you touch anything. Once you've done that and understand what you're facing, then make the calls about what must change to keep the operation running, and write down what you did and why. A notebook with timestamps and a phone full of photos won't satisfy a forensics team, but they will show that you acted reasonably with the resources you had.

The Ransom Question

At some point during a ransomware incident, someone will suggest paying the ransom. Law enforcement guidance is consistent: don't. Payment funds criminal enterprises, encourages future attacks, and guarantees nothing. The FBI, the Cybersecurity and Infrastructure Security Agency (CISA), and their international counterparts all advise against it.[13]

But operational urgency (systems down, product spoiling, a harvest window closing) doesn't always make paying the ransom the right call. Attackers who receive payment often don't deliver working decryption keys. Some deliberately come back to organizations that paid, knowing they'll pay again. Payment may violate sanctions if the attackers have connections to designated entities. Insurance policies increasingly exclude ransom payments or require specific approvals before coverage applies.

But the decision deserves honest analysis, not reflexive refusal. What's the real cost of not paying: spoilage, animal welfare, missed windows, customer relationships, regulatory fallout, and public perception? What are the realistic odds of recovery without payment, and how long would it take? What backups exist? Has legal counsel confirmed payment wouldn't violate sanctions?

For smaller operations, the math is often simpler but cuts closer to the bone. The demand might be $50,000. The year's income if the harvest is missed might be $200,000. The farm might be everything the family has. These aren't abstract business decisions; these are choices that determine whether families keep their land.

There's no universal right answer. But there are wrong ways to get there: deciding in panic, deciding without understanding alternatives, deciding without legal guidance on sanctions. Whatever the choice, make it deliberately and with a cool head.

Farm vs. Factory: Different Contexts, Different Plans

Incident response looks fundamentally different depending on where you sit in the food system.

Agricultural operations usually don't have dedicated IT staff. The person responding to a cyber incident might also be the person who needs to feed livestock in two hours. Geographic isolation makes it worse: technical help may be hours away, and Internet connectivity for cloud-based recovery tools can be unreliable at the end of a gravel road. Seasonal timing adds existential weight: an incident during planting or harvest doesn't just cost money, it can cost the year.

For farms, incident response planning means identifying resources before you need them. Which equipment dealers offer emergency support? Does your precision agriculture provider have a hotline outside business hours? Do neighbors have compatible equipment and the willingness to help finish time-sensitive work? Is there a local IT contractor who knows agricultural systems, or at least one willing to learn? These relationships must exist before 2:00 AM when something breaks.

Modern tractors and combines often have offline modes that maintain basic function when precision features fail, but operators need to know those modes are available and how to activate them before the crisis hits. The manual should be physically accessible, not saved on the computer that isn't working. Neighbor mutual aid, a longstanding agricultural tradition, extends naturally to cyber incidents.

Food production facilities operate under different constraints. Many run around the clock every day. Shutting down involves workforce coordination, shift communication, and sometimes union notification. Regulatory oversight is more intensive. Monday isn't an option when the line runs on Sunday.

Manual procedures must exist and be practiced, not just documented. When Maple Leaf Foods got hit by ransomware in late 2022, the company fell back on what executives called "paper and pencils."[14] That manual capability kept plants running, animals processed, and products shipping. The attackers eventually published stolen data after Maple Leaf refused to pay, but the company had already proven that a sophisticated processor can survive a serious attack without capitulating.

Crisis Communication Under Pressure

Internal communication must work before external communication matters. Most crisis communication failures start inside organizations: departments contradict each other, executives say things that front-line employees can't support, and customers hear different stories depending on whom they talk to.

When Krispy Kreme detected unauthorized activity on November 29, 2024, the day after Thanksgiving, the company faced a test of everything crisis communications is supposed to do. The attack, later attributed to the Play ransomware gang, knocked out online ordering across the United States during one of the busiest periods of the year.[15]

Within two weeks, Krispy Kreme filed an SEC disclosure acknowledging the breach, stating that online ordering was down and that the attack was "reasonably likely to have a material impact" until recovery was complete. At the same time, the company updated its website to warn customers.[16] Physical stores stayed open throughout. Deliveries to retail partners continued. The communication strategy drew a clear line between what was working and what wasn't. Customers frustrated by the website could still walk in and buy doughnuts.

Recovery took weeks. Online ordering came back gradually. The investigation stretched for months, eventually revealing that more than 161,000 people, mostly employees and their families, had their personal information stolen: Social Security numbers, financial accounts, and, in some cases, biometric data.[17] Krispy Kreme refused to pay. The attackers leaked 184 gigabytes on their dark web site.[18] SEC filings put total losses above $11 million, including remediation costs, expert fees, and lost digital revenue.[19] Cyber insurance covered some of it. The long-term brand impact is still unclear.

What stands out: Krispy Kreme disclosed quickly, within two weeks rather than months. The company distinguished between affected and unaffected operations. It filed the required disclosures. And it offered credit monitoring to affected individuals.

For smaller operations without SEC obligations or communications staff, the principle holds: be honest about what's working and what isn't, tell the people who need to know, and don't let silence write a worse story than the truth would.

Food Defense Plans: A Framework Already in Place

Rather than building parallel structures for cybersecurity, food companies can fold cyber considerations into existing Food Defense frameworks. This uses familiar structures, avoids duplication, and treats cyber risk as the food safety issue it is.

FSMA's Intentional Adulteration Rule requires covered facilities to develop Food Defense plans addressing intentional contamination.[20] The rule calls for vulnerability assessment, mitigation strategies, monitoring, and corrective actions when controls fail. That framework was designed around physical threats—the insider who might contaminate product, the intruder accessing critical process points. Cyber wasn't the focus, but the structure accommodates it naturally.

Vulnerability assessment can expand to include digital controls. A pasteurizer running on a networked programmable logic controller is vulnerable to unauthorized access, just from different actors than the pasteurizer with an unlocked door. Network segmentation and authentication become mitigation strategies. Monitoring extends to watching for unauthorized digital access, the same way you watch for unauthorized physical access.

Food safety and quality teams already understand these frameworks. Auditors already assess compliance against them. Adding cyber builds on existing practice rather than creating a separate requirement that operations treats as an IT problem.

Practicing Before the Real Thing

Fire drills work because repetition makes response automatic. Nobody stops to read the evacuation plan when the alarm sounds. Cyber-incident response needs the same conditioning, scaled to whatever makes sense for the operation.

CISA publishes tabletop exercise packages covering ransomware, insider threats, and industrial control system compromise.[21] These adapt readily to food-specific scenarios. For smaller operations, a tabletop can be a kitchen table conversation: What would we do if the tractor's GPS failed mid-planting? If the milk tank monitoring showed no data for

12 hours, how would we know whether the refrigeration temperature held? If the feed system is locked up, how long before animals suffer? Food-sector scenarios need to address constraints that generic exercises miss:

A ransomware attack during harvest: Late September, combines equipped with yield monitoring and GPS guidance, farm management platform encrypted, grain storage monitoring dark. Can you keep harvesting with manual operation? How do you track yield and quality? What do you tell the elevator expecting delivery? The crop keeps maturing whether IT has answers or not.

A cold-chain monitoring gap: Temperature data across three distribution centers shows 15 hours of nothing. Did refrigeration fail? Is product compromised? Do backup records exist? What's the recall threshold, and can you trace which lots moved during the gap?

Credential misuse after termination: A recently departed employee whose access wasn't immediately revoked tries to connect remotely. Did they change anything before you blocked them? Can you verify whether batch records are intact? What about allergen documentation?

After exercises, write down what you learned. What gaps showed up? What would you do differently? For larger organizations, this updates formal plans. For a farm, it might mean writing a phone number on the wall, buying a backup thermometer, or talking to a neighbor about helping each other out.

Reporting and Recovery

The decision to involve law enforcement or regulators paralyzes a lot of organizations. Publicity concerns, regulatory scrutiny, and loss of control lead many to delay or skip reporting entirely. That's usually the wrong decision.

The FBI has increasingly engaged with the food sector in response to significant incidents. Ransomware, suspected nation-state activity, and attacks with major economic impact warrant contact. CISA provides technical assistance for critical infrastructure.[22] These agencies want to help, and reaching out early often gets you assistance before things get worse.

Smaller operations sometimes wonder if their incident is significant enough to bother reporting. When in doubt, report. Law enforcement would rather hear about something that turns out to be minor than miss patterns connecting attacks across the sector.

Insurance claims often require timely notification to law enforcement as a condition of coverage. Wait too long, and you may void the policy right when you need it.

The Analog Fallback Question

If every digital system disappeared tomorrow, could you keep operating?

The answer depends entirely on where you sit. A small farm with computerized records but manually operable equipment could probably muddle through basic operations. A modern dairy with robotic milking and automated feeding would face real problems. A high-speed production line with automated controls wouldn't run at anything close to normal capacity. A cold-chain operation could do manual temperature monitoring, but the labor cost would become unsustainable after a few days.

Building analog capability means documenting manual procedures even when you never expect to use them. Standalone thermometers, paper log sheets, and printed lot tracking forms. Practicing occasionally, and maybe running a "digital-free" shift as part of training.

More than anything, it means being honest about dependencies. Which operations absolutely require digital systems? Which have degraded manual modes? Where would you be completely stuck? Those answers shape incident response planning and technology investment alike.

Documentation for What Comes After

Once the immediate crisis passes, documentation becomes what everyone wants. Insurance claims need contemporaneous records. Regulatory investigations examine response decisions. Legal proceedings pick apart timing and actions. Post-incident reviews require accurate information about what happened versus what should have.

Larger organizations should assign someone specifically to document during incidents. This individual should not make decisions; they should just record who decided what, when, and based on what information. That separation keeps documentation from vanishing into chaos.

Smaller operations where everyone is doing everything still need records. Phone photos with timestamps. Notes when you can grab a minute. Saved texts and emails. After the worst passes, sit down and write out the timeline while your memory is fresh. That record matters later for insurance disputes, regulatory questions, and your own fading recollection of what happened when.

Capture the timeline: detection, key decisions, external notifications. Capture rationale: why production stopped or continued, who made that call, what they knew at the time. Capture communications: what was said, to whom, when. Capture food safety actions: verification steps and whether product is held or released. Track costs from the start: expenses documented in real time are worth far more than reconstructions weeks later.

Moving Forward: Survival as the Measure

Yum! Brands came through their January 2023 incident with one day of closures and no ransom paid. Krispy Kreme weathered weeks of degraded operations without paying. Maple Leaf kept processing and shipping through manual operations while attackers dumped stolen data online. All three recovered. All three continue operating.

That's what success looks like here: survival. Prevention will never be perfect, but it's always worth a try. Response under pressure will never be flawless. Survival means keeping operations running, maintaining food safety, preserving customer relationships, and learning enough from what happened to do better next time.

The organizations that survive well have things in common regardless of size. They thought through what they'd do before anything went wrong. They knew who to call. They treated food safety as the fixed point when everything else was uncertain. And they documented enough to learn from the experience.

The incident is contained, food safety verified, communications handled. But bills are arriving, adjusters want documentation, and lawyers have questions. The next chapter takes up that aftermath: insurance, liability, and the long road from crisis back to continuity.

Notes

1. Kelso, A. (2023). "Yum Brands confirms ransomware attack impacted some U.S. employees." *Nation's Restaurant News.* **https://www.nrn.com/quick-service/yum-brands-confirms-ransomware-attack-impacted-some-u-s-employees**
2. Arghire, I. (2023). "Yum Brands discloses data breach following ransomware attack." *SecurityWeek.* **https://www.securityweek.com/yum-brands-discloses-data-breach-following-ransomware-attack**
3. Yum! Brands, Inc. (2023). "Yum! Brands January 18, 2023 statement." **https://www.yum.com/wps/portal/yumbrands/Yumbrands/news/company-stories/Yum+Brands+January+18+2023+Statement**
4. Gatlan, S. (2023). "KFC, Pizza Hut Owner discloses data breach after ransomware attack." *BleepingComputer.* **https://www.bleepingcomputer.com/news/security/kfc-pizza-hut-owner-discloses-data-breach-after-ransomware-attack**
5. Davis Graham & Stubbs LLP. (2020). "You can cry over excess milk, but don't dump it." **https://davisgraham.com/news-events/dairy-farmers-and-food-producers-must-use-caution-when-disposing-of-excess-milk-and-other-food-waste-in-response-to-coronavirus-driven-demand-reductions**
6. Seely, R. (2020). "There's no use in wasting spilt milk." *GROW Magazine*, University of Wisconsin–Madison College of Agricultural and Life Sciences. **https://grow.cals.wisc.edu/departments/features/theres-no-use-in-wasting-spilt-milk**
7. Jones, D. (2023). "Campbell Soup says summer cyberattack caused limited business impact." *Cybersecurity Dive.* **https://www.cybersecuritydive.com/news/campbell-soup-cyber attack-limited-impact/694758**

8. Nelson, A., Rekhi, S., Souppaya, M., & Scarfone, K. (2025). *Incident Response Recommendations and Considerations for Cyber Risk Management*. NIST Special Publication 800-61 Rev. 3. **https://nvlpubs.nist.gov/nistpubs/SpecialPublications/NIST.SP.800-61r3.pdf**
9. FDA. (2018). *Guidance for Industry: Questions and Answers Regarding the Reportable Food Registry*. **https://www.fda.gov/regulatory-information/search-fda-guidance-documents/guidance-industry-questions-and-answers-regarding-reportable-food-registry-established-food-and-drug**
10. USDA FIS. (2012). "Pathogen reduction; Hazard Analysis and Critical Control Point (HACCP) systems; notification requirements." Code of Federal Regulations, Title 9, Section 418.2. **https://www.ecfr.gov/current/title-9/chapter-III/subchapter-E/part-418/section-418.2**
11. Privacy Rights Clearinghouse. (2026). *Data Breach Notification Laws: A 50-State Survey (2026 Edition)*. **https://privacyrights.org/resources-tools/reports/data-breach-notification-laws-50-state-survey-2026-edition**
12. European Parliament and Council of the European Union. (2022). "Directive (EU) 2022/2555 of the European Parliament and of the Council of 14 December 2022 on measures for a high common level of cybersecurity across the Union, amending Regulation (EU) No 910/2014 and Directive (EU) 2018/1972, and repealing Directive (EU) 2016/1148 (NIS 2 Directive)." *EUR-Lex*. **https://eur-lex.europa.eu/legal-content/EN/TXT/PDF/?uri=CELEX:32022L2555**
13. CISA. (2023). "#StopRansomware guide." **https://www.cisa.gov/stopransomware/ransomware-guide**
14. Maple Leaf Foods investor communications, referenced in *TXOne Networks*. (2025). "Revisiting threats to food & beverage cybersecurity." **https://www.txone.com/blog/revisiting-threats-to-food-beverage-cybersecurity**
15. Greig, J. (2025). "Krispy Kreme: Over 160,000 people had data stolen during November 2024 cyberattack." *The Record*. **https://therecord.media/krispy-kreme-reports-data-breach-from-2024-attack**
16. Franceschi-Bicchierai, L. (2024). "Krispy Kreme discloses cyberattack that is disrupting online orders." *TechCrunch*. **https://techcrunch.com/2024/12/11/krispy-kreme-discloses-cyberattack-that-is-disrupting-online-orders**
17. Kovacs, E. (2025). "161,000 people impacted by Krispy Kreme data breach." *SecurityWeek*. **https://www.securityweek.com/161000-people-impacted-by-krispy-kreme-data-breach**
18. SC Media. (2025). "Ransomware attack hits Krispy Kreme systems." **https://www.scworld.com/brief/ransomware-attack-hits-krispy-kreme-systems**
19. Hill, E. (2025). "New details emerge on Krispy Kreme's 2024 data breach." *National CIO Review*. **https://nationalcioreview.com/articles-insights/extra-bytes/new-details-emerge-on-krispy-kremes-2024-data-breach**
20. FDA. (2016). "FSMA final rule on mitigation strategies to protect food against intentional adulteration." **https://www.fda.gov/food/food-safety-modernization-act-fsma/fsma-final-rule-mitigation-strategies-protect-food-against-intentional-adulteration**
21. CISA. (n.d.). "CISA tabletop exercise packages." **https://www.cisa.gov/resources-tools/services/cisa-tabletop-exercise-packages**
22. CISA. (n.d.). "Reporting a cyber incident." **https://www.cisa.gov/reporting-cyber-incident**

CHAPTER 23

Insurance, Liability, and Business Continuity

When the Policy Meets the Pasteurizer

In October 2022, after four years of litigation, Mondelez International and Zurich American Insurance Company reached a confidential settlement over a claim that had rattled the entire insurance industry. The dispute stemmed from the 2017 NotPetya cyberattack, which had crippled Mondelez facilities worldwide, disrupting chocolate and cookie production across multiple continents. Zurich had denied the claim under its policy's war exclusion, arguing that because the U.S. and UK governments attributed NotPetya to the Russian military, the attack constituted an act of war. Mondelez countered that war exclusions were never meant to cover cyberattacks against private companies, regardless of who launched them.[1]

As we examined in Chapter 8, the operational impact of NotPetya on food production was devastating. But the insurance angle reveals something different: the collision between century-old insurance concepts and 21st-century risks. War exclusions date back to the age of battleships and trenches. They were designed to protect insurers from the catastrophic, correlated losses of military conflict. Nobody drafting those exclusions in the early 20th century imagined a scenario in which a cyberattack launched from Moscow could simultaneously disable cookie factories in Australia, cheese plants in Wisconsin, and chocolate production in Belgium.

The Mondelez–Zurich dispute forced the insurance industry to confront a fundamental question: what counts as war in cyberspace? The answer matters enormously for food and agriculture, a sector increasingly dependent on connected systems yet historically underserved by cyber insurance markets. If nation-state attacks are excluded from coverage, what about the ransomware groups that operate with tacit state approval? What about hacktivists targeting farms over animal welfare concerns? What about domestic extremists who view industrial agriculture as an ideological enemy?

This chapter examines the financial architecture of cyber risk in food systems: how insurance works, where it fails, what liability exposure looks like after an incident, and how to build business continuity plans that account for the reality that food production operates on biological timelines that don't pause recovery efforts.

The State of Cyber Insurance in Food and Agriculture

The cyber insurance market has grown rapidly over the past decade, but coverage for food and agriculture remains uneven. According to the Coalition's 2025 Cyber Claims Report, manufacturing and consumer-facing industries, categories that include food processing and agriculture, filed the highest volume of cyber insurance claims in 2024. Yet the same report noted that many businesses in these sectors lack adequate coverage, often because they underestimate their exposure or cannot afford premiums that reflect actual risk.

The numbers paint a sobering picture. Ransomware demands, although they declined slightly from their 2023 peak, averaged approximately $ 1.1 million per incident in 2024. Business email compromise and funds-transfer fraud accounted for 60 percent of all claims. Food companies, with their complex vendor relationships and high transaction volumes, face exposure to these social engineering attacks. A single compromised email account in a procurement department can redirect hundreds of thousands of dollars to criminal accounts before anyone notices.[2]

The scale of losses in agriculture specifically has become staggering. One insurance provider reported that its customers in agriculture alone lost $10 million in 2024 on cyber claims, primarily due to invoice manipulation, fraudulent phone calls that tricked employees into revealing banking information, and hacked emails that delivered fake wiring instructions.[3] These attacks are happening to feedyards, cattle operations, and grain elevators, sometimes costing them not only their own money but their customers' money as well.

What Policies Actually Cover

Cyber insurance policies generally cover two broad categories: first-party losses, meaning direct costs to the policyholder, and third-party losses, meaning liability to others. First-party coverage typically includes incident response costs such as forensic investigation, legal guidance, and public relations support. It may also cover business interruption losses, data recovery expenses, and ransom payments, although the latter has become increasingly contentious as some insurers restrict or exclude ransom coverage entirely.[4]

Third-party coverage addresses liability when a breach affects others: customers whose data was exposed, business partners whose operations were disrupted, or individuals harmed by downstream effects of a security failure. For food companies, this third-party exposure is particularly significant. If a cyberattack compromises food safety systems, leading to contaminated products reaching consumers, the liability implications extend far beyond typical data breach scenarios.

Business interruption coverage deserves particular attention because it is frequently misunderstood. A forensic accountant specializing in cyber claims explained a common misconception: many policyholders believe that wages paid to idle workers during a cyber incident are separately indemnifiable. In reality, business interruption policies indemnify for lost gross profit, which already includes fixed costs like wages. If sales are ultimately recovered after systems are restored, the idle labor costs are recouped through those recovered sales. Only additional labor costs incurred specifically to recover lost sales, such as overtime or temporary workers brought in to catch up on delayed orders, are separately indemnifiable.[5]

This distinction matters enormously for food operations. A four-week shutdown of a dairy processing facility might involve substantial wages paid to workers who cannot process milk while systems are down. But if the facility operates double shifts after recovery to catch up on orders, the business interruption claim covers the net loss, not the full four weeks of idle wages. Failing to understand this calculation leads to unrealistic expectations during claims, often creating friction between policyholders and insurers at precisely the moment when cooperation matters most.

When Insurance Actually Works

The headlines tend to feature disputed claims and denied coverage, but insurance does work, sometimes even when carriers initially refuse to pay. Understanding what successful claims look like helps set realistic expectations and provides a template for documentation and persistence.

In March 2021, Yoshida Foods International, an Oregon-based manufacturer of beverages and sauces, discovered that ransomware had encrypted its computer systems. The company faced a familiar dilemma: paying the ransom or attempting to rebuild from backups. When its IT consultant couldn't arrange payment quickly enough, company president Junki Yoshida made an unusual decision. He paid the ransom, approximately $107,000, from his personal cryptocurrency holdings. Within days, the company recovered its data and restored operations.[6]

Over a year later, Yoshida Foods reimbursed its president and filed a claim under the computer fraud coverage of its policy with Federal Insurance Company, a Chubb subsidiary. The insurer denied the claim, arguing that the company hadn't suffered a direct loss because the president, not the company, had made the initial payment. Federal also invoked the fraudulent-instructions exclusion, suggesting that because someone at the company had approved the payment, it wasn't truly unauthorized.[7]

Yoshida Foods sued. In December 2022, the United States District Court for Oregon granted summary judgment in the company's favor. The court found that both the ransom payment and the subsequent reimbursement were proximately caused by the hacker's intrusion. No intervening event broke the chain of causation. The ransom payment, the company's reimbursement of its president, and the underlying attack were all part of one unbroken sequence. Whether the president or the company initially transferred the cryptocurrency was, in the court's words, irrelevant.[8]

The court also rejected the fraudulent-instructions exclusion. Paying a ransom under duress, the court reasoned, is not the same as approving a fraudulent transfer. The president didn't authorize the payment because he wanted to; he did so because criminals had locked the company out of its own systems. Coerced transactions don't become authorized merely because someone clicked a button.[9]

The Yoshida case matters for several reasons. It demonstrates that initial claim denials are not final answers. It establishes useful precedents about causation and coercion in ransomware scenarios. And it provides a roadmap: document everything, understand your policy language, and be prepared to push back when denials don't align with the facts.

The Exclusion Minefield

Every insurance policy contains exclusions: circumstances under which coverage does not apply. Cyber policies are no exception, and the exclusions they contain have become increasingly consequential as attack patterns evolve. War exclusions, as the Mondelez case demonstrated, present the most dramatic example. But several other exclusions are relevant to food and agriculture.

In March 2023, Lloyd's of London issued a bulletin requiring its syndicates to include explicit exclusions for nation-state cyberattacks in stand-alone cyber policies. Lloyd's directive specified that policies must exclude losses arising from cyber operations conducted during war, retaliatory attacks between states, and attacks that significantly impair the functioning of a state. Importantly, the directive requires attribution, meaning the exclusion applies only if the attack can be affirmatively attributed to a nation-state actor.[10]

Attribution, however, is notoriously difficult in cyberspace. Sophisticated attackers routinely obscure their origins, route attacks through compromised infrastructure in third countries, and adopt tactics associated with other groups to confuse investigators. Even when government agencies issue formal attributions, these often come months or years after an incident, long after insurance claims have been filed and disputed.

Many cyber policies attempt to address this ambiguity by including terrorism carve-backs: provisions that restore coverage for attacks meeting certain definitions of terrorism. A typical carve-back might cover attacks committed with the intent to further state social, ideological, religious, economic, or political objectives. This language appears designed to maintain coverage for ransomware groups and hacktivists while still excluding traditional state military operations.[11]

For food and agriculture, the terrorism carve-back raises uncomfortable questions. Animal rights activists have targeted agricultural facilities for decades, from releasing mink from fur farms to vandalizing equipment at livestock operations. Environmental extremists have sabotaged logging equipment and agricultural machinery. These actors typically pursue social and political objectives, potentially qualifying for the terrorism carve-back. But some policies contain further exclusions for activities in support of military actions or committed by groups simultaneously engaged in armed conflict. The boundaries become blurry, and blurry boundaries favor insurers at claim time.[12]

Consider a hypothetical scenario: a hacktivist group with loose ties to animal liberation movements launches ransomware against a large pork processing facility, explicitly citing factory farming practices as motivation. Does the terrorism carve-back apply? What if the same group has previously claimed attacks against government facilities in other countries? What if intelligence agencies suspect but cannot prove state sponsorship? These questions have no clear answers, and the uncertainty itself becomes a risk factor that food operations must account for.[13]

Claims That Went Sideways

Understanding how insurance claims fail illuminates the gaps between what policyholders expect and what policies deliver. In May 2022, CiCi Enterprises, the Texas-based pizza restaurant chain, suffered a ransomware attack that encrypted its systems. The attackers demanded $2 million dollars; CiCi's negotiators got the demand down to $400,000. Total losses exceeded $1.2 million dollars. CiCi's held a cyber insurance policy with a $3 million aggregate limit, and the company had done everything right: it notified its insurer promptly, retained approved counsel and incident response firms, and documented its losses carefully.[14]

Then the insurer invoked a ransomware sublimit endorsement, arguing that the maximum payout was capped at $250,000 regardless of the policy's stated limits. CiCi's sued. In February 2026, a federal court in Texas ruled that the sublimit endorsement didn't apply because the policy language failed to clearly state how the cap worked or that it modified the cyber extortion coverage. The court noted that if the insurer "wanted the $250,000 sub-limit to apply across the board. . . it was incumbent upon [them] to expressly and clearly state that." The case continues, but the ruling illustrates a troubling pattern: insurers adding coverage limitations through endorsements whose implications become clear only when a claim is filed.[15]

Claim denials have become increasingly common. Industry data indicates that nearly 40 percent of cyber insurance claims faced denial or coverage limitations in 2024. The most frequent grounds for denial involve failure to maintain security controls represented in policy applications. Cyber insurance applications typically include detailed questionnaires about security practices: whether multifactor authentication (MFA) is implemented, how often systems are patched, and whether backups are tested regularly. These representations create warranties, binding commitments that the policyholder will maintain stated security practices.[16]

If an incident reveals that a represented control was not actually in place, insurers may void coverage entirely. One widely discussed case involved a company that represented on its application that it had implemented MFA across all remote access points. The post-incident investigation revealed that several legacy systems lacked the stated protection. The insurer denied the claim, arguing that the misrepresentation was material to its underwriting decision. The policyholder contended that the application questionnaire was ambiguous and that most access points were indeed protected. Litigation followed.[17]

Food and agricultural operations face challenges with security representation. Legacy operational technology systems often cannot support modern security controls. A pasteurization controller from the 1990s may have no mechanism for MFA. A grain elevator monitoring system may run on software that cannot be patched without voiding the manufacturer's support contract. Accurately representing security posture on an insurance application requires careful attention to these nuances, often more technical expertise than small operations possess.

Insurance Requirements as a Security Lever

For cybersecurity professionals working in food and agriculture, insurance requirements can become powerful allies in the perpetual struggle for security investment. When an insurer's questionnaire demands MFA, or when premiums triple because backup systems aren't properly segmented, that external pressure often accomplishes what internal advocacy could not.

The dynamic works because insurance requirements translate technical risks into financial consequences that executives and boards understand. Explaining that unpatched systems create an attack surface is one thing. Showing that the insurance carrier will either deny coverage or increase premiums by six figures unless those systems are addressed is another. The conversation shifts from abstract security concerns to concrete budget impact.

Many carriers now require specific controls before they write coverage. MFA on remote access. Endpoint detection and response tools. Tested backup and recovery capabilities. Network segmentation between IT and OT environments. Regular vulnerability scanning. Employee security awareness training. Each requirement that appears on an insurance application becomes leverage for the security team.[18]

The strategy requires timing. Security professionals should be involved in insurance renewal discussions months before deadlines. When the broker presents application questionnaires, the security team can identify gaps between the current state and the required state and then frame the necessary investments as insurance prerequisites rather than security wish lists. The argument changes from "we should implement MFA because it's best practice" to "we must implement MFA, or we lose coverage."

This approach also helps prioritize investments. Not every security improvement carries equal weight with insurers. Understanding which controls most affect premiums and coverage availability helps focus on limited resources where they deliver both security value and insurance benefit. A control that reduces premiums enough to offset its cost becomes self-funding from a budget perspective.

Liability Beyond Insurance

Insurance addresses some, but not all, of the financial exposure from cyber incidents. Liability to injured parties may exceed policy limits. Regulatory penalties may fall outside coverage. Contractual obligations to business partners may create exposure that insurance cannot address. And reputational damage, although sometimes partially covered under crisis management provisions, often represents the most significant long-term cost.

Food safety incidents carry liability weight. Under product liability law in most jurisdictions, manufacturers and sellers of defective products that cause injury face strict liability, meaning plaintiffs need not prove negligence or intent, only that the product was defective and caused harm. If a cyberattack compromises allergen labeling systems, leading to mislabeled products that trigger allergic reactions, the legal framework for liability is well established, even if the cyber component is novel.[19]

Consider the implications: a ransomware attack disables a facility's quality control systems. Under pressure to resume production and avoid spoilage of raw materials, management decides to ship products based on manual checks rather than wait for full system restoration. If those manual checks miss something, a contaminated product reaching consumers could trigger recall costs, personal injury claims, and regulatory action. The decision to ship during degraded operations, even if understandable given business pressures, may be characterized as negligent in subsequent litigation.

Supply chain liability adds another dimension. The Food Safety Modernization Act requires food facilities to implement preventive controls and supply chain programs that hold them accountable for the safety of ingredients they receive. If a cyber incident at a supplier compromises the integrity of ingredients and those ingredients cause problems in finished products, liability may flow to the receiving facility even though the attack occurred elsewhere. Contracts between supply chain partners typically address these scenarios, but contract language drafted before cyber risks were well understood may not provide the protection that parties assume.[20]

Business Continuity When Biology Won't Wait

Business continuity planning (BCP) for food and agriculture must account for a fundamental reality that distinguishes the sector from most others: biological processes do not pause for system recovery.

The November 2023 incident at Vital Bircher's Swiss dairy farm, which we examined at the beginning of the book, illustrated this reality with tragic clarity. When ransomware locked the milking parlor's control system, the immediate crisis was not data loss but animal welfare. The robot could continue basic milking operations in degraded mode, but the health-monitoring data that would have flagged a cow in distress was inaccessible. A pregnant cow showing subtle signs of trouble went unnoticed. The calf died, and the cow had to be euthanized. The incident cost approximately 11,000 Swiss francs in direct losses, but the harder cost was measured in animal lives.[21]

Effective BCP for agricultural operations must explicitly address these biological imperatives. The Cyber Security Agriculture Canada framework identifies several elements that agricultural business continuity plans should address: critical digital assets and their dependencies, manual fallback procedures for essential operations, communication protocols when normal channels are unavailable, and relationships with vendors that can provide emergency support.[22]

Manual fallback procedures deserve particular emphasis. Modern dairy operations may have staff who have never performed manual milking. Greenhouse operations may rely entirely on automated climate control. Feed mills may have formulation processes that exist only in software. Documenting manual procedures, training staff to execute them, and periodically testing that capability are not merely good practices. They are the difference between a bad week and a catastrophe.

The question of what constitutes acceptable reduced operations varies by facility and season. A grain elevator might continue receiving harvest deliveries with paper documentation while systems are down, reconciling records later. A meat processing plant might pause slaughter operations but continue processing already-slaughtered carcasses that would otherwise spoil. A dairy might shift to bulk tank storage that doesn't require individual cow identification. Each of these decisions involves trade-offs between operational continuity, food safety, regulatory compliance, and financial loss.

Building a Plan That Actually Works

Business continuity plans gather dust when they read like compliance documents rather than operational guides. The test of a good plan is whether someone grabbing it during a crisis can follow it. This means concrete actions, named responsibilities, and current contact information, not abstract frameworks and aspirational statements.

For food operations, a functional business continuity plan should address several distinct phases. The first phase covers immediate response in the first minutes to hours: how to detect that an incident has occurred, who has authority to make emergency decisions, how to communicate when normal systems may be compromised, and how to preserve evidence for later investigation without impeding immediate recovery. Designating decision-making authorities in advance prevents dangerous delays when systems go down and the people who normally approve of major decisions cannot be reached.

The second phase covers stabilization, typically the first hours to days: implementing manual workarounds, engaging incident response resources, communicating with employees and business partners, and triggering which systems to restore first. Food operations must weigh system restoration priorities against biological timelines. Restoring email can wait; restoring refrigeration monitoring cannot.

The third phase covers recovery, which may extend for weeks or months: rebuilding compromised systems, implementing improvements to prevent recurrence, managing ongoing communications, and addressing the financial aftermath, including insurance claims. Recovery is when the boring but essential work happens: documenting losses, coordinating with insurers, updating security controls, and training staff on whatever changes result from the incident.

Testing the plan matters as much as writing it. Tabletop exercises, during which key personnel walk through hypothetical scenarios without disrupting operations, reveal gaps that look obvious in hindsight but are invisible on paper. Can the procurement team reach backup suppliers if primary communication channels are down? Does the plant manager know how to switch fermentation tanks to manual control? Has anyone tested whether the backup generator will power critical refrigeration for the duration it might be needed? These questions are better answered in a conference room than during an actual incident.

When Size Determines Options

The insurance and continuity strategies available to a multinational food corporation bear little resemblance to those accessible by a family farm or small regional processor. Large enterprises maintain dedicated risk management teams, carry policies with limits measured in hundreds of millions of dollars, and can retain specialized incident response firms on a standing contract. A 50-cow dairy operation might struggle to justify the premium for any cyber coverage at all.

This disparity creates structural vulnerability in the food system. Large operations can absorb and recover from incidents that would bankrupt smaller competitors. The resulting consolidation pressure runs counter to the food system resilience that diverse, distributed production provides. A food system dominated by a few large producers may appear efficient, but it contains hidden fragility: there are fewer points of failure, but each failure carries greater consequences.

Small operations have options, even if those options require different approaches than enterprise risk management. Cooperative structures allow smaller farms and processors to pool resources for shared security services and group insurance programs. Some agricultural associations have negotiated group cyber insurance programs that offer coverage at lower rates than individual policies. Trade associations and extension services increasingly offer cybersecurity resources tailored to agricultural contexts.

The most fundamental protection available to operations of any size is maintaining relationships that can provide support during a crisis. Knowing a neighboring farm that could temporarily house livestock if barn systems fail. Having a backup veterinarian's cell phone number written somewhere other than on the computer. Understanding which local IT consultant has worked with agricultural equipment versus generic office systems. These relationships cannot be established during an incident; they must exist beforehand.

International Perspectives

Cyber insurance markets and regulatory expectations vary significantly across jurisdictions, creating complexity for food operations that cross borders. The European Union's Network and Information Security Directive, commonly known as NIS2, which took effect in October 2024, imposes cybersecurity requirements on food production and processing entities that meet certain thresholds. These requirements include risk management measures, incident reporting obligations, and supply chain security programs. Failure to comply can result in significant penalties.[23]

Regulatory pressure in Europe has begun to influence insurance markets. Insurers increasingly inquire about NIS2 compliance during underwriting, and demonstrated compliance may improve policy terms. Conversely, facilities that cannot demonstrate adequate security measures may find coverage difficult to obtain at any price.[24]

Australia has seen significant agricultural cyber incidents that have shaped regional awareness. The 2022 attack affecting GrainCorp disrupted grain handling and logistics across the country.[25] The Lion Dairy and Drinks ransomware attack in June 2020 halted dairy and beverage production at one of Australia and New Zealand's largest food and beverage companies, forcing the company to implement manual processes just as pandemic restrictions were easing and demand was recovering.[26] These incidents demonstrated that agricultural cyber risk is not uniquely American; it follows connectivity wherever it spreads.

Canadian agricultural operations face their own regulatory environment, with provincial variations adding complexity. Cyber Security Agriculture Canada has developed resources specifically for agricultural BCP, recognizing that generic corporate frameworks do not translate well to farming contexts. The organization emphasizes that BCP is not merely a compliance exercise but a survival skill for operations that depend on digital systems they may not fully control.[27]

What It Teaches Us

The insurance industry's approach to cyber risk continues evolving, sometimes in ways that benefit policyholders and sometimes in ways that do not. Premium increases that peaked in 2022 have moderated, with many buyers seeing rate decreases in 2024 and 2025. Competition among insurers has driven improvements in coverage and additional services, including proactive security tools offered directly by carriers.[28]

At the same time, exclusions have become more refined and more consequential. The gap between what policyholders believe they have purchased and what their policies cover remains a persistent problem. Reading policy language carefully, understanding exclusions, and asking pointed questions before incidents occur is more important than ever.

For food and agricultural operations, the path forward requires integrating cyber risk into the same frameworks used to manage operational, food safety, and financial risks. Insurance is one tool, not a complete solution. Understanding liability exposure, both direct and through the supply chain, helps inform decisions about what risks to accept, what risks to mitigate, and what risks to transfer.

The coming chapter turns from financial risk to cultural risk: how to build a food safety culture that incorporates cybersecurity, and how to build a cybersecurity culture that understands food. The intersection is where real protection happens, when people understand not just that security matters, but why it matters for what they care about: the safety of the food they produce and the people who consume it.

Notes

1. Satariano, A., & Perlroth, N. (2019). "Big companies thought insurance covered a cyberattack. They may be wrong." *New York Times*. **https://www.nytimes.com/2019/04/15/technology/cyberinsurance-notpetya-attack.html**; Adriano, L. (2022). "Zurich, Mondelez settle long-standing lawsuit over $100 million claim." *Insurance Business*. **https://www.insurancebusinessmag.com/us/news/cyber/zurich-mondelez-settle-longstanding-lawsuit-over-100-million-claim-426741.aspx**
2. Coalition, Inc. (2025). "2025 Cyber Claims Report finds ransomware stabilized but remains costly for businesses." **https://www.coalitioninc.com/announcements/2025-cyber-claims-report**
3. American Cattlemen. (2026). "Why cyber insurance is critical for feedyards & cattle producers." **https://americancattlemen.com/why-cyber-insurance-is-critical-for-feedyards-cattle-producers**
4. FTC. (2021). "Cyber insurance." **https://www.ftc.gov/business-guidance/small-businesses/cybersecurity/cyber-insurance**
5. Heiss, M. (2025). "Understanding business interruption following a cyber attack." *MDD*. **https://www.mdd.com/forensic-accounting-articles/business-interruption-after-cyber-attack**
6. Greenwald, J. (2022). "Chubb unit ordered to cover ransomware reimbursement: Yoshida Foods International." *Business Insurance*. **https://www.businessinsurance.com/chubb-unit-ordered-to-cover-ransomware-reimbursement-yoshida-foods-internationa**
7. Law Street Media. (2022). "Food company sues insurer after rejected claim for cyberattack ransom." **https://lawstreetmedia.com/news/agriculture/food-company-sues-insurer-after-rejected-claim-for-cyberattack-ransom**
8. Keeley, M. (2022). "Direct Means Proximate? Oregon District Court Holds Ransomware Payment Is a Direct Loss." *Clark Hill PLC*. **https://www.clarkhill.com/news-events/news/direct-means-proximate-oregon-district-court-holds-ransomware-payment-is-a-direct-loss**
9. ibid.
10. Lloyd's of London. (2022). "Market bulletin Y5381: cyber-attack exclusions." **https://assets.lloyds.com/media/35926dc8-c885-497b-aed8-6d2f87c1415d/Y5381%20Market%20Bulletin%20-%20Cyber-attack%20exclusions.pdf**
11. Global Business Alliance Insurance. (2022). "Assessing cyber insurers' war exclusions & terrorism carvebacks." **https://www.gbainsurance.com/cyber_war_222**
12. Gear, D. (2025). "Property insurance exclusions: lessons from recent high-profile incidents." *Sedgwick*. **https://www.sedgwick.com/blog/property-insurance-exclusions-lessons-from-recent-high-profile-incidents**
13. Bateman, J. (2020). "War, terrorism, and catastrophe in cyber insurance: understanding and reforming exclusions." *Carnegie Endowment for International Peace*. **https://carnegieendowment.org/research/2020/10/war-terrorism-and-catastrophe-in-cyber-insurance-understanding-and-reforming-exclusions**
14. Justia. (2026). "CiCi Enterprises LP v. HSB Specialty Insurance Company." **https://law.justia.com/cases/federal/district-courts/texas/txndce/3:2023cv02155/381618/74**
15. Bongat, C. (2026). "Court blocks HSB's ransomware sublimit in first-of-its-kind cyber ruling." *Insurance Business Magazine*. **https://www.insurancebusinessmag.com/us/news/cyber/court-blocks-hsbs-ransomware-sublimit-in-firstofitskind-cyber-ruling-567006.aspx**
16. Jain, S. (2026). "160 Cyber Insurance Statistics 2026." *Astra Security*. **https://www.getastra.com/blog/security-audit/cyber-security-statistics**

17. American Bar Association, Commercial & Business Litigation Committee. (2023). "Travelers v. International Complaint." **https://www.americanbar.org/content/dam/aba/publications/litigation_committees/commercial/cases/2023/travelers-v-international-complaint.pdf**
18. LawFinity. (2025). "Cyber insurance and IT compliance: what underwriters look for." **https://www.ebcgroup.co.uk/lawfinity-news-insights/cyber-insurance-and-it-compliance?utm_source=chatgpt.com**
19. Said, Z. K. (2022). "Product liability based on dangerous or defective condition." In *Tort Law: A 21st-Century Approach* (CALI eLangdell). **https://saidtorts2d.lawbooks.cali.org/chapter/product-liability-based-on-dangerous-or-defective-condition**
20. Code of Federal Regulations. (2016). Title 21, chapter I, subchapter B, part 117, subpart G, supply-chain program. **https://www.ecfr.gov/current/title-21/chapter-I/subchapter-B/part-117/subpart-G**
21. CSO. (2025). "Ransomware attack paralyzes milking robots, cow dead." **https://www.csoonline.com/article/3484349/ransomware-attack-paralyzes-milking-robots-cow-dead.html**
22. Agriculture and Agri-Food Canada. (2026). "Cyber security and your farming business: prepare." **https://agriculture.canada.ca/en/programs/tools-manage-farm-risk-and-finance/cyber-security-and-your-farming-business#prepare**
23. European Union. (2022). "Directive (EU) 2022/2555 of the European Parliament and of the Council of 14 December 2022 on measures for a high common level of cybersecurity across the Union, amending Regulation (EU) No. 910/2014 and Directive (EU) 2018/1972, and repealing Directive (EU) 2016/1148 (NIS 2 Directive)." *EUR-Lex.* **https://eur-lex.europa.eu/eli/dir/2022/2555/oj**
24. Zappaterra, G. (2024). "The NIS2 Directive – will it affect insurance companies?" *DLA Piper.* **https://www.dlapiper.com/en-us/insights/publications/derisk-newsletter/2024/the-nis2-directive-will-it-affect-insurance-companies**
25. Kirby, I. (2022). "Incident: ransomware attack on Australian agricultural company GrainCorp's payroll provider Kronos." *kirbyidau.com.* **https://kirbyidau.com/2022/02/23/incident-ransomware-attack-on-australian-agricultural-company-graincorps-payroll-provider-kronos**
26. Corfield, G. (2020). "Cyber attack targeted Australian brewery 'n' dairy biz Lion." *The Register.* **https://www.theregister.com/2020/06/11/australia_brewery_lion_cyber_attack**
27. Agriculture and Agri-Food Canada. (2026). "Cyber security and your farming business: prepare."
28. Risk & Insurance. (2025). "Global cyber insurance market reaches $16.6 billion in 2024." **https://riskandinsurance.com/global-cyber-insurance-market-reaches-16-6-billion-in-2024**

CHAPTER 24

Food Safety Culture and Cybersecurity

When What You Don't See Can Kill

Three people died eating ice cream in 2015. *Listeria monocytogenes* had been lurking in Blue Bell Creameries' Oklahoma plant for at least two years before anyone connected the dots.[1] The outbreak that eventually triggered a full recall of all Blue Bell products wasn't sudden. Investigators traced it back to March 2013, when the company first detected *Listeria* on equipment and surfaces. Then the company found it again. Seventeen separate positive tests over two years, and product kept rolling off the lines.[2]

What makes Blue Bell's story relevant here is the systemic blindness that allowed the problem to persist. The Delaware Supreme Court, in a landmark ruling about the company's board, noted there was no committee addressing food safety, no regular process requiring management to report compliance practices to directors, and no schedule for considering food safety risks. When third-party auditors raised what the court called "red, or at least yellow, flags," those warnings never reached the boardroom. The company had procedures on paper, but the culture of acting on them wasn't there.[3]

Digital monitoring systems could have automatically escalated repeated positive tests. Robust traceability could have enabled surgical recalls instead of a company-wide shutdown. These technologies exist. The question is whether any organization's culture is ready to use them and whether that culture recognizes that protecting digital systems has become part of protecting food.

What Food Safety Culture Means in Practice

The Global Food Safety Initiative (GFSI) defines *food safety culture* as "shared values, beliefs and norms that affect mindset and behavior toward food safety in, across and throughout an organization."[4] Stripped of the formality, it's whether the newest person on the sanitation crew will stop the line if something looks wrong, confident that leadership will back them up.

GFSI identifies five dimensions: *vision and mission* (is food safety embedded in why the company exists?), *people* (how leadership behaves, how employees are trained), *consistency* (are resources actually allocated?), *adaptability* (can the organization respond when conditions change?), and *hazard awareness* (does everyone understand what can go wrong?).[5]

Swap "cybersecurity" for "food safety," and you'd have a reasonable framework for any security program. Both disciplines protect people from harm. Both require cultures in which doing the right thing is easier than cutting corners. The problem is that in most food companies, these cultures developed separately. Food safety came from microbiology, Hazard Analysis and Critical Control Points (HACCP) plans, and regulatory inspections. Cybersecurity came from IT, network architecture, data, and the ongoing fight with threat actors. The people in these fields report to different executives and use different vocabularies.

That separation made sense when production was mostly mechanical and cybersecurity meant email servers. But not anymore! A ransomware attack can shut down pasteurization. A compromised credential can alter allergen labels. The same sensors monitoring food safety data connect to networks that cross oceans.

When Digital Coordination Works and When It Doesn't

The multinational *Listeria* outbreak linked to enoki mushrooms between 2016 and 2020 showed what's possible. Forty-eight illnesses across the United States and Canada were eventually traced to a single manufacturer in South Korea, a connection that emerged only through whole-genome sequencing data shared internationally.[6] When the CDC posted genomic data to shared databases, French authorities responded within days with related isolates from Korean-origin mushrooms collected in 2017.[7] Australian officials found matching cases going back three years.[8]

But that same connectivity creates fragility. The INFOSAN network, which coordinates food safety responses among WHO member states, depends on digital platforms for rapid alerts.[9] What happens when those platforms go down? When ransomware locks genomic sequencing equipment during an emerging outbreak?

JBS, Schreiber Foods, Dole; the food sector has experienced its share of cyberattacks. We haven't yet seen one that compromises food safety systems during an active contamination event, but that absence won't last forever; sadly, it's only a matter of time.

When the Cultures Actually Talk

Most of this chapter documents failures. But some organizations are building something different.

Nestlé, the world's largest food and beverage company, has digitalized core elements of its food safety management system, including early-warning systems, predictive analytics for supplier risk assessment, and AI-enabled consumer feedback analysis.[10] The company's approach treats cybersecurity not as a separate IT function but as infrastructure that food safety depends on. When digital tools flag a potential contamination signal, the integrity of those tools matters as much as the signal itself. The company works with Microsoft on PhishScreener, a custom tool that identifies phishing attempts before employees click.[11]

What makes this relevant isn't the size of Nestlé's security budget; it's integration. The company's global head of food safety describes HACCP as "the engine room" of the food safety management system, and Nestlé has become an early adopter of HACCP digitalization.[12] That digitalization creates dependencies on networks, databases, and authentication systems. Nestlé appears to understand this, working those dependencies into its food safety thinking rather than treating them as someone else's problem.

Arla Foods, the European dairy cooperative, has built a framework of integrated tools specifically designed around the recognition that cyberattacks could disrupt food operations. According to GlobalData's analysis, Arla Foods is actively developing contingency plans for scenarios in which digital systems fail. And Ferrero has established dedicated teams coordinating security and compliance across its operations.[13]

These aren't perfect organizations. They face the same pressures everyone else does: margin compression, talent shortages, legacy equipment. What distinguishes them is that someone at a senior level decided that cybersecurity belonged in conversations about food safety culture, not just IT strategy. That's the starting point.

The Organizational Divide

In most food companies, food safety reports up through operations or quality. Security reports through IT or, less commonly, a dedicated risk function. These reporting chains rarely intersect below the C-suite. Both teams have their own budgets, their own metrics, their own languages. Telling them to collaborate doesn't address the structural reality that they're often competing for resources and executive attention.

What does work: finding existing forums where both perspectives belong. IT typically runs a change advisory board (CAB) that reviews modifications to production systems before implementation. Food safety should have a seat at that table, not to become IT experts, but to flag when proposed changes could affect food safety systems. If a patch requires rebooting a programmable logic controller (PLC) that controls pasteurization temperatures, someone who understands pasteurization needs to be in that conversation.

The same applies in reverse. Food safety investigations often convene cross-functional teams when contamination is suspected. If the investigation might reveal a compromised sensor or altered data, IT or OT security should be present from the beginning, not called in later when forensic evidence may have been overwritten.

Joint meetings don't require budget approvals. They require one person to decide to send an invitation.

For Food Safety Teams

If you work in food safety, quality, or Food Defense, cybersecurity probably feels like someone else's job. You've got HACCP plans to maintain, audits coming, sanitation schedules to verify, and maybe a food fraud assessment that's been sitting in your inbox for weeks. But you're already doing related work. Your Food Defenseplan, developed under FSMA's Intentional Adulteration rule or as part of a Threat Assessment Critical Control Point (TACCP) exercise, considers scenarios in which someone deliberately contaminates product. Does "someone" include an attacker at a keyboard in another country, manipulating systems that control your processing parameters? If your threat assessment only pictures a disgruntled employee near a mixer, you've got a gap.

As an exercise, walk your production floor with different eyes. Every piece of equipment with a digital interface, every sensor feeding a supervisory control and data acquisition (SCADA) system, every login credential granting access to processing controls: these are places where food safety and cybersecurity overlap. You don't need to become a security expert. Ask basic questions: Who has access? When was the software updated, and how is it updated? What happens if this controller receives wrong instructions? What's the backup if this display goes blank?

The methodological parallels to standards like IEC 62443 are significant.[14] Here's how that parallel works in practice:

HACCP concept	IEC 62443 equivalent
Hazard Analysis	Threat Assessment / Risk Analysis
Critical Control Point (CCP)	Security Zone / Conduit
Critical Limits	Security-Level Targets
Monitoring Procedures	Continuous Security Monitoring
Corrective Actions	Incident Response
Verification	Security Audits / Penetration Testing
Documentation	Asset Inventory / Configuration Management

HACCP asks, what biological, chemical, or physical hazards could enter this process, and where must we place controls to prevent harm? IEC 62443 asks the same question about cyber threats: what attack vectors exist, and where must we implement defenses? Both frameworks segment complex systems into zones, identify points where failures would be catastrophic, establish thresholds for acceptable operation, and require continuous monitoring and documented response when those thresholds are crossed.

The difference is that HACCP has decades of regulatory enforcement behind it. Food safety teams know exactly what auditors will ask because inspectors have been asking those questions for years. Cybersecurity in food production lacks regulatory maturity, which is why food safety's documentation discipline could become a model for OT security programs.

The regulatory landscape is evolving faster than most practitioners realize. No single regulation currently mandates cybersecurity specifically for food production or agriculture in the way that other critical infrastructure sectors are regulated. But the gap is filling from multiple directions outside of cybersecurity.

BRCGS Issue 9 explicitly extends food defense to include cybercrime, stating that "food defense can encompass measures to protect against cybercrime" and requiring organizations to address this within their threat assessments. The standard requires systems that "protect products, premises and brands from malicious actions," with cybersecurity now falling within that definition.[15]

FSSC 22000 Version 6 introduced Information and Communication Technology (ICT) requirements, creating audit expectations around how organizations manage digital systems that connect to food safety management. Auditors are asking: how is your FSMS protected from digital threats? What happens to your traceability system if the network goes down? Can you demonstrate that critical food safety data hasn't been tampered with?[16]

In the UK, PAS 96:2017 includes Annex D, which specifically covers "10 steps to cyber security" as a board-level responsibility within Food Defense.[17] It's guidance rather than regulation, "should" rather than "shall," but it establishes the expectation that organizations applying TACCP consider digital attack vectors alongside physical ones.

Even FSMA's Intentional Adulteration rule, although focused on protecting against deliberate contamination by insiders, creates a framework that can accommodate cyber threats. The rule requires vulnerability assessments and mitigation strategies for "actionable process steps."[18] An attacker who gains remote access to a control system to contaminate product is still an attacker engaged in intentional adulteration. The mechanism is digital, but the outcome of wide-scale public health harms is exactly what the rule addresses.

What auditors will increasingly ask: have you conducted a vulnerability assessment that includes cyberattack vectors? Do your mitigation strategies address remote access risks? Can you demonstrate monitoring and verification for digital systems that affect food safety? Do your personnel understand how cyber threats could be used for intentional adulteration?

Try this: invite someone from the IT and OT teams to your next HACCP team meeting. Not to present on their topics, but to listen. Let them hear how you think about hazard analysis and verification procedures.

For Security Teams

If you work in cybersecurity, IT, or OT within a food company, the production floor probably feels foreign. You understand endpoints and encryption, not pasteurization and allergen controls. The plant manager speaks a different language and likely views your patch requirements as threats to uptime.

The food safety team is doing work that looks familiar. They assess hazards, implement controls, monitor continuously, respond to deviations, and document everything. They talk about biological hazards instead of threat actors, CCPs instead of network segmentation. The underlying logic and mission are similar.

What's different in food production is that availability and integrity often matter more than confidentiality. When the quality manager can't access the lab information system during a product-hold decision, that's a food-safety delay. When process control data gets altered, the result could be food that looks normal but isn't safe.

Learn what a CCP means in your facility. Walk the floor with Food Safety, and have them explain why certain measurements matter, why certain processes can't be interrupted, and what happens downstream when something fails. You'll probably find systems you didn't know existed—aging PLCs on legacy operating systems, sensors transmitting over protocols you've never heard of, remote access set up years ago by vendors and never properly secured.

I once worked on a risk assessment at a media factory: CDs, DVDs, and console games. It's a heavy chemical process. We were walking the production floor when the tour stopped in a space that looked like somewhere villains fight in action movies: industrial pipes, harsh lighting, the works. Several engineers with long tenures stood there staring at me like I'd lost my mind, which, to be fair, was a reasonable reaction to a cybersecurity auditor beaming at their wastewater treatment equipment. I was grinning under my hard hat. This room was where they stripped copper from production wastewater before returning it to the municipal supply. I thought it was fascinating and was perfectly happy to stand there listening to their stories about the process. While they talked, my partner noticed a machine near the process with a network cable running out of it. He asked about it. No one seemed to know that it was connected to the network. The Environmental Protection Agency (EPA) had installed it, they said, and nobody had bothered with it after that. If we hadn't walked that floor, and if I weren't such a nerd about industrial processes, we'd never have noticed it. It's remarkable what you find when you let curiosity lead you through your own facility.

You'll likely discover something else: the documentation gap runs in one direction. Food safety teams have been documenting everything for decades—regulatory requirements made that nonnegotiable. Every temperature check, every sanitation verification, and every corrective action gets logged and retained for audits. That discipline exists because inspectors show up and ask to see records. OT security in food environments often has almost none of this. No complete asset inventory. No patch records. No access logs. The cultural habits that make food safety auditable largely don't exist for the systems on which food safety now depends. That's an opportunity. The documentation rigor that food safety teams already practice could become a model for OT security programs.

Consider joint tabletop exercises. Food safety teams run mock recalls and contamination scenarios; your team runs incident response exercises. Combine them. A ransomware attack that encrypts your SCADA system during production isn't just a cybersecurity incident; it's potentially a food safety incident if you can't verify that CCPs were maintained. Working through these scenarios together builds understanding that policy documents can't.

There's a philosophical gap worth acknowledging. Food safety culture often operates from a zero-tolerance mindset: any positive pathogen test is a failure; any temperature deviation triggers corrective action. Cybersecurity culture typically accepts residual risk when you prioritize threats based on likelihood and impact, knowing you can't eliminate every vulnerability. When these teams try to work together, they can talk past each other. The food safety professional asking, "Is this system secure?" wants certainty. The security professional knows that absolute certainty doesn't exist and instead offers probability assessments.

Here's a bridging principle: when cybersecurity controls protect food safety systems, adopt the Food Safety team's risk tolerance. Zero-tolerance makes sense when the failure mode is "someone gets sick or could die." The pasteurization control system gets treated differently from the marketing department's file server. Security teams already practice these critical systems and receive more protection than peripheral ones. The translation is recognizing which systems food safety considers critical, which may not match technical intuition.

And when a joint incident eventually occurs, a cyberattack that creates a food safety question, the teams need to have worked through tensions in advance. Food Safety's instinct will be to clean equipment and resume production; Security's instinct will be to preserve evidence and understand what happened. Both instincts are right. The conflict emerges when no one has discussed how to balance them before the crisis. Joint business continuity and disaster recovery planning isn't about creating more documents. It's about ensuring that when the head of cybersecurity and the head of food safety are in the same room during a real incident, they've already built a working relationship to make decisions together.

And when you're designing controls, remember the constraints. You can't always patch during operating hours, because lines can't stop. Biometrics access doesn't work on equipment that operators access with gloved hands dozens of times per shift. Production staff focused on moving 10,000 pounds of product through a complex process aren't going to catch phishing attempts. Security that ignores operational reality becomes a reason for workarounds and increased risk.

What It Teaches Us: The Cost of Waiting

Budgets are tight. Margins in food production are thin. Every dollar spent on integrated programs is a dollar not spent elsewhere.

Consider these numbers. The USDA estimates that foodborne illness costs the United States roughly $75 billion annually in medical expenses, lost productivity, and premature death.[19] That's conservative—it doesn't include recalls or cascading effects on supply chains. Blue Bell's outbreak cost an estimated $175 million, forced layoffs of more than a third of the workforce, and required private equity investment to avoid collapse.[20] Chipotle's outbreaks contributed to a $6 billion drop in market value in 60 days.[21]

Those were food safety failures. But JBS paid $11 million in ransom after a ransomware attack halted meat operations across North America and Australia.[22] It wasn't just paying to recover files. It was preventing the food safety consequences of prolonged shutdowns, spoiled inventory, supply chain chaos, and products that couldn't be verified as safe.

And there's trust. Food is personal. When consumers learn that a company failed to protect them, something breaks that takes years to repair. Blue Bell was a beloved regional brand before 2015. It survived. But the unconditional trust customers once had took much longer to return; some would say it hasn't fully.

The next chapter examines how this integration works in practice: the specific programs where food safety and cybersecurity must merge, and how traditional frameworks can be strengthened to address digital vulnerabilities alongside physical ones.

The framework exists. Standards are evolving. What remains is the work of translation: taking principles both teams already understand and weaving them together. That starts with conversations. The Food Safety and Security managers should have coffee or tea to understand how the other thinks. Small steps that don't require budget approvals—just recognizing that the person on the other side of the building is working on the same problem from a different angle.

Notes

1. CDC. (2015). "Multistate outbreak of listeriosis linked to Blue Bell Creameries ice cream products." **https://archive.cdc.gov/www_cdc_gov/listeria/outbreaks/ice-cream-03-15/index.html**
2. Ibid.
3. Justia. (2019). *Marchand v. Barnhill, et al.* **https://law.justia.com/cases/delaware/supreme-court/2019/533-2018.html**
4. GFSI. (2019). *A Culture of Food Safety: A Position Paper from the Global Food Safety Initiative.* **https://mygfsi.com/wp-content/uploads/2019/09/GFSI-Food-Safety-Culture-Full.pdf**
5. Ibid.
6. Pereira, E., Conrad, A., Tesfai, A., Palacios, A., Kandar, R., Kearney, A., Locas, A., Jamieson, F., Elliot, E., Otto, M., Kurdilla, K., Tijerina, M., Son, I., Pettengill, J. B., Chen, Y., Fox, T., Lane, C., Aguillon, R., Huffman, J., Low, M. S. F., Wise, M., Edwards, L., Bidol, S., Blankenship, H. M., Rosen, H. E., Leclercq, A., Lecuit, M., Tourdjman, M., Herber, H., Singleton, L. S., Viazis, S., & Bazaco, M. C. (2023). "Multinational outbreak of listeria monocytogenes infections linked to enoki mushrooms imported from the Republic of Korea 2016–2020." *Journal of Food Protection*, 86(7), 100101. **https://doi.org/10.1016/j.jfp.2023.100101**
7. Pettengill, J. B., Markell, A., Conrad, A., Carleton, H. A., Beal, J., Rand, H., Musser, S., Brown, E. W., Allard, M. W., Huffman, J., Harris, S., Wise, M., & Locas, A. (2020). "A multinational listeriosis outbreak and the importance of sharing genomic data." *The Lancet: Microbe*, 1(6), e233. **https://doi.org/10.1016/S2666-5247(20)30122-1**
8. WHO. (2020). "INFOSAN quarterly summary, 2020 #2." **https://www.who.int/news/item/29-07-2020-infosan-quarterly-summary-2020-2**
9. WHO. (n.d.). "International Food Safety Authorities Network (INFOSAN)." **https://www.who.int/groups/fao-who-international-food-safety-authorities-network-infosan/about**
10. Robinson, N. (2025). "Nestlé's global head of food safety on digital and AI." *FoodNavigator.* **https://www.foodnavigator.com/Article/2025/03/07/how-does-nestle-tackle-food-safety-on-a-global-scale**
11. DeepLearning.AI. (2021). "24/7 Phish fry: how Nestlé built an AI-powered phishing detector with Azure." **https://www.deeplearning.ai/the-batch/24-7-phish-fry**
12. Robinson, N. (2025). "Nestlé's global head of food safety on digital and AI."
13. Just Food. (2022). "Revealed: the food manufacturers leading the way on cybersecurity." **https://www.just-food.com/features/revealed-the-food-manufacturers-leading-the-way-on-cybersecurity**

14. International Society of Automation Global Cybersecurity Alliance. (2022). "ISA/IEC 62443." **https://isagca.org/isa-iec-62443-standards**
15. BRCGS. (2025). "Food defence." **https://www.brcgs.com/about-brcgs/news/2026/public-consultation-food-safety-issue-9/#:~:text=BRCGS%20is%20commencing%20the%20revision,was%20launched%20in%20August%202022**
16. FSSC 22000. (2023). *FSSC 22000 Scheme Version 6.* **https://www.fssc.com/wp-content/uploads/2023/03/FSSC-22000-Scheme-Version-6-Final.pdf**
17. British Standards Institution. (2017). *PAS 96:2017: Guide to Protecting and Defending Food and Drink from Deliberate Attack.* **https://www.food.gov.uk/sites/default/files/media/document/pas962017_0.pdf**
18. FDA. "FSMA final rule for mitigation strategies to protect food against intentional adulteration." 21 CFR Part 121.
19. USDA, Economic Research Service. (2025). "Cost estimates of foodborne illnesses." **https://www.ers.usda.gov/data-products/cost-estimates-of-foodborne-illnesses**
20. Food Safety News. (2019). "The Blue Bell ice cream Listeria outbreak and its fallout." **https://www.foodsafetynews.com/2019/08/the-blue-bell-ice-cream-listeria-outbreak-and-its-fallout**
21. Tuttle, H. (2016). "Día de la crisis: The Chipotle outbreaks highlight supply chain risks." *Risk Management Magazine.* **https://www.rmmagazine.com/articles/article/2016/03/01/-Dia-de-la-Crisis-The-Chipotle-Outbreaks-Highlight-Supply-Chain-Risks-**
22. Associated Press. (2021). "REvil, a notorious ransomware gang, was behind JBS cyberattack, the FBI says." *NPR.* **https://www.npr.org/2021/06/03/1002819883/revil-a-notorious-ransomware-gang-was-behind-jbs-cyberattack-the-fbi-says**

PART IV

Strategic & Societal Themes

CHAPTER 25

Food Protection: Food Safety, Defense, and Quality Control

When Testing Stops

On the weekend of June 1–2, 2019, ransomware hit Eurofins Scientific. The Luxembourg-based company is the world's largest provider of food and environmental testing services, operating more than 800 laboratories across 47 countries and employing roughly 45,000 people. Its scientists test food for pathogens and verify pharmaceutical purity; its forensics division analyzes evidence for law enforcement agencies. Within hours, systems went down across multiple countries.[1]

Certificates of analysis (CoAs), the documents food companies require before shipping product, couldn't be generated. Pathogen testing ground to a halt, and quality data that regulatory agencies demand became inaccessible. One week later, the company acknowledged the financial impact would be "material."[2] Reports later suggested Eurofins paid a ransom, although the company wouldn't confirm it.[3]

In the UK alone, Eurofins handled more than 70,000 criminal forensics cases annually and processed an estimated half of all outsourced forensic work for the British police.[4] The ripple effects lasted months. But what deserves more attention is the food safety dimension. Every day, laboratories like Eurofins stand between potentially contaminated products and grocery store shelves. They test Salmonella in chicken and *E. coli* in spinach. They confirm allergen labels match what's in the package. When those laboratories go offline, the infrastructure keeping our food supply trustworthy goes with them.

This chapter examines what happens when digital systems underpinning food safety fail. The verification principles we've applied to physical food safety for decades now need to extend into digital systems, and we're behind on that work.

The Digital Architecture of Food Safety

Documentation in food production has always been sacred. Every temperature reading, sanitation step, and batch release decision gets logged somewhere. Paper clipboards gave way to software: specifically, laboratory information management systems (LIMSs). A LIMS tracks samples from arrival through analysis to final reporting: sample registration, test assignment, instrument data capture, results review, and certificate generation. Modern platforms connect to laboratory instruments, enterprise resource planning systems, supply chain databases, and regulatory reporting portals.[5]

For food companies that don't run their own labs, the concept might seem abstract, but think about what a CoA represents. When you receive a shipment of flour with a CoA stating that it's free of specific allergens and within microbial limits, you're trusting an entire digital chain. The LIMS managed the sample. Instruments generated data. Software calculated results. Another system produced the certificate. Compromise any link, and that document means nothing.

This architecture extends beyond testing labs. Modern food production facilities rely on digitally monitored critical control point (CCPs): specific steps where applying a control measure prevents or reduces food safety hazards. Temperature logging during pasteurization. Time monitoring during thermal processing. Metal detection before packaging. These CCPs increasingly use automated sensors feeding data into electronic systems rather than workers with thermometers and clipboards. A thermometer can't be hacked. A paper log can be falsified, but it requires physical access and leaves physical evidence. A networked sensor and the database storing its readings? Those can be manipulated remotely, and the manipulation might never leave a trace visible to human review.

Digital tools enable more precise, continuous, automated monitoring, but they create new vulnerabilities. A walk-in cooler logging temperature continuously to the tenth of a degree provides better documentation than hourly manual checks. Unless, of course, someone manipulates those records to hide a process deviation.

Integrating Cyber Threats into HACCP

I'm sure you're asking yourself, what does it look like, practically, to incorporate cyber threats into a Hazard Analysis and Critical Control Points (HACCP) plan? When you're filling out your hazard analysis worksheet, what do you write?

Start by mapping what might be called "digital critical control points" for each traditional CCP in your process, and identify the corresponding digital infrastructure. What sensors monitor that control point? What programmable logic controllers (PLCs) control the equipment? Where does the data go? What historian or LIMS stores the records? For each of these digital touchpoints, ask one question: what's the digital failure mode that could hide this hazard or falsify its control?

Consider some concrete scenarios:

Clean-in-place (CIP) cycle spoofing: A CIP system on a bottling line reports that caustic concentration, temperature, and contact time all met specifications, when the cycle ran with diluted chemicals at lower temperatures. The equipment appears sanitized in your records, but biofilm persists in the filler heads. The next production run picks up Listeria.

Historian tampering: Someone deletes a week of environmental monitoring data. Without actual values, you can't verify preventive controls for that period. That triggers hold-and-test protocols, potential recall, and regulatory scrutiny.

Supplier portal breach: A compromised supplier quality system injects falsified CoAs, allowing ingredients with undeclared allergens into your production line.

These aren't exotic scenarios; they're the digital equivalents of hazards HACCP was designed to address. The difference is that your auditor probably hasn't asked about them yet. That doesn't mean they won't, and it certainly doesn't mean the risks aren't real.

When documenting cyber-related hazards in your HACCP plan, treat them as what they are: potential causes of biological, chemical, or physical hazards that could reach consumers. A manipulated temperature sensor isn't an IT problem, it's a pathogen survival problem. Frame it that way, and your food safety team will understand why it belongs in their documentation.

When Falsified Data Kills

If you're wondering whether falsified quality data has ever actually caused deaths, whether this is a theoretical concern or a documented reality, let's revisit the actions of the Peanut Corporation of America (PCA) that we discussed in Chapter 9. In late 2008 and early 2009, 9 people died and at least 714 fell ill from Salmonella poisoning traced to PCA's peanut processing plant in Blakely, Georgia. The outbreak triggered the largest food recall in U.S. history at the time: more than 3,900 products across 46 states, from crackers to ice cream to pet treats.[6]

What made PCA different from other outbreaks was the documentation. Internal tests came up positive for Salmonella at least 12 times between 2007 and 2008. Executives responded by repeatedly running samples until one came back clean and then shipping as if the positive test results had never existed. In other cases, they shipped products before receiving lab results at all. They sent falsified CoAs to customers, assuring them that the peanuts were safe.[7]

One email became a centerpiece of the federal prosecution. When told that a customer order would be delayed pending Salmonella test results, CEO Stewart Parnell replied, "Just ship it. I can't afford to lose another customer." On January 12, 2009, while the outbreak was killing people, Parnell wrote to employees, "We have never found any salmonella at all." The company had actually documented at least 12 positive results.[8]

Stewart Parnell received a 28-year federal prison sentence, the harshest ever imposed for a food safety crime. His brother, Michael, was sentenced to 20 years. The quality assurance manager who falsified records got five years. The company declared bankruptcy.[9]

PCA's data falsification was analog—selective retesting, paper certificates, emails instructing staff to ship anyway. But here's the point: the same outcome could be achieved digitally with less effort and less evidence. Change a result in a LIMS database. Alter a certificate template. Modify an automated report. No emails saying, "just ship it." No paper trail of positive tests that someone chose to ignore. The digital version of PCA could be harder to detect and harder to prosecute, but no less deadly.

Quality Data as a Target

In food production, data can matter as much as physical products. A batch of cheese in cold storage is wheels of cheddar, yes, but it's also records proving proper pasteurization, test results confirming absence of Listeria, and temperature logs demonstrating safe storage conditions. Without that data, the cheese might be perfectly safe, but you can't prove it. And in food safety, what you can't prove might as well not have happened.

The FDA's 21 CFR Part 11, established in 1997, sets requirements for electronic records and signatures to ensure that they're as trustworthy as paper equivalents.[10] Although it was developed primarily for pharmaceutical manufacturing, its principles increasingly apply to food production as electronic systems replace paper. Systems must be validated. Access must be controlled. Audit trails must be maintained. Records must be preserved accurately.

The food industry uses ALCOA+ as shorthand for data integrity principles: *Attributable* (every entry traces to a specific person), *Legible* (records are permanent and readable), *Contemporaneous* (data is recorded at the time of activity), *Original* (records are preserved as created), *Accurate* (information is error-free). The plus (+) adds *Complete*, *Consistent*, *Enduring*, and *Available*. Anyone working in cybersecurity will recognize that an attack could undermine every one of these properties.[11]

Batch-release systems serve as gates between production and distribution. Before product ships, these systems verify that specifications were met and tests were completed. Manipulate data—change a pathogen test from positive to negative, adjust an allergen test result—and contaminated product moves into the supply chain with documentation that looks entirely proper. Equipment calibration records prove that analytical instruments functioned correctly. If a pH meter reads 4.2 when the actual value is 4.8, every product measured with it carries incorrect data. Calibration records exist to catch these deviations. Corrupt those records, and systematic measurement errors become invisible.

And then there's the instrument itself. Laboratory equipment, mass spectrometers, PCR machines, even digital microscopes are computers with firmware. In 2017, researchers at the University of Washington demonstrated that they could encode malware into synthetic DNA that, when sequenced, took control of the computer analyzing the data.[12]

The attack surface isn't limited to databases and networks. If someone compromises the analytical instrument, independent verification becomes impossible because the original measurement is wrong. The number the LIMS records is wrong from the moment it's generated.

What makes quality data attacks particularly dangerous is their invisibility. Manipulated data looks like legitimate data. Without independent verification, and in interconnected systems, true independence requires deliberate architecture: falsified records can persist undetected for a long time.

What Independent Verification Actually Looks Like

If your LIMS connects to your instruments, and your instruments are networked, and your backup pulls from the same system, where's the independence? It's a fair question, and the honest answer is that true independence requires intentional design, not default configuration.

Some options to consider: periodic manual spot-checks using stand-alone equipment that isn't connected to your production network. A thermometer and a stopwatch aren't glamorous, but they provide a ground-truth reference that's hard to spoof. Sending duplicate samples to a second laboratory, preferably one using different systems for critical parameters. Maintaining offline backups that aren't just for disaster recovery but serve as independent verification sources. For your highest-risk CCPs, consider keeping redundant paper records that can't be altered remotely.

None of this is convenient. It adds cost and friction to operations that already run on tight margins. But the alternative is trusting systems that may have been compromised without any way to detect the compromise. The inconvenience of periodic manual verification is considerably less than the inconvenience of shipping contaminated products because someone manipulated your digital records.

Finding Anomalies: Tools You Might Already Have

If you work at a larger organization, your company likely has anomaly-detection tools deployed for operational technology security. These systems monitor network traffic on the plant floor, establish baselines of normal behavior, and flag deviations. They exist because OT traffic is more predictable than IT traffic: OT devices do the same things repeatedly, which makes it unusually easy to spot a problem.[13] The question is whether your food safety team knows that these tools exist.

If you're on the food safety side, consider reaching out to your IT or OT security colleagues. Ask whether the company has network monitoring or anomaly detection deployed on production systems. Ask whether you can see the dashboards or receive alerts relevant to food safety equipment. You're not asking them to do more work; you're asking to be included in information they're already collecting.

The same applies to LIMS and quality system data. Patterns in test results, sudden changes in failure rates by supplier, unusual data entry times, and access from unexpected users can indicate problems. Some LIMS platforms now include AI-driven anomaly detection for exactly this purpose. If yours does, is anyone reviewing those alerts? If it doesn't, could your quality team develop informal monitoring practices to spot unusual patterns?

For smaller organizations without dedicated security teams, the options are more limited but not nonexistent. A basic audit trail review that looks at who accessed what and when catches more than you'd expect. Periodic comparison of electronic records against physical spot-checks provides a sanity check. These aren't sophisticated detection systems, but they're better than assuming everything is fine because no one's complained.

Who Owns the Risk

This question comes up in every discussion about cybersecurity: who owns the risk? When a quality system is compromised, does the Food Safety and Quality Assurance (FSQA) director own it or the Chief Information Security Officer (CISO)? In most food companies, no one explicitly owns this intersection. It falls between job descriptions.

The legal answer is increasingly clear: the board owns it. As we discussed in the previous chapter, the Delaware Supreme Court's Marchand v. Barnhill decision established that for "mission-critical" compliance issues, boards must implement board-level monitoring and reporting systems.[14] Food safety is mission-critical for a food company. When cybersecurity becomes a food safety risk, it becomes a board-level oversight issue.

For small and medium operations without formal boards, this rolls up to ownership and leadership. But the principle holds: food safety cyber risk isn't something that can fall between job descriptions and have nobody accountable. Just because the FSQA director and the IT manager both think it's the other's responsibility doesn't mean neither will be called to answer when something goes wrong. They both will.

This is another reason joint assessments matter. When food safety and cybersecurity teams report together, leadership gets a complete picture. The conversation shifts from "Whose problem is this?" to "How do we address this together?" Enterprise risk management frameworks, where they exist, should incorporate food safety cyber risks alongside other operational risks. Where they don't exist, the joint assessment creates a foundation for building that integrated view.

When Verification Fails

Food safety runs on verification. We trust that the metal detector caught contaminants because the system logged every package and flagged rejects. We trust that pathogen testing happened because laboratory systems documented results. We trust traceability because databases tracked product movement. When digital systems fail, we lose the ability to verify, and verification is what allows trust to be justified rather than blind.

A testing laboratory outage, like the one Eurofins experienced, means samples can't be processed. Products sit in warehouses while their shelf life ticks away. Companies face ugly choices between shipping without testing and holding product while absorbing losses.

LIMS corruption creates a different nightmare. Historical test results become unreliable. Auditors ask for documentation of past compliance, and you can't provide it with confidence. Regulatory submissions based on compromised data face scrutiny. Even if current products are safe, the inability to demonstrate past compliance creates legal exposure that can persist for years.

Process control manipulation is perhaps the most insidious failure mode. The monitoring system shows that pasteurization reached 72 °C for 15 seconds, the minimum required for milk safety, even though the actual temperature never exceeded 65 °C. Product ships with documentation indicating proper treatment. The pathogen survives. People get sick. The records, if anyone bothers to check them, show nothing wrong.

CoA fraud works at the supply chain level. A supplier provides documentation showing that ingredients meet specifications. The documentation is falsified. Substandard or contaminated ingredients enter production carrying certificates that guarantee nothing except that someone typed the right numbers into a form.

Building Cyber-Aware Food Safety Systems

Food safety professionals conducting HACCP evaluations should include digital threats among the hazards considered. What happens if this control point's monitoring system is manipulated? What if the data logging temperature is wrong? What if the batch-release system approves products that should be held? The hazards are food safety hazards; they just happen to have IT dimensions.

Data integrity belongs in food safety management, not just documentation management. Implement audit trails capturing who accessed data, what changed, and when. Validate electronic systems to ensure that they perform as intended. Control access so only authorized personnel can modify safety-critical records. None of this is new thinking; it extends principles that governed paper documentation for decades.

Quality control teams need to understand data flows. Where does information originate? How does it move through systems? Where could manipulation occur? You don't need to become an IT or cybersecurity expert, but you need enough understanding to recognize when something doesn't add up.

Laboratory operations, whether internal or contracted, deserve particular attention. Maintain offline backups of critical data, not just for disaster recovery but as independent verification sources. Segment LIMS from general business networks so ransomware spreading through email doesn't automatically reach laboratory infrastructure. Implement multifactor authentication for systems that control or document food safety activities. These measures add friction to daily operations, but that friction serves a protective function.

The intersection of food safety and cybersecurity requires skills that few professionals currently possess in combination. Building that capacity takes time. But the alternative is waiting for incidents to force change, and those incidents carry consequences that ultimately fall on the people eating the food.

The Foundation of Verified Trust

Food safety has always been about building justified trust. The entire apparatus of modern food safety regulations—certifications, audits, documentation—exists so that our trust in products we can't individually verify isn't misplaced. The digital infrastructure underlying that trust has become invisible to most people. We don't think about the LIMS managing laboratory samples or the sensors monitoring pasteurization temperatures. They work, so we don't notice them.

Systems we don't notice tend to be systems we don't protect adequately. When ransomware locks a testing laboratory or an insider manipulates allergen records, we lose more than information; we lose the ability to verify. Food safety culture has emphasized "trust but verify" for decades. In a digital environment, that principle needs to be extended: we also need to protect the systems that make verification possible in the first place.

Threats to food trust don't end with malware and system intrusion, though. In the next chapter, we turn to a different kind of attack: one that doesn't require touching food or food safety systems at all. A false recall notice spreading on social media. A manipulated video claiming contamination. A coordinated campaign to undermine confidence in a brand. When the weapon is misinformation rather than malware, the defenses look different, and so do the UK Law Enforcement Data Put at Risk in "Highly Sophisticated" Eurofins Scientific Attack.

Notes

1. Tech Monitor. (2019). "UK law enforcement data put at risk in "highly sophisticated" Eurofins Scientific attack." **https://www.techmonitor.ai/technology/cybersecurity/eurofins-scientific-hacked**
2. Finanznachrichten.de. (2019). "Eurofins Scientific: update on the cyber-attack announced on June 3, 2019." **https://www.finanznachrichten.de/nachrichten-2019-06/46911952-eurofins-scientific-update-on-the-cyber-attack-announced-on-june-3-2019-004.htm**

3. Kovacs, E. (2019). "Eurofins Scientific paid up in response to ransomware attack: report." *SecurityWeek*. **https://www.securityweek.com/eurofins-scientific-paid-response-ransomware-attack-report**
4. Cyber Security Intelligence. (2019). "British Police forensics attacked." **https://www.cybersecurityintelligence.com/blog/british-police-forensics-attacked-4360.html**
5. LabKey. (2025). "What is a LIMS?" **https://www.labkey.com/what-is-a-lims**
6. Charles, D. (2015). "Peanut exec gets 28 years in prison for deadly salmonella outbreak." *NPR*. **https://www.npr.org/sections/thesalt/2015/09/21/442335132/peanut-exec-gets-28-years-in-prison-for-deadly-salmonella-outbreak#:~:text=Peanut%20Corporation%20of%20America's%20then,deadlier%2C%20outbreaks%20of%20foodborne%20illness**
7. U.S. House Energy and Commerce Committee. (2009). "The salmonella outbreak: the continued failure to protect the food supply." **https://www.congress.gov/111/chrg/CHRG-111hhrg63824/CHRG-111hhrg63824.pdf**
8. Wagner, L. (2015). "Former Peanut Corp. exec gets 28 years for role in deadly salmonella outbreak." *NPR*. **https://www.npr.org/sections/thetwo-way/2015/09/21/442273854/former-peanut-company-exec-could-get-life-sentence-in-deadly-salmonella-outbreak**
9. Kim, S. (2015). "Peanut exec Stewart Parnell sentenced to 28 years in prison for salmonella outbreak." *ABC News*. **https://abcnews.com/Business/peanut-exec-face-prison-sentence-803-years-salmonella/story?id=33920375**
10. FDA. (2003). *21 CFR Part 11: Electronic Records; Electronic Signatures*. **https://www.fda.gov/regulatory-information/search-fda-guidance-documents/part-11-electronic-records-electronic-signatures-scope-and-application**
11. FDA. (2018). *Data Integrity and Compliance with Drug CGMP: Questions and Answers*. **https://www.fda.gov/regulatory-information/search-fda-guidance-documents/data-integrity-and-compliance-drug-cgmp-questions-and-answers**
12. Regalado, A. (2017). "Scientists hack a computer using DNA." *MIT Technology Review*. **https://www.technologyreview.com/2017/08/10/150013/scientists-hack-a-computer-using-dna**
13. McCarthy, J., Powell, M., Stouffer, K., Tang, C., Zimmerman, T., Barker, W., Ogunyale, T., Wynne, D., & Wiltberger, J. (2020). "Securing Manufacturing Industrial Control Systems: Behavioral Anomaly Detection." NIST. **https://csrc.nist.gov/pubs/ir/8219/final**
14. Justia. (2019). Marchand v. Barnhill, et al. **https://law.justia.com/cases/delaware/supreme-court/2019/533-2018.html**

CHAPTER 26

Misinformation, Disinformation, and Narrative Warfare

In September 2024, a video circulated on social media showing what appeared to be live fish swimming inside packets of instant ramen noodles. The clip spread rapidly, accumulating millions of views as people expressed horror at the thought of eating contaminated food. Within days, it had been debunked as a hoax—the fish were digitally added, and no such product existed.[1]

No one was poisoned, and no product was recalled. Yet the damage rippled outward anyway: consumer confidence shaken, brand reputations questioned, and hours of corporate communications time spent responding to something that never happened. This is the arithmetic of modern information warfare: the cost of fabrication is nearly zero, and the cost of correction remains stubbornly high.

For food and agriculture, this imbalance creates a particular kind of vulnerability. The sector operates under conditions that amplify false information: biological timelines that cannot wait for fact-checkers, supply chains too complex for most consumers to trace, and a cultural weight that makes food-related fears especially contagious. When someone claims your dinner might kill you, you do not pause to verify sources.

Understanding the Terms

The terminology matters because different problems require different responses.

Misinformation refers to false or inaccurate information shared without intent to deceive. Your grandmother forwarding a warning about contaminated fruit because she genuinely believes it will protect her family would be considered misinformation. *Disinformation* involves the deliberate creation and spread of false content designed to mislead.

A competitor fabricating quality concerns to damage a rival, or a nation-state spreading food safety rumors to destabilize markets, would be disinformation.[2]

Narrative warfare operates at a higher level of abstraction. Rather than spreading specific false claims, it seeks to shape how people interpret events and whom they trust. A narrative warfare campaign might not claim that a particular shipment is contaminated; instead, it works to erode confidence in food safety systems generally, so that when real or fabricated incidents occur, people are primed to believe the worst.[3]

For cybersecurity professionals, these categories connect directly to familiar concepts. Misinformation functions like an unintentional insider threat: someone with access causing harm without malicious intent. Disinformation resembles targeted social engineering: attackers crafting messages designed to manipulate specific behaviors. And narrative warfare parallels advanced persistent threats: patient, strategic campaigns that reshape the operating environment over time.

Food defense professionals will also recognize a similar framework. The distinction between unintentional and intentional adulteration—accidental contamination versus deliberate tampering—maps neatly onto the misinformation–disinformation divide. In both cases, the harm may look identical from the outside, but the response differs fundamentally. You do not investigate an accident the same way you investigate sabotage, and you do not counter a grandmother's forwarded warning the same way you counter a coordinated influence campaign, right?

The Hybrid Threat: Where Cyber Meets Disinformation

False information and cyberattacks increasingly operate as force multipliers for each other. A ransomware attack on a food processor becomes more damaging when social media amplifies panic about contamination. A fabricated outbreak report gains credibility when it coincides with genuine system outages. This convergence creates what security researchers call *hybrid threat operations* that combine technical intrusion with information manipulation to achieve effects neither could accomplish alone.[4]

The numbers tell the story of an accelerating problem. Ransomware attacks against the food and agriculture sector reached 212 documented incidents in 2024, up from 167 the previous year. The first quarter of 2025 alone saw 84 attacks, more than double the number in the same period in 2024.[5] Remember when JBS, the world's largest meat processor, suffered a ransomware attack in May 2021? The financial impact extended far beyond the ransom: cattle slaughter dropped 22 percent, 7,000 Australian employees were stood down, and wholesale beef prices spiked as the USDA scrambled to encourage other processors to increase capacity.[6]

What the statistics cannot capture is how disinformation amplified the JBS disruption. Social media speculation about meat safety, conspiracy theories about deliberate sabotage, and rumors of extended shortages all contributed to consumer anxiety that outlasted the actual outage. Security researchers at one firm found that one in five food processing and distribution companies have known vulnerabilities in their Internet-facing systems, technical weaknesses that become exponentially more dangerous when combined with information operations designed to exploit public fear.[7]

Historical Roots, Modern Acceleration

Agricultural misinformation predates the Internet by decades. In the 1930s, Soviet agronomist Trofim Lysenko promoted theories that rejected genetics in favor of environmental adaptation, claiming that simply exposing plants to cold would make them cold-resistant. The Soviet government adopted Lysenkoism as official policy, forcing farmers to plant crops in unsuitable conditions. The resulting agricultural failures contributed to famines that killed millions.[8]

In 1958, China launched the Four Pests Campaign, ordering farmers to kill sparrows based on the mistaken belief that the birds were consuming too much grain. With sparrow populations decimated, locusts lost a key predator. The resulting insect swarms devastated crops and contributed to what became known as the Great Chinese Famine, one of the deadliest in human history.[9]

The Villejuif Leaflet, which circulated through Europe in the 1970s and 1980s, spread false claims about carcinogenic additives in common food products. Distributed through photocopied brochures rather than social media, it nonetheless triggered widespread panic and changed purchasing behavior across multiple countries.[10]

What distinguishes the current moment is both pace and volume. The same false claims that once took weeks to spread through photocopies now achieve global reach in hours. Between 2020 and 2025, documented cyber incidents against the food and agriculture sector increased more than threefold, with ransomware attacks alone jumping from fewer than 60 incidents annually to more than 200. Research has found that approximately 10 percent of all media coverage of agricultural biotechnology contains inaccuracies that diverge from scientific consensus: a rate that exceeds vaccine misinformation and matches the politicization of climate change discussions. The potential global audience exposed to such misinformation exceeds a quarter of a billion people.[11]

Why Food Systems Are Particularly Vulnerable

Biological times, emotional connection, speed, and amount of information make the sector very vulnerable. A rumor about contaminated produce spreads in minutes; a thorough investigation takes days or weeks. By the time the truth emerges, the perishable product may already be in landfills, and consumer confidence may not recover for months. During the 2017 plastic seaweed hoax in China, false claims that seaweed was being manufactured from black plastic cost the industry more than 100 million yuan in losses. Chinese authorities eventually arrested 18 people, including several who used the viral panic to extort payments from seaweed producers.[12]

Modern food supply chains are too intricate for most people to trace. This complexity creates space for false narratives to take root. When consumers cannot verify claims about how their food is produced, processed, and transported, they become susceptible to stories that exploit that uncertainty. A plausible-sounding claim that a particular additive is harmful, or that a processing method is unsafe, can spread unchecked because most people lack the technical knowledge to evaluate it.

Food systems face genuine challenges: real contamination events, actual safety lapses, documented corporate misconduct. This history of legitimate concerns provides cover for fabricated ones. Disinformation campaigns can exploit past incidents, citing real recalls to lend credibility to false claims about current products or referencing genuine regulatory failures to justify blanket distrust of the entire system.

The economics overwhelmingly favor attackers. A targeted disinformation campaign can be launched for under $100 with a few hours of effort, perhaps some basic video-editing software or AI generation, and a handful of social media accounts. Defending against it requires crisis communications teams, legal review, regulatory engagement, and sometimes laboratory testing to prove a negative. Response costs routinely exceed a million dollars for even moderate incidents. For small agricultural operations, this asymmetry is especially punishing. A family farm cannot afford the response infrastructure that a false viral claim might require.

Global Dimensions: Developing Regions and Informal Channels

The challenge of agricultural misinformation takes distinct forms across different regions. In parts of Africa and South Asia, where formal information channels often reach rural communities less effectively than word-of-mouth networks, false claims about seeds, fertilizers, and farming techniques spread through pathways that are difficult to monitor or counter.[13]

Research on agricultural communication shows how misinformation affects farmer decision-making across multiple countries. Food misinformation creates anxiety, uncertainty, and confusion among both farmers and consumers, reducing their ability to make science-based decisions. Controversial topics including genetically modified crops, organic farming practices, and animal welfare are commonly surrounded by competing claims: some representing legitimate scientific debates, others reflecting deliberate distortion.[14]

In Kenya, misinformation delayed the commercial approval of Bt maize and Bt cotton (crops engineered to carry a natural pest-fighting protein from the soil bacterium *Bacillus thuringiensis*, giving the plant the ability to repel insects without spraying) for years, resulting in substantial agricultural losses as farmers continued using less effective pest management approaches.[15] The 2021 Sri Lankan organic farming mandate provides a cautionary example of policy shaped by misinformation. Government campaigns discouraged the use of chemical fertilizers, partly based on exaggerated claims about organic alternatives. When the abrupt transition to organic methods failed to produce promised yields, crop production collapsed, food prices spiked, and the resulting crisis contributed to broader economic and political instability.[16]

Case Studies: The Cyber–Disinformation Intersection

The following incidents illustrate how technical attacks and information operations increasingly converge:

Case	Cyber element	Disinformation amplifier	Impact
Canadian Buttergate (2021)	None (pure disinformation)	Food bloggers' claim about palm oil in cow feed went viral	Measurable decline in butter sales
Ukraine grain corridor (2023)	Port infrastructure attacks, DDoS on logistics	Russian officials threatened food weapon use	Millions of tons delayed; global price volatility
UNFI ransomware (2025)	Ransomware on organic distributor systems	Social media speculation about organic supply chain integrity	Whole Foods shortages; consumer anxiety
BrucePac recall (2024)	None (legitimate recall)	Conspiracy narratives about deliberate contamination	11.7 million pounds recalled; extended distrust

Not all agricultural misinformation involves fear; some spreads through novelty and apparent whimsy. In late 2019, images circulated widely on social media showing cows wearing virtual reality headsets, supposedly to reduce stress and increase milk production by watching peaceful green pastures. The claim spread not because it frightened people but because it seemed innovative and shareable, a glimpse of agricultural technology's quirky future.[17]

The images came from a small pilot project by Russia's Ministry of Agriculture involving a handful of cattle at a single farm near Moscow. A Turkish farmer conducted a similar experiment in 2022. These were isolated tests, not widespread practice, and certainly not a technology deployed across American dairy operations, as many posts implied. The misinformation spread because it was entertaining, not alarming, but the underlying dynamic was the same. People shared without verification, building false assumptions about how modern agriculture operates. I've recently been sent posts about it, and it's 2026 now.

The Deepfake Horizon

The emergence of synthetic media adds another dimension to information threats. In early 2024, an employee at engineering firm Arup transferred $25.6 million to criminals after a video call with what appeared to be the company's chief financial officer and other executives, all of whom were deepfake re-creations.[18] The technology that enabled that fraud is becoming increasingly accessible and increasingly difficult to detect—you guessed it, AI. By late 2025, security researchers were documenting deep-fake audio clips impersonating food safety officials in at least three countries, used to spread false contamination warnings. The food sector has yet to experience its defining deepfake crisis, but the trajectory suggests it is only a matter of time.

The implications extend beyond corporate fraud in food and agriculture. Synthetic video could fabricate food safety violations, create false footage of contamination events, or impersonate regulatory officials. A convincing deepfake of a USDA spokesperson announcing an outbreak that never occurred could trigger market panic before anyone verified the source. Ferrari's security team recently demonstrated one defensive approach: when an executive received calls from someone impersonating their CEO, they asked a verification question only the real CEO could answer. The caller hung up.[19]

Such verification protocols will become increasingly important across the food sector. But they require awareness that the threat exists and systems for confirming identity through channels that synthetic media cannot compromise.

Geopolitics and Food as Gray-Zone Warfare

Food has always been a tool of statecraft, but what has changed is the integration of information operations into food security strategy. In July 2023, former Russian President Dmitry Medvedev described food as a quiet weapon that could achieve political objectives without military confrontation. Russia controls approximately 22.5 percent of global wheat exports, giving such statements practical weight.[20]

Gray-zone warfare is the term security analysts use for competition between nations that stays below the threshold of actual military conflict. For food systems, this means adversaries do not need to bomb a grain elevator to achieve their objectives. They can spread doubt about food safety, amplify arguments about farming practices that already divide communities, or take a real contamination incident and pour gasoline on it through social media. The point is not necessarily to make anyone sick; it is to make people stop trusting the systems that feed them, to make good decisions harder to reach, and to create chaos that serves the attacker's broader goals.[21] The July 2023 attacks on Ukrainian port infrastructure illustrated this convergence. Missile strikes and cyberattacks on grain export facilities coincided with propaganda questioning whether Ukraine's grain was safe for global consumption. The combination of physical disruption and information operations created greater effects than either could achieve alone.[22]

What This Looks Like on the Ground

Away from the geopolitical arena, the day-to-day reality of agricultural misinformation often plays out on a more personal scale. I have watched small operators turn down documentary crews and media inquiries because they'd seen what happened when coverage was edited to tell a predetermined story, or they thought it would put them on the radar of radical animal extremist groups. The economic margins in agriculture are thin enough that reputational damage can end a family operation.

What frustrates me about discussions on this topic is how often they treat farmers as either victims to be protected or problems to be managed. The reality is more complicated. The people who grow and raise our food are navigating an information environment that feels increasingly hostile, where their practices are scrutinized by audiences with little agricultural context, and where any misstep can become a permanent digital record. Many have developed sophisticated informal networks for sharing information about which journalists can be trusted, which activist groups operate in good faith, and which claims circulating online require a response.

When activist campaigns cross into manipulation, selectively editing footage, staging incidents, or misrepresenting standard practices, the damage extends beyond the targeted operation. The Animal Agriculture Alliance documented 189 direct actions against U.S. agricultural facilities in 2024 alone.[23] Some of these represent legitimate advocacy; others involve deception that undermines public understanding of food production. Distinguishing between them requires more agricultural literacy than most consumers possess.

Building Defenses: Framework and Practice

The DISARM framework, developed to analyze and counter disinformation, has been endorsed by organizations including NATO and the European Union Agency for Cybersecurity (ENISA). It provides a structured approach for mapping how false narratives are created and spread, identifying intervention points where countermeasures might be effective.[24] For food and agriculture operations, the framework can inform both defensive monitoring and response planning.

DISARM helps with the information side, but hybrid threats require hybrid defenses. If a ransomware attack hits your facility and a disinformation campaign amplifies the panic, you need to address both vectors. The IEC 62443 series provides guidance for securing the operational technology that attackers might target, the systems whose failure gives disinformation campaigns something real to exploit. CISA has published sector-specific guidance and conducted exercises focused on food and agriculture incident response, including the 2024 Cyber Storm exercise that exposed critical gaps in how the federal government shares threat information with industry operators.[25] Organizations like CGIAR are building networks for sharing intelligence across borders, recognizing that neither cyberattacks nor false narratives respect national boundaries. The goal is layered defense: make the systems harder to breach, and make the breaches harder to weaponize through information operations.

For smaller operations without dedicated security resources, practical defenses begin with relationships. Building connections with local extension services, commodity organizations, and peer networks creates early warning systems for emerging threats. Documenting standard practices through video and photography provides evidence to counter false claims. Establishing relationships with local media before crises occur means having trusted channels for response when needed.

Social engineering attacks by phishing emails or text messages crafted to look like supplier communications, or phone calls impersonating inspectors, often provide the entry point for both cyber intrusions and information gathering that enables later manipulation. Teaching staff to verify unusual requests through independent channels protects against both technical and social vectors. Also, a good rule of thumb is to stay curious and trust your instincts. If you feel like a message "isn't quite right," delete it. If that message was real and important, the sender will find a way to reach you, and then you can educate them on their delivery, too.

Ethical Boundaries for Security Professionals

Those of us working in security have our own responsibilities here. The first 48 hours after an incident are when speculation does the most damage and when the temptation to weigh in is strongest. Social media rewards speed and hot takes, but jumping into an active situation often just adds noise. Sometimes the most useful thing you can do is shut up until you know something. We are supposed to be helping, not farming engagement.

There is also a boundary around how we discuss attribution. Naming threat actors without solid evidence can complicate response efforts, create legal liability, and potentially trigger geopolitical consequences. The urge to appear authoritative by offering confident assessments is worth resisting when confidence is not actually warranted. You do not have to be first to say it; what you need to do is wait for people to do their job and report the facts so you can learn from or help with the incident. As security professionals, we cannot be part of the problem.

For Individual Consumers

I would like to offer a set of reliable tools for navigating food-related information, but honesty compels a more modest conclusion. The patchwork of CGIAR technical reports, university extensions, and agency alerts aids verification but lacks unified, real-time coverage for food/agriculture misinformation crises. They were not designed for the speed and scale of modern information flows.[26]

What I can suggest is this: verify through primary sources. When you see claims about food safety incidents, check whether official agencies have issued statements. The USDA,

FDA, and their counterparts in other countries maintain public databases of recalls and advisories. Consider emotional temperature: content designed to provoke strong reactions such as fear, disgust, and outrage often sacrifices accuracy for engagement. Information that makes you want to share immediately deserves extra scrutiny before you do. Trace the source: anonymous claims, screenshots of deleted posts, and content with unclear origins warrant skepticism. Legitimate food safety information generally comes from identifiable sources that can be held accountable for accuracy. Recognize that uncertainty is normal, and real food safety situations often involve evolving information. Claims of absolute certainty that something is definitely safe or definitely dangerous early in an incident should prompt caution.

Policy Outlook

In August 2025, the Africa Biennial Biosciences Communication symposium in Lusaka, Zambia, brought together researchers, policymakers, and industry representatives to address misinformation as a barrier to agricultural innovation. The gathering reflected a growing recognition that the hybrid threat problem we have been discussing is not just a Western concern; false narratives about seeds, fertilizers, and biotechnology are costing African farmers real yields and real income. Participants pushed for stronger media literacy programs, better coordination between scientists and communicators, and regulatory frameworks that can move faster than rumors spread.[27] Whether those recommendations translate into action remains to be seen. But the fact that disinformation now ranks as the second greatest global risk in the World Economic Forum's assessment, trailing only extreme weather, suggests that the conversation is overdue.[28]

Policy recommendations from such gatherings tend to follow predictable patterns: harmonize regulations, invest in media literacy, strengthen public–private collaboration. These are not wrong, but their implementation remains uneven. The COMESA (Common Market for Eastern and Southern Africa) effort to harmonize biosafety laws across 21 member states represents meaningful progress, but such coordination takes years while misinformation operates at internet speed.[29]

More promising may be the emergence of youth-led AI initiatives that use chatbots and real-time monitoring to track misinformation and provide farmers with verified information in local languages. Early pilot programs in Kenya and Uganda have shown measurable improvements: farmers with access to AI-assisted fact-checking tools made demonstrably better decisions about seed selection and pest management compared to control groups relying on traditional information channels.[30] Such tools cannot solve the underlying problem of trust deficits, but they can reduce the delay between false claims and accurate responses.

For regulatory frameworks, the challenge is building requirements that address information integrity without stifling legitimate speech or creating compliance burdens that smaller operations cannot bear. The European Union's NIS2 Directive, which imposes cybersecurity requirements on critical infrastructure, including food production, represents one model, although its effectiveness in addressing hybrid threats remains to be demonstrated.

Throughout this chapter, we have examined threats that operate primarily through deception: false claims, manipulated media, and strategic narratives designed to erode trust. The damage this causes is real, but it differs in kind from what comes next. The following chapter examines what happens when intent shifts from deception toward direct physical harm: when bad actors target food systems not to shake confidence but to cause injury, illness, or death. The threat landscape of agroterrorism and ecoterrorism operates by different rules, with different defensive requirements.

Notes

1. Ciceronie, J. (2024). "CDC hasn't warned about deadly ramen noodles despite viral video." *PolitiFact*. **https://www.politifact.com/factchecks/2024/sep/18/instagram-posts/cdc-hasnt-warned-about-deadly-ramen-noodles-despit**
2. Čechmánek, K. (2024). "Disinformation, misinformation and the agri-food sector." *EU Agrarian Law*, 13(1), 21–27. **https://doi.org/10.2478/eual-2024-0003**
3. Chowdhury, A., Kabir, K. H., Abdulai, A., & Alam, M. F. (2023). "Systematic review of misinformation in social and online media for the development of an analytical framework for agri-food sector." *IDEAS*. **https://ideas.repec.org/a/gam/jsusta/v15y2023i6p4753-d1090342.html**
4. Ibid.
5. Freed, A. (2025). "Ransomware attacks targeting agriculture and food production doubled in 2025." *Halcyon*. **https://www.halcyon.ai/blog/ransomware-attacks-targeting-agriculture-and-food-production-doubled-in-2025**; Yang, J. (2025). "Cybercriminals targeting U.S. food and agriculture sector." *Foundation for Defense of Democracies*. **https://www.fdd.org/analysis/2025/02/13/cybercriminals-targeting-u-s-food-and-agriculture-sector-now-more-than-ever**
6. USDA. (2021). "Statement on JBS USA ransomware attack." **https://www.usda.gov/about-usda/news/press-releases/2021/06/01/statement-us-department-agriculture-jbs-usa-ransomware-attack**; Rosenbaum, E. (2021). "JBS cyberattack: from gas to meat, hackers are hitting the nation where it hurts." *CNBC*. **https://www.cnbc.com/2021/06/02/from-gas-to-burgers-hackers-hit-consumers-where-it-hurts.html**
7. SecurityScorecard. (2021). "JBS ransomware attack started in March and much larger in scope than previously identified." **https://securityscorecard.com/blog/jbs-ransomware-attack-started-in-march**
8. Kolchinsky, E. I., Kutschera, U., Hossfeld, U., & Levit, G. S. (2017). "Russia's new Lysenkoism." *Current Biology*, 27(19), R1042–R1047. **https://doi.org/10.1016/j.cub.2017.07.045**
9. Chowdhury, A. (2023). "Separating fact from fiction in agricultural misinformation." *Futurum Careers*. **https://futurumcareers.com/Ataharul-Chowdhury-Article.pdf**
10. Kapferer, J.-N. (2004). *Managing Brands through Rumors*. *ResearchGate*. **https://www.researchgate.net/publication/278812817_Managing_brands_through_rumors**
11. Lynas, M., Adams, J., & Conrow, J. (2025). "Misinformation in the media: global coverage of GMOs 2019-2021." *GM Crops & Food*, 16(1), 18–27. **https://doi.org/10.1080/21645698.2022.2140568**
12. Wenyu, S. (2017). "18 people arrested in China for spreading false information about 'plastic seaweed.'" *People's Daily Online*. **https://en.people.cn/n3/2017/0607/c90000-9225531.html**; Qiqing, L. (2017). "Plastic seaweed rumors lead to 18 arrests." *Sixth Tone*. **https://www.sixthtone.com/news/1000309/plastic-seaweed-rumors-lead-to-18-arrests**

13. Annang, R. (2025). "MSc perspective: misinformation in African agritech." *The Sainsbury Laboratory.* **https://www.tsl.ac.uk/news/msc-perspective-misinformation-in-african-agritech**
14. Chowdhury, A. (2020). "Separating fact from fiction in agricultural misinformation."
15. Meeme, V. (2025). "Misinformation is delaying Africa's leap forward in agriculture and health." *African Agricultural Technology Foundation.* **https://www.aatf-africa.org/misinformation-is-delaying-africas-leap-forward-in-agriculture-and-health**
16. Wijerathna-Yapa, A., Henry, R. J., Dunn, M., & Beveridge, C. A. (2023). "Science and opinion in decision making: a case study of the food security collapse in Sri Lanka." *Modern Agriculture*, 1(2), 142–151. **https://doi.org/10.1002/moda.18**
17. Cornall, J. (2019). "Russian VR takes dairy cows to the moovies." *Dairy Reporter.* **https://www.dairyreporter.com/Article/2019/11/27/Russian-VR-takes-dairy-cows-to-the-moovies**; BBC News. (2019). "Russian cows get VR headsets to 'reduce anxiety'." **https://www.bbc.com/news/world-europe-50571010**; ExtremeTech. (2022). "Farmer uses VR to make cows think they're in a green pasture." **https://www.extremetech.com/extreme/330440-farmer-uses-vr-to-make-cows-think-theyre-in-a-green-pasture**
18. Chen, H., & Magramo, K. (2024). "Finance worker pays out $25 million after video call with deepfake CFO." *CNN.* **https://www.cnn.com/2024/02/04/asia/deepfake-cfo-scam-hong-kong-intl-hnk**
19. Lepido, D. (2024). "Ferrari exec foils deepfake plot by asking a question only real CEO could answer." *Fortune.* **https://fortune.com/2024/07/27/ferrari-deepfake-attempt-scammer-security-question-ceo-benedetto-vigna-cybersecurity-ai**
20 Gus, C. (2022). "Russia's former president Medvedev warns agricultural supplies restricted to 'friendly countries'." *Politico Europe.* **https://www.politico.eu/article/russias-former-president-medvedev-warns-agricultural-supplies-restricted-to-friendly-countries**
21. Azad, T. M., Haider, M. W., & Sadiq, M. (2022). "Understanding gray zone warfare from multiple perspectives." *World Affairs*, 186(1), 81–104. **https://doi.org/10.1177/00438200221141101**
22. Reuters. (2023). "Russia strikes Ukraine grain port after pulling out export deal." **https://www.reuters.com/world/europe/russia-strikes-ukraine-grain-port-after-pulling-out-export-deal-2023-07-18**
23. AgDaily. (2024). "Industry report reveals new layers to animal rights movement." **https://www.agdaily.com/news/industry-report-reveals-new-layers-to-animal-rights-movement**
24. DISARM Foundation. Framework for countering disinformation. **https://www.disarm.foundation**
25. CISA. (2024). *Cyber Storm IX: After-Action Report.* **https://www.cisa.gov/sites/default/files/2024-10/Cyber%20Storm%20IX%20After-Action%20Report%20v00%2020241001_508.pdf**
26. CGIAR. (2025). *2024 CGIAR Technical Report.* **https://www.cgiar.org/2024-cgiar-technical-report**
27. Africa Biennial Biosciences Communication Symposium. (2025). **https://abbcsymposium.org/program-overview**
28. Elsner, M., Atkinson, G., & Zahidi, S. (2025). *Global Risks Report 2025* (World Economic Forum). **https://www.weforum.org/publications/global-risks-report-2025**
29. ISAAA AfriCenter. (2024). "Efforts towards harmonisation of regional biosafety policies in COMESA revived." **https://africenter.isaaa.org/efforts-towards-harmonisation-regional-biosafety-policies-comesa-revived**
30. iSDA Africa. (2026). "10,000 farmers: next generation virtual agronomist launches in Kenya and Uganda." **https://www.isda-africa.com/posts/10000-farmers-next-generation-virtual-agronomist-launches-in-kenya-and-uganda**

CHAPTER 27

Agroterrorism and Ecoterrorism

When the Threat Arrives on Six Legs

In September 2014, a Pennsylvania Department of Agriculture inspector found an insect clinging to a tree of heaven in Berks County that nobody in the Western Hemisphere had seen before. The spotted lanternfly, *Lycorma delicatula*, had likely hitchhiked on stone imports from Asia. Within months, it established itself across southeastern Pennsylvania. By 2024, 17 states were fighting it, with Pennsylvania alone suffering an estimated $324 million in annual agricultural losses.[1]

The arrival of the spotted lanternfly was accidental, but as agricultural security professionals watched state and federal agencies scramble to contain an invasive species nobody anticipated, they confronted an uncomfortable question: if one unintended introduction could cause this much damage, what could someone accomplish on purpose? That question runs through both agroterrorism and its ideologically driven cousin, ecoterrorism. Neither aims to kill people directly, but both target the biological and economic systems that feed nations. As a Congressional Research Service report put it, the goal is to "cripple the biological infrastructure underlying agriculture: i.e., its livestock and its crops."[2]

What We're Talking About

Agroterrorism is the deliberate introduction of an animal or plant disease to generate fear, cause economic losses, or undermine social stability.[3] Unlike conventional *bioterrorism* (which targets people directly, such as Anthrax), agricultural attacks aim at economic stability rather than immediate casualties.[4] Agricultural attacks work through what researchers call a "systems disruption model": "sluggish onset, delayed discovery, and cascade failures across economic and social systems."[5] The delay is strategic and often looks like natural outbreaks, which complicates both attribution and response.

Ecoterrorism overlaps but differs; it involves the use or threatened use of criminal violence against people or property by environmentally-oriented groups for political reasons.[6] By 2005, ecoterrorism had become the number-one domestic terrorism challenge originating within U.S. borders.[7]

The broader category of *agro-crime* encompasses everything from food fraud to animal smuggling to the deliberate release of pathogens.[8] The threat landscape of agro-crime breaks down further: biological agroterrorism (pathogens and pests), chemical (contamination), digital (cyberattacks on farming systems), ecological (invasive species, genetic sabotage), and psychological (threats designed to instill fear without physical attack).[9] Most modern agricultural operations face some combination of these vulnerabilities.

The Historical Weaponization of Food and Agriculture

Using hunger as a weapon predates modern terrorism by millennia. Siege warfare in ancient Mesopotamia around 2,500 BCE employed starvation systematically. The Holodomor in Ukraine from 1932 to 1933 demonstrated what a state could accomplish with food policy alone—Soviet grain quotas that exceeded available supply killed between 3 and 7 million people.[10] Nazi Germany's Hunger Plan of 1941–1944, designed by Food Minister Herbert Backe, aimed to starve Ukrainians, Russians, and Slavs to feed German soldiers during Operation Barbarossa.[11]

Twentieth-century governments developed agricultural bioweapons with alarming sophistication. Swedish and German researchers conducted the most rigorous historical analysis and documented programs across multiple nations: the Soviet Union studied African swine fever, foot-and-mouth disease, wheat stem rust, and rice blast from 1928 to 1992; the United States weaponized similar agents from 1943 to 1969; during World War II, the United Kingdom produced 5 million anthrax-laden cattle cakes for potential use against German livestock, although the program was never executed.[12]

German forces during World War I may have conducted the first documented state-sponsored agroterrorism campaign, clandestinely inoculating horses and mules at U.S. ports with anthrax and glanders before shipment to Allied forces. Japan reportedly used animal and plant pathogens against Russia and Mongolia in 1940.[13]

Non-state actors, by contrast, have rarely attacked agriculture deliberately. The most comprehensive analysis of the period from 1945 to 2012 identified only four documented cases.[14] In 1952, Mau Mau insurgents in Kenya poisoned 33 cattle by cutting their hides and introducing African milk bush latex, killing 8. In 1984, followers of Bhagwan Shree Rajneesh contaminated Oregon salad bars with *Salmonella typhimurium*, sickening 751 people to suppress voter turnout, although this attack targeted humans rather than agriculture directly.[15]

The researchers' conclusion matters: "Agroterrorism has not been a serious problem in the period from 1945 to 2012." But they also found that perpetrators exploited the difficulty of distinguishing attacks from natural outbreaks, facing minimal detection risk. Proving deliberate intent remains difficult decades later.[16]

The Economics of Agricultural Attack

Agroterrorism's strategic logic rests on economics rather than casualties. In the 1970s, Israeli citrus exports dropped 40 percent following Palestinian threats to contaminate Jaffa oranges with mercury. Then, in 1989, cyanide was discovered in two Chilean grapes destined for American markets, triggering temporary import bans and costing Chile more than $200 million.[17] A 2007–2008 Japanese frozen food contamination that poisoned 10 people caused economic and psychological damage far exceeding the direct harm.

Natural disease outbreaks illustrate the potential scale. The 2001 foot-and-mouth epidemic in the United Kingdom—accidental, not deliberate—cost between £25 billion and £30 billion and required destroying more than 6 million animals. Taiwan's 1997 foot-and-mouth outbreak caused $7 billion in losses.[18] South Korea's 2010–2011 epidemic cost $2.6 billion and required culling 3.48 million animals and deploying 33,000 military personnel.[19]

Analysts project that a deliberate livestock attack in the United States could generate between $10 billion and $30 billion in damage. The Congressional Research Service outlined how damage cascades: direct losses from containment and destruction, multiplier effects through compensation and related industry losses, and international costs from trade embargoes.[20]

Political consequences can arise even without an actual attack. The 2008 Korean beef riots—demonstrations against American beef imports over food safety concerns—mobilized thousands and helped undermine South Korea's newly elected administration. Urban supermarkets typically stock about a week's worth of food. Researchers warn that significant disruption could trigger "severe panic and rioting."[21]

Terrorist Interest

Evidence of terrorist organizations exploring agricultural capabilities accumulated after 2001. When American forces overran al-Qaeda sanctuaries in Afghanistan in 2002, captured materials included U.S. agricultural documents and training manuals.[22] The 2012 edition of *Inspire*, al-Qaeda's English-language magazine, made the interest explicit. Anwar al-Awlaki wrote that "the use of poisons or chemical and biological weapons against population centers is allowed and is strongly recommended due to its great effect on the enemy."[23] Al-Qaeda and the Taliban reportedly considered weaponizing wheat rust. Iraq under Saddam Hussein researched agricultural bioweapons during the late 1980s, including wheat-smut fungi and aflatoxin-producing Aspergillus strains, with reports that Hussein actually deployed wheat smut over Iranian fields during the Iran-Iraq War.[24]

Economics favor biological over nuclear weapons for resource-constrained actors. A nuclear weapon costs roughly $1 billion to develop; a biological capability requires approximately $10 million with a small laboratory and trained personnel.[25] Agricultural pathogens lower the barrier further: many exist naturally or can be cultured with modest expertise, and they can escape conventional detection technologies and allow delayed symptom onset that lets perpetrators disappear before anyone realizes an attack occurred.[26]

European researchers assessed 50 agents as potential agroterrorism candidates: 35 fungal, 9 bacterial, and 6 viral, with 32 capable of damaging staple crops. Fungal pathogens concern specialists, particularly because spores disperse naturally through the air, can incubate undetected, and can travel enormous distances.[27]

The Domestic Threat

Although international terrorism captures attention, the more documented threat to American agriculture comes from domestic animal rights extremism. The Animal Agriculture Alliance has monitored adversarial groups since 1987, tracking organizations "that just don't believe there's a way to ethically and responsibly raise animals for food."[28]

Direct Action Everywhere (DxE) released a 149-page investigation manual after its leader's criminal conviction, detailing drone selection for farm surveillance, heat-sensing cameras to detect building occupancy, supply chain tracking methods, and techniques for cutting plastic barn curtains to avoid detection.[29] Hannah Thompson-Weeman, president of the Animal Agriculture Alliance, describes coordinated actions involving hundreds of people bused from activist conferences. Smaller groups receive specialized training for trespassing and interacting with law enforcement. Participants wear color-coded armbands indicating training level and then claim in court that they didn't know they were breaking the law.[30]

The California Department of Food and Agriculture documented a probable connection between DxE farm visits and subsequent highly pathogenic avian influenza (HPAI) outbreaks at those same facilities. During an ongoing trial, activists returned to farms involved in the proceedings; those operations then experienced HPAI outbreaks. Earlier incidents forced the complete depopulation of poultry barns after activist intrusions raised biosecurity concerns about introduced diseases.[31]

Harassment extends to personal targeting, with activists burying dead piglets in executives' yards, allegedly dumping manure lagoon contents on residential property, and arriving at an Iowa executive's home while his grandchildren were visiting. Jonathan Lawler, an Indiana farmer known as the "Punk Rock Farmer," received 800 death threats over three months for publicly defending animal agriculture. Vehicles appeared on his property at 6:30 AM; strangers drove around locked gates toward his barn.[32]

The Animal Liberation Front (ALF) and Earth Liberation Front (ELF) have committed more than 600 criminal acts in the United States since 1996, resulting in damages in excess of $43 million. The RAND Corporation estimated that ELF alone has caused $35 million to $45 million in direct property damage across multiple states.[33]

Congress enacted the Animal Enterprise Terrorism Act in 2006 to address these threats, criminalizing conduct that damages animal enterprise operations or creates "reasonable fear of injury."[34] Prosecution remains difficult because economic damage must meet statutory thresholds, and the agricultural supply chain's resilience often prevents disruptions from reaching those levels.

Distinguishing Attack from Outbreak

One of agricultural security's hardest problems is telling deliberate sabotage from natural disease. Perpetrators exploit this ambiguity, knowing that biological attacks can look like bad luck.

Experienced farmers hold irreplaceable baseline knowledge. They know their animals' normal behaviors, seasonal disease patterns, and local risks. That familiarity lets them recognize anomalies: mass die-offs without gradual fever progression, uniform lesions across animals regardless of age or immune history, symptoms from pathogens that don't exist in their region.

Red flags suggesting human intervention can include sudden outbreaks of non-endemic diseases without natural vectors (foot-and-mouth appearing where no imported livestock arrived, avian influenza in facilities isolated from migratory birds) or disease that spreads faster than biology allows; something that hits a peak overnight or jumps quarantine zones warrants heightened scrutiny. When losses exceed 10 percent mortality without an identifiable cause, coincidence becomes statistically implausible.

Forensic markers can confirm sabotage when investigators find them. Genetic fingerprints from lab-engineered strains are detectable through PCR testing. Delivery leaves traces of aerosol residues, and fragments of dispersal devices occasionally survive. Non-agricultural chemicals in feed or water suggest contamination. Physical evidence of tampering speaks for itself. Ideological markers like graffiti or claims of responsibility sometimes provide direct attribution, although sophisticated actors know how to avoid them.

The transition from disease response to criminal investigation typically occurs within 72 hours if diagnostics find no natural explanation, surveillance detects parallel threats, or physical evidence of tampering emerges. Standard protocols continue quarantine, testing, depopulation if necessary, but evidence preservation becomes critical.

Farmers confronting potential incidents should contact their local sheriff or state veterinarian first. They activate the response chain under the National Animal Health Emergency Response Plan (NAHERP), notifying the USDA Animal and Plant Health Inspection Service (APHIS) for quarantine and testing.[35] Provide photographs, symptom descriptions, timelines, animal counts, and access logs documenting property visitors. Authorities loop in the FBI through established Agroterrorism Annex protocols if criminality is suspected. The layered approach preserves evidence while ensuring that the biological response to the incident architecture that Chapter 22 examines applies with particular urgency here.[36]

Cyber as Force Multiplier

As Chapter 18 explored, cyber-physical attacks exploit critical digital systems and physical processes. For agroterrorism, cyber intrusions serve as force multipliers, amplifying biological attacks, helping conceal sabotage, or obstructing the investigation that follows.

Animal rights extremists have already demonstrated interest. Thompson-Weeman describes activists using fake online personas, what law enforcement calls "romance fraud," to build relationships with farm and plant workers, extracting operational information and mapping supply chains. Activists who gained physical access to the facility have installed spyware to monitor operations and identify the optimal timing for subsequent actions.

More concerning: when feed mills and JBS processing facilities suffered ransomware attacks, activist channels celebrated. Some explicitly expressed interest in partnering with hacking groups.

Beyond activist threats, specific attack chains illustrate how cyber enables agroterrorism:

- Modern poultry and swine facilities use networked sensors to monitor temperature, humidity, gases, and ventilation. Attackers who compromise these systems could spoof readings to mask early disease symptoms such as respiratory distress, abnormal clustering, and fever indicators, delaying awareness until containment becomes impossible. Air-gapped backup monitoring helps, as do anomaly detection systems that flag discrepancies between sensor data and observable conditions. Manual welfare checks that can't be digitally manipulated remain essential.
- Feed mill operations present different vulnerabilities. SCADA systems control batch recipes, ingredient ratios, and additive concentrations. Remote manipulation could introduce low-dose contaminants across regional supply chains in quantities that mimic natural spoilage rather than triggering acute poisoning. Contaminated feed reaching multiple farms simultaneously would generate diffuse illness patterns that are hard to trace back to a common source. Defenses here rely on blockchain ingredient tracing, laboratory testing when shipments arrive, and network microsegmentation that prevents lateral movement from business systems into operational technology.
- Refrigerated transport creates another opening. Ransomware targeting telematics could cause spoilage attributed to equipment failure rather than sabotage. Attackers selecting specific shipments—vaccines, breeding stock, high-value products—inflict damage while maintaining deniability. Redundant thermal logging and forensic-ready backups provide protection.
- Epidemiological tracing depends on veterinary health records that link animal movements, treatment histories, and disease patterns across facilities. Corrupting or deleting these records obstructs the investigation of any outbreak, deliberate or otherwise. Immutable logging architectures become critical infrastructure for response capability.

These scenarios exploit agriculture's OT legacy systems, which are designed for reliability rather than security and often run protocols like Modbus without authentication. The Purdue Model's hierarchical network zoning and CISA's sector-specific guidance provide defensive frameworks,[37] but implementation remains uneven.

Globalized agricultural supply chains concentrate vulnerabilities in unexpected ways. Network theory reveals that modern systems exhibit "scale-free" structures where a few hubs—seed companies, cold storage facilities, export terminals—hold disproportionate importance. Targeting these nodes can generate cascading failures without affecting the entire network. Agriculture functions as what systems theorists call a "complex adaptive system": interdependent subsystems where disruption reverberates throughout. The combination makes agriculture what security theorists term a "soft target": critically important, geographically distributed, and genuinely difficult to defend at every point.

International Governance

Global governance for agricultural security remains fragmented. The Biological Weapons Convention (BWC) of 1972 implicitly covers agroterrorism but lacks enforcement mechanisms. The Food and Agriculture Organization (FAO) coordinates through the International Plant Protection Convention. The WHO conducts food safety surveillance under International Health Regulations. The UN Office on Drugs and Crime (UNODC) addresses terrorism financing. The WTO's Sanitary and Phytosanitary Agreement provides trade frameworks. None of these was designed specifically for agricultural attack.

In October 2018, WOAH, FAO, and INTERPOL launched "Building Resilience Against Agro-Crime and Agro-Terrorism," backed by Global Affairs Canada. The initiative fostered collaboration between animal health and security sectors, assessed emergency capabilities, and conducted an international simulation exercise by 2022.[38]

U.S. domestic policy centers on Homeland Security Presidential Directive 9 (HSPD-9), issued in 2004, establishing "a national policy to defend the agriculture and food system against terrorist attacks, major disasters, and other emergencies." The directive established coordination frameworks among the DHS, USDA, FDA, and FBI through the Strategic Partnership Program for Agroterrorism (SPPA).[39] In October 2019, Congress introduced the Protecting America's Food and Agriculture Act, acknowledging that border inspection capabilities had not kept pace with import volumes.[40]

India's framework evolved through successive terrorism legislation to the current Unlawful Activities Prevention Act (UAPA), which covers "disruption of any supplies or services essential to the life of the community." With agriculture representing 20 percent of GDP and employing two-thirds of the population, agricultural security becomes a matter of national survival.[41]

Practical Measures

Protecting distributed agricultural systems requires layered approaches. The Animal Agriculture Alliance recommends motion-sensor lighting and surveillance cameras, signage, functional locks that people use, visitor verification processes, and background checks for employees.

The FBI emphasizes that "identifying threats of agroterrorism and stopping them before they happen" requires partnerships among "the local sheriff and farmers, ranchers, meatpackers, truckers, feedlot owners, and other critical members of the food supply chain." Early warning indicators include theft of vaccines, medicines, and agricultural equipment.[42] Veterinarians, plant pathologists, and extension agents serve as first responders who need both agricultural and security training.

Classical epidemiological methods remain vital for distinguishing natural outbreaks from attacks. Research facilities face tensions between scientific openness and biosecurity; recommendations include molecular detection strategies and nationwide diagnostic networks.

Insurance deserves more attention than it gets. Standard property policies can include terrorism coverage endorsements for $0–200 annually under the Terrorism Risk Insurance Act. Cyber liability policies protect against business interruption from digital incidents, and claims from foreign-source attacks are currently being paid.[43]

Effective preparedness requires multisectoral coordination. Many pathogens considered for bioterrorism are zoonotic. Response requires joint assessments, contingency plans, and surveillance systems connecting animal health, public health, law enforcement, and intelligence.[44]

The Human Cost

Behind the frameworks stand people bearing consequences. A turkey farmer told Lawler about finding six dead birds one morning, watching half his flock die within 12 hours, and culling the rest. "He agreed with the decision to cull and understood why it needed to be done," Lawler said. "He's not some conspiracy guy. He's just a guy trying to raise turkeys."

The mental health toll compounds practical challenges. Thompson-Weeman observes that when you're "doing such a demanding job for a public that can be very skeptical of you," while extremists "make it worse," the weight can fall heavily on producers, especially those who are personally targeted.

Disconnection between agricultural reality and public perception creates its own problems. Lawler recounts someone calling animal control because cattle were lying down, which the caller interpreted as death or distress but which was simply normal bovine behavior. A delivery driver also reported concerns about visible ribs on Jersey cattle, a breed characteristic. These misunderstandings consume resources and erode trust between agricultural communities and the public.

What the Rest of Us Can Do

Most readers aren't farmers, veterinarians, or FBI agents. So, what do we do?

It always starts with awareness. Understand that food systems face deliberate threats, not just accidents and weather. Challenge how you interpret disruptions. When outbreaks happen or prices spike, resist the urge to amplify panic or spread unverified claims about causes. Misinformation during agricultural crises can cause as much economic damage as the event itself.

Support regional food systems where practical. The more concentrated our supply chains become, the more attractive they are as targets and the more catastrophic single points of failure become. Buying from local producers when possible doesn't just support farmers, it builds redundancy into the system. A community with relationships with nearby farms weathers disruption better than one entirely dependent on distant supply chains.

Respect biosecurity! Farm tourism, agritourism, and educational visits benefit everyone, but follow the protocols. Those boot-wash stations and restricted areas exist because diseases hitch rides on shoes, clothing, and vehicles. The spotted lanternfly spread partly because people didn't take "check your car before traveling" advisories seriously.

If you see something genuinely suspicious (not cattle lying down or ribs on a Jersey cow), report it to local authorities rather than social media. Farmers deal with enough false alarms from well-meaning but uninformed observers. But actual surveillance, tampering, or threats deserve official attention.

Finally, recognize that the people producing your food operate under pressures most consumers never see. Tight margins, unpredictable weather, disease risks, injury risks, regulatory burdens, and now deliberate threats from actors who want their industry to disappear. A little understanding goes further than you might think. Thank the farmers you interact with in your community and beyond.

Looking Ahead

Agroterrorism and ecoterrorism occupy awkward territories serious enough to warrant attention, but not dramatic enough to command resources devoted to other threats. Major non-state agricultural attacks remain rare. But natural outbreaks, accidental introductions, and the few deliberate incidents that have occurred demonstrate how agricultural disruptions cascade beyond the initial damage, affecting economies and political systems.

Precision agriculture and concentrated supply chains create new vulnerabilities alongside productivity gains. Physical activism converging with cyber capabilities means threat actors will continue adapting. Meanwhile, international coordination fragments even as pathogens and ideologies cross borders without difficulty.

Agricultural security warrants the attention given to other critical infrastructure sectors. Defense depends on partnerships that connect farming communities with law enforcement, with veterinary and plant science professionals, and with intelligence services. Resilience—the capacity to absorb disruption and continue functioning when prevention fails—matters more than perfect prevention.

Spotted lanternflies keeps spreading, and we will keep stomping on them. Agricultural systems face constant pressure from threats anticipated and unanticipated alike. Building defenses means getting good enough at distinguishing deliberate attacks from natural disasters, robust enough to respond when either occurs, and honest about what security can accomplish. The alternative is a food system caught unprepared when adversity arrives.

Notes

1. Cornell Integrated Pest Management. (n.d). "Spotted lanternfly." **https://cals.cornell.edu/integrated-pest-management/outreach-education/whats-bugging-you/spotted-lanternfly**
2. Segarra, A. E. (2001). *Agroterrorism: Options in Congress* (Congressional Research Service). **https://digital.library.unt.edu/ark:/67531/metacrs1384**
3. Monke, J. (2007). *Agroterrorism: Threats and Preparedness* (Congressional Research Service). **https://www.congress.gov/crs-product/RL32521**
4. Chalk, P., Hoffman, B., Reville, R. T., & Kasupski, A.-B. (2004). *Hitting America's Soft Underbelly: The Potential Threat of Deliberate Biological Attacks Against the U.S. Agricultural and Food Industry* (RAND Corporation). **https://www.rand.org/pubs/monographs/MG135.html**
5. Nayak, D., Mukherjee, S., Chakraborty, S., & Amin, D. (2025). "Agroterrorism: A theoretical exploration of global threats to agriculture and food security." *International Journal of Latest Technology in Engineering, Management & Applied Science*, 14(8), 1392. **https://www.researchgate.net/publication/395699647_Agroterrorism_A_Theoretical_Exploration_of_Global_Threats_to_Agriculture_and_Food_Security**
6. Jarboe, J. F. (2002). "The threat of eco-terrorism." Testimony before the U.S. House Resources Committee, Subcommittee on Forests and Forest Health. *FBI*. **https://archives.fbi.gov/archives/news/testimony/the-threat-of-eco-terrorism**
7. Chalk, P., et al. (2004). *Trends in Terrorism*.
8. Kishore, S. M., Police Patil, C. B., Shalini, K. S., Chinmayi, S., & Vinay Chander, A. (2024). "Agroterrorism: Assessing the growing threat to global food security and economic stability." *International Journal of Biochemistry Research & Review*, 33(6), 208. **https://www.researchgate.net/publication/384142435_Agroterrorism_Assessing_the_Growing_Threat_to_Global_Food_Security_and_Economic_Stability**
9. Nayak, D., et al. (2025). "Agroterrorism."
10. Álvarez, J. D. R. (2024). "The weaponization of hunger: An analysis of food security in conflict and post-conflict scenarios." *Journal of Human Security and Global Law*, 3(2024), 1–24. **https://revistes.uab.cat/jhsgl/article/view/v3-rodriguez/42-pdf-en**
11. Runge, C. F., & Graham, L. (2020). "Viewpoint: hunger as a weapon of war: Hitler's hunger plan, native American resettlement and starvation in Yemen." *Food Policy*, 92, 101835. **https://doi.org/10.1016/j.foodpol.2020.101835**
12. Keremidis, H., Appel, B., Menrath, A., Tomuzia, K., Normark, M., Roffey, R., Knutsson, R. (2013). "Historical perspective on agroterrorism: lessons learned from 1945 to 2012." *Biosecurity and Bioterrorism*, 11(Suppl 1), S17–24. **https://doi.org/10.1089/bsp.2012.0080**. **https://pubmed.ncbi.nlm.nih.gov/23971803**
13. Govern, K. (2009). "Agroterrorism and ecoterrorism: a survey of Indo-American approaches under law and policy to prevent and defend against these potential threats ahead." *ResearchGate*. **https://www.researchgate.net/publication/237501594_agroterrorism_and_ecoterrorism_a_survey_of_indo-american_approaches_under_law_and_policy_to_prevent_and_defend_against_these_potential_threats_ahead**
14. Keremidis, H., et al. (2013). "Historical perspective on agroterrorism."
15. Kishore, S. M., et al. (2024). "Agroterrorism."
16. Manuel, F. Z. (2017). "Agro terrorism: a global perspective." *Journal of Political Sciences & Public Affairs*, 5(1), 3–4. **https://www.longdom.org/open-access/agro-terrorism-a-global-perspective-36433.html**; Yeh, J. Y., Seo, H. J., Park, J. Y., Cho, Y. S., Cho, I. S., Lee, J. H., Hwang,

J. M., & Choi, I. S. (2012). "Livestock agroterrorism: the deliberate introduction of a highly infectious animal pathogen." *Foodborne Pathogens and Disease,* 9(10), 869–877. **https://doi.org/10.1089/fpd.2012.1146**
17. Kishore, S. M., et al. (2024). "Agroterrorism."
18. Manuel, F. Z. (2017). "Agro terrorism."
19. Yeh, J. Y., et al. (2012). "Livestock agroterrorism."
20. Monke, J. (2007). "Agroterrorism."
21. Kishore, S. M., et al. (2024). "Agroterrorism."
22. Olson, D. (2012). "Agroterrorism: threats to America's economy and food supply." FBI law enforcement bulletin. *LEB.* **https://leb.fbi.gov/articles/featured-articles/agroterrorism-threats-to-americas-economy-and-food-supply**
23. Keremidis, H., et al. (2013). "Historical perspective on agroterrorism."
24. Kishore, S. M., et al. (2024). "Agroterrorism."
25. Keremidis, H., et al. (2013). "Historical perspective on agroterrorism."
26. Kishore, S. M., et al. (2024). "Agroterrorism."
27. Suffert, F., Latxague, É., & Sache, I. (2009). "Plant pathogens as agroterrorist weapons: assessment of the threat for European agriculture and forestry." *Food Security,* 1, 221–232. **https://www.researchgate.net/publication/226121097_Plant_pathogens_as_agroterrorist_weapons_Assessment_of_the_threat_for_European_agriculture_and_forestry**
28. Johnson, P. (2017). "Celebrating 30 years of bridging the gap between farm and fork." *Animal Agriculture Alliance.* **https://animalagalliance.org/celebrating-30-years-of-bridging-the-gap-between-farm-and-fork**
29. Greenberg, A. (2023). "This is the ops manual for the most tech-savvy animal liberation group in the US." *Wired.* **https://www.wired.com/story/dxe-animal-agriculture-investigation-guide**
30. Thompson-Weeman, H., Rose, A., & Urbanik, M. (2025). "Ep. 028 - Agroterrorism & cyber threats: how farms are under attack." Interviews by Kristin King (Demoranville). *Bites and Bytes Podcast.* **https://www.bitesandbytespodcast.com/episodes/agroterrorism-cyber-threats-how-farms-are-under-attack**
31. California Department of Food and Agriculture. (2024). "UPDATE (HPAI Confirmed): CDFA statement on possible introduction of highly pathogenic avian influenza at California dairy farms." *UC Davis Weill, School of Veterinary Medicine.* **https://www.vetmed.ucdavis.edu/news/cdfa-statement-possible-introduction-highly-pathogenic-avian-influenza-california-dairy-farms**
32. Lawler, J. (2025). "Ep. 036 - Farming without apologies: Jonathan Lawler on bulls, bullsh*t, and being real." Interview by Kristin King (Demoranville). *Bites and Bytes Podcast.* **https://www.bitesandbytespodcast.com/episodes/farming-without-apologies-jonathan-lawler-on-bulls-bullsht-and-being-real**
33. U.S. Senate Committee on Environment and Public Works. (2005). "Oversight on eco-terrorism specifically examining the Earth Liberation Front (ELF) and the Animal Liberation Front (ALF)." **https://www.epw.senate.gov/public/index.cfm/2005/5/full-committeeoversight-on-eco-terrorism-specifically-examining-the-earth-liberation-front-elf-and-the-animal-liberation-front-alf**
34. United States Code. (2000). "Animal Enterprise Terrorism Act of 2006." 18 U.S.C. section 43. **https://uscode.house.gov/view.xhtml?req=granuleid:USC-1999-title18-section43&num=0&edition=1999#:~:text=18%20USC%20section%2043%20of%20the%20Animal,production%20*%20Agriculture%20*%20Research%20*%20Testing**

35. USDA APHIS. "National Animal Disease Preparedness and Response Program." **https://www.aphis.usda.gov/funding/nadprp**
36. Congressional Research Service. (2007). "Agroterrorism: threats and preparedness." **https://www.congress.gov/crs_external_products/RL/PDF/RL32521/RL32521.7.pdf**
37. CISA. (n.d.). "Food and agriculture cybersecurity checklist and resources." **https://www.cisa.gov/resources-tools/resources/food-and-agriculture-cybersecurity-checklist-and-resources**
38. Kishore, S. M., et al. (2024). "Agroterrorism."
39. Department of Homeland Security. (2007). "The Department of Homeland Security's role in food defense and critical infrastructure protection." **https://www.oig.dhs.gov/sites/default/files/assets/Mgmt/OIG_07-33_Feb07.pdf**
40. Kishore, S. M., et al. (2024). "Agroterrorism."
41. Govern, K. (2009). "Agroterrorism and ecoterrorism."
42. Schmitt, G. R. (2007). "Agroterrorism--Why we're not ready: a look at the role of law enforcement." *National Institute of Justice Journal*, 257. **https://nij.ojp.gov/topics/articles/agroterrorism-why-were-not-ready-look-role-law-enforcement**
43. California Department of Insurance. *Terrorism Risk Insurance Act.* **https://www.insurance.ca.gov/0250-insurers/0800-rate-filings/upload/TRIA_for_posting_to_Internet.pdf**
44. Kishore, S. M., et al. (2024). "Agroterrorism."

CHAPTER 28

Regulations and Compliance

The Two Factories

Picture two nearly identical food processing facilities, one outside Munich, Germany, and the other in rural Ohio, United States; both make frozen prepared meals. Both run similar equipment, employ comparable workforces, and ship to grocery chains across their respective continents. And in the fall of 2024, both experienced ransomware attacks that locked their production systems for the better part of a week.

The operational impact was nearly identical. Production lines stopped and refrigeration monitoring went offline, forcing emergency protocols to prevent spoilage. Workers stood idle while IT teams scrambled, or, in the United States, were most likely sent home without pay. Product already in distribution had to be tracked manually, and retailers wondered whether their next deliveries would arrive.

What happened next diverged sharply.

The German facility was required to notify regulators within 24 hours under the European Union's Network and Information Security Directive 2 (NIS2).[1] Its leadership faced questions about whether the company's risk management measures had been adequate. Auditors scheduled reviews. The incident became part of a national database informing future policy. Board members understood that they could face personal liability if the company's cybersecurity preparations were found to be lacking.

The Ohio facility filed an insurance claim. Management considered whether to mention the incident to the FDA, but because no contamination had occurred and no products were adulterated, there was no clear obligation to do so. The company issued a press release acknowledging "operational disruptions" without mentioning the word "cyber." Within a month, the incident had faded from public view. No regulator asked questions. No database recorded it, and no policymaker learned from it.

Why does geography determine whether a cyberattack on food becomes a matter of public record or a private inconvenience? And what does that gap tell us about how seriously different societies take the digital vulnerabilities in their food supply?

This chapter maps the regulatory terrain: the rules that exist, the gaps that remain, and why compliance matters even when enforcement is uneven.

Why Regulations Exist (and Why They Often Don't)

Regulations rarely emerge from careful foresight, because they follow disasters. Upton Sinclair's *The Jungle* didn't prompt the Meat Inspection Act because Congress suddenly developed an interest in slaughterhouse sanitation. It happened because the public recoiled at descriptions of what was going on in the making of their sausages.[2] The 2006 *E. coli* outbreak in spinach didn't create the Food Safety Modernization Act (FSMA) overnight, but it built momentum that eventually pushed the legislation through.[3] The Colonial Pipeline attack in 2021 triggered executive orders on critical infrastructure cybersecurity not because the risk was newly discovered, but because a gas shortage on the East Coast made the abstract suddenly concrete.[4] Food safety regulation evolved over more than a century of tragic lessons. Cybersecurity in food is being written in real time, often without waiting for the equivalent body count. The question is whether we're moving fast enough or waiting for a disaster that could have been prevented.

The governance gap in food and agriculture is striking when you step back and look at it. Food safety regulation is well-established, mandatory, and enforced. If a processor fails to maintain proper temperatures or sanitation, inspectors have the authority to shut them down. Cybersecurity regulation, by contrast, remains fragmented, often voluntary, and unevenly enforced. The result is a sector with mature safety cultures but immature security governance, even though cyber incidents can cause the same harm, including contamination, spoilage, and supply disruption, that food safety regulations were designed to prevent.

Consider what happens when ransomware disables a pasteurization monitoring system. The milk may not actually be contaminated, but the processor can't prove it meets safety standards. The processor faces a choice: ship products it can't verify, or halt distribution and absorb losses. Either way, the cyber incident has created a food safety problem. Yet only the traditional safety failure—say, an actual contamination event—would trigger regulatory scrutiny. The digital cause often goes unexamined.

The United States: A Patchwork with Plenty of Gaps

American food regulation is split between two primary agencies. The FDA oversees most food products, seafood, produce, packaged goods, beverages, and dairy.[5] The USDA handles meat, poultry, and egg products.[6] This division creates its own complexities. A pepperoni pizza, for instance, involves FDA jurisdiction over the crust and cheese and USDA jurisdiction over the pepperoni. Neither agency has comprehensive cybersecurity requirements, although both have begun to acknowledge the risk in various advisories and guidance documents and at conferences.

FSMA, signed into law in January 2011, represented a fundamental shift in American food safety from reactive to preventive. Rather than waiting for outbreaks and responding after people get sick, the law requires facilities to identify hazards before they cause harm and implement controls to prevent them. Seven major rules emerged from FSMA, but two matter most for understanding cybersecurity's place in this framework[7]:

The Preventive Controls for Human Food rule requires facilities to conduct hazard analyses and implement risk-based preventive controls.[8] The logic is straightforward: identify what could go wrong, figure out how to prevent it, monitor your controls, and document everything. Cyber vulnerabilities fit awkwardly into this framework. If a digital system controls a critical food safety function—temperature monitoring, for instance, or sanitation chemical dosing—then a cyber failure could compromise that function. But the rule doesn't explicitly address cyber threats, leaving companies to interpret whether and how digital risks should be incorporated into their hazard analyses.

The Intentional Adulteration rule, finalized in May 2016 with compliance dates phased through 2020, addresses acts intended to cause wide-scale harm.[9] The rule focuses on "actionable process steps" in food production where someone could introduce contaminants with maximum impact. Facilities must conduct vulnerability assessments, implement mitigation strategies, and maintain food defense plans. The rule's primary concern is the insider threat: someone with legitimate access who decides to cause harm.[10] FDA guidance emphasizes that the greatest risk comes from individuals who already have the keys to the building.

Here's where interpretation gets complicated. The Intentional Adulteration rule doesn't mention cybersecurity explicitly either. But if a digital system controls an actionable process step, and that system can be compromised remotely, then cyber vulnerability becomes a food defense concern. An attacker who gains access to temperature controls or mixing systems could potentially cause the same harm as someone physically tampering with product. Some food safety professionals interpret the rule to include cyber threats at vulnerable process points; others maintain a narrower reading. The FDA has not issued definitive guidance resolving this ambiguity. If your organization maintains cybersecurity policies, those policies should explicitly address how cyber risks at actionable process steps are identified, assessed, and documented within your food defense plan. Don't wait for the FDA to clarify.

One FDA regulation directly addresses electronic systems, although it's not a cybersecurity rule per se. Title 21 of the Code of Federal Regulations, Part 11, effective since 1997, establishes requirements for electronic records and electronic signatures.[11] The regulation requires that electronic records be trustworthy, reliable, and equivalent to paper records. For food facilities using electronic systems for production records, batch documentation, or quality data, Part 11 creates requirements around access controls, audit trails, and system validation. Interestingly, the FDA has specifically exempted records maintained under the FSMA Preventive Controls rule from Part 11 requirements, but records that satisfy FSMA requirements and are also required under other regulations remain subject to Part 11.[12] The regulation's emphasis on access controls and audit trails overlaps with basic cybersecurity hygiene, making it relevant for cybersecurity professionals working in FDA-regulated food environments.

The FBI sounded a significant warning in September 2021 with a Private Industry Notification specifically targeting the food and agriculture sector.[13] The notification documented ransomware attacks against food sector businesses, including a bakery that lost access to servers, files, and applications, halting production, shipping, and receiving. The FBI warned that larger businesses are targeted based on their perceived ability to pay higher ransom demands, whereas smaller entities are seen as soft targets, particularly those in earlier stages of digitizing their processes. The notification was prescient: within weeks, grain cooperatives NEW Cooperative and Crystal Valley were hit by ransomware attacks.[14] The FBI followed up in April 2022 with another notification warning that ransomware attacks on agricultural cooperatives might be timed to coincide with critical planting and harvest seasons, when pressure to restore operations would be highest.[15]

The Cybersecurity and Infrastructure Security Agency (CISA) designates food and agriculture as one of 16 critical infrastructure sectors under Presidential Policy Directive 21.[16] This designation sounds significant, but it primarily means coordination and voluntary guidance rather than enforceable requirements. Compare food and agriculture to the energy sector, where mandatory standards under the North American Electric Reliability Corporation (NERC) include specific cybersecurity requirements with real audit consequences. Or compare it to financial services, where multiple regulatory bodies conduct regular examinations of institutions' cybersecurity practices. Food and agriculture have no equivalent.

A practical point worth emphasizing is that in the United States, there are no mandatory routine cybersecurity inspections for food and agriculture facilities. CISA and the FDA do not show up unannounced to audit your cybersecurity practices the way food safety inspectors audit sanitation. Unless there's an incident that triggers an investigation or you voluntarily engage with CISA's services, government agencies aren't examining your cyber defenses. Most cybersecurity assessments in food and agriculture are contracted by the organizations themselves, to meet either customer requirements, insurance conditions, or internal risk management goals.

Change is coming, although the timeline remains uncertain. The Cyber Incident Reporting for Critical Infrastructure Act of 2022 (CIRCIA) will eventually require entities in critical infrastructure sectors, including food and agriculture, to report substantial cyber incidents to CISA within 72 hours and ransom payments within 24 hours.[17] The proposed rule, published in April 2024, would apply to entities exceeding Small Business Administration (SBA) size thresholds or meeting sector-specific criteria.[18] CISA estimates that more than 316,000 entities across all critical infrastructure sectors will be covered.[19]

Here's something worth noting: CISA did not propose additional sector-specific criteria for food and agriculture. Unlike sectors such as energy or healthcare, where specific facility types or activities could pull smaller entities into scope regardless of size, food and agriculture coverage will rely solely on size-based criteria.[20] This means small food operations generally won't face CIRCIA reporting requirements, but it also means the regulation provides no special recognition of food system criticality beyond what size thresholds capture.

One compliance wrinkle deserves attention: CIRCIA allows entities to avoid duplicate reporting if they've already reported to another federal agency under "substantially similar" requirements. But the proposed rule explicitly states that state reporting requirements will not be considered substantially similar.[21] If you experience an incident requiring both state breach notification and CIRCIA reporting, you'll likely need to file both. The final rule isn't expected until sometime in 2026, with requirements taking effect sometime after that. Until then, reporting remains voluntary.

Publicly traded food companies face an additional layer of disclosure requirements. In July 2023, the SEC finalized rules requiring public companies to disclose material cybersecurity incidents within four business days of determining the incident is material.[22] The disclosure must describe the material aspects of the incident's nature, scope, and timing, as well as its material impact or reasonably likely material impact on the company. The SEC also requires annual disclosure of cybersecurity risk management, strategy, and governance in Form 10-K filings.[23]

Materiality remains the key question. The SEC uses the standard from federal securities law: information is material if there's a substantial likelihood that a reasonable investor would consider it important in making an investment decision. A ransomware attack that shuts down production for a week might be material. A phishing attempt that was caught before causing damage probably isn't. But the determination requires judgment, and the four-business-day clock starts ticking once the company determines materiality, not when the incident occurs. Getting that assessment process right matters.

State-level requirements add another layer of complexity. California's Consumer Privacy Act (CCPA) and its successor, the California Privacy Rights Act, impose requirements on businesses that collect personal information from California residents, including customer data, employee data, and potentially data from loyalty programs or direct sales.[24] New York's SHIELD Act requires businesses that hold private information about New York residents to implement reasonable safeguards, including cybersecurity measures appropriate to the size and complexity of the business.[25] A food company operating in multiple states may face different breach notification requirements, different definitions of protected information, and different timelines for reporting. This patchwork means compliance teams need to track not just federal developments but potentially dozens of state-level requirements.

The European Union: NIS2 Changes the Game

The European Union has gone further than the United States in mandating cybersecurity requirements for food. The Network and Information Security Directive 2 (NIS2), adopted in December 2022 with a transposition deadline of October 2024, significantly expands cybersecurity obligations across the EU.[26] The original NIS Directive from 2016 focused narrowly on operators of essential services and certain digital service providers. NIS2 dramatically widens the scope, categorizing entities as either "essential" or "important" based on sector and size.

Food production, processing, and distribution fall under the "important entity" classification. Medium and large enterprises in the food sector, generally those with 50 or more employees or annual revenue exceeding €10 million, now face mandatory cybersecurity requirements. The directive's supply chain provisions extend accountability further:

even smaller entities may be affected if they supply regulated companies, because those companies must assess and address supplier security risks.[27]

The requirements are substantial. Covered entities must implement risk management measures addressing technical, operational, and organizational aspects of security. They must notify regulators of significant incidents within 24 hours of becoming aware, with follow-up reports due within 72 hours and final reports within 1 month.[28] They must address supply chain security, maintain business continuity capabilities, and ensure that management bodies approve and oversee cybersecurity measures. This last point matters: NIS2 places accountability at the board level, not just with IT departments.

Penalties for noncompliance can reach €7 million or 1.4 percent of global annual turnover, whichever is higher.[29] These numbers approach the levels associated with the General Data Protection Regulation (GDPR), signaling that the EU takes cyber-physical security as seriously as data protection. Personal liability provisions mean that management can be held responsible, not just the corporate entity.

National implementation varies. Germany's law is expected to affect approximately 30,000 companies nationwide.[30] Italy has incorporated NIS2 into national legislation, with the National Cybersecurity Agency overseeing compliance. Some member states have expanded the scope beyond the directive's minimum requirements. Italy, for instance, now includes legal services for large grocery retailers within its covered entities.[31]

The EU Cyber Resilience Act (CRA), adopted in late 2024 with main obligations applying from December 2027, represents a different approach to improving food sector cybersecurity, one that could matter more than operator-focused regulations for many food companies.[32] Rather than placing the primary burden on entities operating digital systems, the CRA shifts it to manufacturers of products with digital elements. This includes agricultural equipment, food processing machinery, IoT sensors, and the software that runs them.

Why does this matter for food and agriculture? Consider the situation many food processors and farms face today. They're told they need to secure their OT environments, but they're staring at equipment with proprietary systems they can't modify, outdated operating systems they can't patch without voiding warranties, and a lack of expertise to assess what security measures are even possible. They didn't design these systems. They often can't get security updates from manufacturers who designed equipment for durability and functionality rather than ongoing security maintenance. The CRA addresses this upstream problem.

Under the CRA, manufacturers must ensure that products are designed with security in mind from the beginning, a principle known as *secure by design*. They must provide security updates throughout the product's expected lifecycle, which the regulation sets at a minimum of five years.[33] They must report actively exploited vulnerabilities to authorities within 24 hours. Products must bear CE markings, demonstrating that they meet cybersecurity requirements before being sold in EU markets. The European Agricultural Machinery Association has already published implementation guidance recognizing the substantial impact on agricultural equipment design.[28]

For food processors and agricultural operations that have struggled to secure equipment they didn't design and can't modify, the CRA represents hope for an upstream fix. It won't solve legacy system problems overnight—existing equipment isn't covered—but it should mean that equipment purchased after 2027 comes with security built in rather than bolted on, with manufacturers responsible for ongoing vulnerability management rather than leaving operators to fend for themselves.

Which Regulations Apply to Me?

Once you understand the specific regulations, you need to know which ones affect your operation. The answer depends primarily on your organization's size, where in the world you operate, and who your customers are:

Small operations: Family farms, local processors, and small cooperatives often fall below the thresholds of cybersecurity regulations. NIS2 generally applies to medium and large enterprises. CIRCIA's proposed rule focuses on entities that exceed the SBA size standards.[34] But "exempt from regulation" doesn't mean "exempt from risk." Small operations increasingly use the same connected technology as large ones. And if you supply larger companies, their compliance requirements may flow down to you through contracts. Customer-driven requirements can matter more than government mandates.

Mid-sized operations: Regional processors, multilocation food service companies, and agricultural cooperatives sit in an interesting position. You may be large enough to fall under regulations like NIS2 if you operate in Europe (generally 50 or more employees or €10 million in revenue) but not large enough to have dedicated compliance or cybersecurity staff. If you're a publicly traded company, SEC cybersecurity disclosure requirements apply regardless of your primary industry.[35] This is often the hardest position: subject to requirements without the resources larger competitors deploy to meet them.

Large operations: National or multinational food companies, major agricultural enterprises, face the full complement of applicable regulations. If you operate in the EU, NIS2 applies. If you're publicly traded in the United States, SEC rules apply. CIRCIA will apply once it is finalized. Customer certifications like British Retail Consortium Global Standards (BRCGS) and Safe Quality Food (SQF) require demonstrable cybersecurity practices. The good news: you likely have legal, compliance, IT, and cybersecurity departments to help navigate these regulations. The challenge is coordinating across jurisdictions with different requirements.

One critical point for American companies: even if U.S. regulations don't require specific cybersecurity practices, exporting to Europe or supplying European customers may pull you into NIS2's orbit. The directive requires covered entities to assess and address supplier security risks. A European food company buying ingredients from an American supplier will increasingly ask about that supplier's cybersecurity practices, not because American law requires it, but because European law requires the buyer to assess supply chain risk. Commercial contracts are becoming compliance mechanisms.[36]

Beyond Europe and America: A Global Patchwork

Australia moved earlier than many countries to address critical infrastructure cybersecurity. The Security of Critical Infrastructure Act 2018, substantially amended in 2021 and 2022, covers 11 sectors, including food and grocery.[37] Critical food and grocery assets include distribution and supply networks owned or operated by critical supermarket retailers, food wholesalers, or grocery wholesalers. Responsible entities must register critical assets, report cyber incidents, and maintain written risk management programs addressing cybersecurity alongside physical, personnel, and supply chain hazards.[38] This approach includes a notable feature: government assistance powers that allow intervention in response to serious cyberattacks when entities are unable or unwilling to respond adequately.[39] The Australian Signals Directorate can provide technical assistance, and in extreme circumstances, the government can issue directions to affected entities. This interventionist model contrasts with the American preference for voluntary cooperation.

The United Kingdom, having transposed the original NIS Directive before Brexit, maintains similar requirements under the NIS Regulations 2018, which is now being updated independently of EU developments.

Canada lacks specific food sector cyber regulation but is beginning to address agricultural technology risks through various departmental initiatives. Integrated North American supply chains mean Canadian and American food system vulnerabilities are deeply intertwined regardless of regulatory boundaries.

International harmonization remains elusive. The Codex Alimentarius Commission, the joint body of the Food and Agriculture Organization and WHO that sets international food standards and has been adopted by 189 countries, focuses on traditional food safety concerns.[40] Cyber threats are not explicitly addressed, creating a gap between international food safety harmonization and digital system integrity. No global framework exists to address cyber-related food safety incidents, leaving supply chains spanning dozens of jurisdictions to navigate inconsistent requirements.

The Private Standards That Often Matter More

For many food companies, private certification schemes have more immediate impact than government regulations. Major retailers require suppliers to hold certifications such as BRCGS, SQF, or Food Safety System Certification 22000 (FSSC 22000). Losing certification can mean losing access to major customers, a more tangible consequence than the uncertain prospect of regulatory enforcement.

These standards increasingly incorporate cybersecurity requirements. The BRCGS Global Standard for Food Safety, Version 9, released in 2022, explicitly includes IT systems within its site security requirements.[41] Companies pursuing certification must demonstrate that they have documented procedures to protect information technology systems. FSSC 22000 Version 6 enhanced food defense requirements, with implications for cyber. The trend is clear: private standards are filling gaps left by incomplete government regulation.

Information security standards designed for other contexts are also finding application in food. ISO/IEC 27001, the international standard for information security management,[42] is increasingly adopted by food companies seeking to demonstrate security practices to customers. The NIST Cybersecurity Framework provides a voluntary structure organized around five core functions: Identify, Protect, Detect, Respond, and Recover.[43] Its scalability makes it adaptable to organizations of varying sizes and levels of sophistication.

For operational technology environments common in food processing, IEC 62443 provides a comprehensive standard for the security of industrial automation and control systems.[44] The challenge is that IEC 62443 is complex, expensive to implement, and unfamiliar to most food industry personnel. The gap between what standards require and what smaller food companies can realistically achieve remains substantial.

What Regulations Don't Address

The most striking regulatory gap lies in agricultural technology. Farm equipment safety is regulated. The Environmental Protection Agency (EPA) oversees pesticide application equipment. The Occupational Safety and Health Administration (OSHA) sets standards for workplace safety. But cybersecurity in agricultural equipment, precision agriculture software, IoT sensors, and cloud platforms for farm data? It's essentially unregulated in the United States. The EU's Cyber Resilience Act will address this for equipment sold in European markets after 2027, but American farmers and food producers buying equipment domestically have no equivalent protection.

It isn't an accident of timing but reflects how agricultural technology has evolved. Equipment manufacturers design for durability and harsh operating conditions, not for security against deliberate attack. Farm equipment was traditionally sold as durable goods, not as connected systems requiring ongoing security maintenance. Regulatory agencies lacked technical expertise in agricultural operational technology. And frankly, tractors weren't originally conceived as critical infrastructure in the cyber sense.

The Ukraine incident we discussed in earlier chapters revealed what this means in practice. John Deere's ability to remotely disable stolen equipment demonstrated capabilities that, in different hands or different circumstances, could be exploited to disable legitimate operations. Security researchers examining agricultural equipment have found outdated operating systems, unencrypted data transmission, and default passwords that should have been changed years ago. Agricultural equipment manufacturers built for durability and field performance because those were the requirements. Nobody in the procurement conversation asked about patch management schedules, and manufacturers weren't volunteering information about network vulnerabilities. Starting in 2027, the EU's Cyber Resilience Act will require something different for equipment sold in European markets. American farmers buying from domestic manufacturers have no equivalent rule coming their way.

Small operators face another gap. Regulations often include size thresholds. NIS2 generally applies to medium and large enterprises. CIRCIA focuses on entities that exceed the SBA size standards. But small farms and processors use the same technology as large farms and processors. The family dairy with robotic milkers faces the same cyber risks as a corporate operation but has none of the resources to address them. Regulatory compliance becomes a burden only for those large enough to be covered, while smaller operators remain invisible to both enforcement and protection.

Legacy systems present yet another challenge. Many food processing facilities operate equipment that's been running for decades. Regulations often don't apply retroactively. Upgrading requires capital that thin-margin operations can't spare. The result is a two-tier system: new installations may meet modern standards, whereas older facilities operate with equipment designed before security was a consideration.

The Economics of Compliance

Compliance costs money. Security technology, personnel, training, audits, and remediation each add to the burden. For food companies operating on margins measured in single digits, cybersecurity spending competes with every other investment. The cost of ISO 27001 certification, the fees for BRCGS audits with cyber components, and the expense of hiring or contracting cybersecurity expertise aren't trivial for organizations already stretched thin.

But the cost of noncompliance can exceed the cost of compliance. NIS2 penalties can reach €7 million euros. Loss of food safety certifications can mean loss of major customer contracts. A ransomware incident can cost millions in recovery, downtime, spoiled product, and reputational damage. Insurance increasingly requires baseline security measures, and post-incident discovery of inadequate controls can complicate claims.

The cost of doing nothing compounds over time. Each unpatched vulnerability, each unchanged default password, each deferred security investment increases exposure. Attackers specifically target organizations that appear weak. The "hope for the best" strategy offers short-term savings but accumulates long-term risk that eventually comes due.

Looking Ahead

The regulatory landscape is moving, even if it's not moving as fast as threats evolve. In the United States, CIRCIA will eventually require cyber incident reporting from food and agriculture entities above size thresholds. CISA's secure by design initiative encourages manufacturers to build security into products from the beginning. The FDA continues to evolve its approach to data integrity in food safety systems. Congressional attention to agricultural supply chain security suggests potential legislative action in future Farm Bills.

In Europe, NIS2 implementation is setting precedents that will shape global expectations. The Cyber Resilience Act will fundamentally change the security expectations for agricultural and food-processing equipment, shifting the burden from operators who can't secure equipment they didn't design to manufacturers who can.

Private standards will continue tightening. Within five years, meaningful cybersecurity requirements will likely be standard in major food safety certifications. Companies that invest early will find compliance easier; those who delay will face rushed implementations when customer contracts suddenly require what they don't have.

Beyond the Audit

The fundamental tension in food cybersecurity regulation is this: it imposes costs on an industry with thin margins, but the absence of regulation allows risks to accumulate. The question isn't whether to regulate, but how to regulate effectively, in ways that improve security without crushing the operators who can least afford compliance burdens.

Compliance is the floor, not the ceiling. Attackers don't check your certification status before deploying ransomware. Security requires continuous adaptation, not annual audit cycles. The mindset shifts matter: moving from "Are we compliant?" to "Are we actually secure?"

The regulatory framework we've built so far covers the part of the food system that shows up in corporate filings. Those are organizations large enough to audit and hold to account. But the supply chain doesn't stop at the gate of a publicly traded food company. It runs through packing sheds and small cooperatives and family operations that have been digitizing their processes without anyone asking whether they've thought about what that connectivity costs them in terms of exposure.

Eighty-nine percent of American farms are classified as small. A family operation adopting precision agriculture because some sustainability certification now requires data reporting. They're facing the same ransomware gangs that target multinational processors. The budget for cybersecurity consultation doesn't exist, and neither does the in-house expertise to evaluate whether that new irrigation controller just opened a hole in their network. A three-person cooperative trying to help members modernize can barely keep up with the technology itself, let alone the threat landscape around it. Regulations designed for enterprises with compliance departments don't reach operators who've never heard of NIS2.

The next chapter looks at this gap, what digital equity means in food and agriculture, and whether there's any realistic path forward for the small operators who grow most of what we eat.

Notes

1. European Parliament and Council of the European Union. (2024). "Regulation (EU) 2024/2847 of the European Parliament and of the Council of 23 October 2024 on horizontal cybersecurity requirements for products with digital elements and amending regulations (EU) No. 168/2013 and (EU) 2019/1020 and Directive (EU) 2020/1828 (Cyber Resilience Act)." *EUR-Lex*. **https://eur-lex.europa.eu/eli/reg/2024/2847/oj/eng**
2. Sinclair, U. (1906). *The Jungle* (Doubleday); for the Act, U.S. Congress. (1906). Federal Meat Inspection Act. Pub. L. No. 59-392, 34 Stat. 674.

3. FDA. (2018). "Background on the FDA Food Safety Modernization Act (FSMA)." **https://www.fda.gov/food/food-safety-modernization-act-fsma/background-fda-food-safety-modernization-act-fsma**
4. Biden, J. R., President. (2021). "Executive order 14028: Improving the Nation's cybersecurity." *Federal Register* 86(93), 26633–26649. **https://www.federalregister.gov/documents/2021/05/17/2021-10,460/improving-the-nations-cybersecurity**
5. FDA. (2023). "What we do." **https://www.fda.gov/about-fda/what-we-do**
6. USDA. (n.d.). "Mission areas." **https://www.usda.gov/about-usda/general-information/mission-areas**
7. FDA. (2017). "Full text of the Food Safety Modernization Act (FSMA)." **https://www.fda.gov/food/food-safety-modernization-act-fsma/full-text-food-safety-modernization-act-fsma**
8. FDA. (2025). "FSMA final rule for preventive controls for human food." **https://www.fda.gov/food/food-safety-modernization-act-fsma/fsma-final-rule-preventive-controls-human-food**
9. FDA. (2022). "FSMA final rule on mitigation strategies to protect food against intentional adulteration." **https://www.fda.gov/food/food-safety-modernization-act-fsma/fsma-final-rule-mitigation-strategies-protect-food-against-intentional-adulteration#:~:text**
10. FDA. (2020). "Protecting the food supply from intentional adulteration, such as acts of terrorism." **https://www.fda.gov/food/conversations-experts-food-topics/protecting-food-supply-intentional-adulteration-such-acts-terrorism**
11. Code of Federal Regulations. "Electronic records; electronic signatures." Title 21, Chapter I, Subchapter A, Part 11.
12. Code of Federal Regulations. "Scope." Title 21, Chapter I, Subchapter A, Part 11, Subpart A, § 11.1. **https://www.ecfr.gov/current/title-21/chapter-I/subchapter-A/part-11/subpart-A/section-11.1**
13. FBI. (2021). "Cyber criminal actors targeting the food and agriculture sector with ransomware attacks." Joint Cybersecurity Advisory (210907). **https://www.ic3.gov/CSA/2021/210907.pdf**
14. McVan, M. (2021). "FBI says ransomware attacks on food and agriculture industry are increasing." *Investigate Midwest.* **https://investigatemidwest.org/2021/10/13/fbi-says-ransomware-attacks-on-food-and-agriculture-industry-are-increasing**
15. FBI. (2022). "Ransomware attacks on agricultural cooperatives." Joint Cybersecurity Advisory (220420-2). **https://www.ic3.gov/CSA/2022/220420-2.pdf**
16. U.S. Department of Homeland Security. (2013). *NIPP 2013: Partnering for Critical Infrastructure Security and Resilience.* **https://www.cisa.gov/sites/default/files/2022-11/national-infrastructure-protection-plan-2013-508.pdf**
17. United States Congress. (2022). "H.R.2471 - Consolidated Appropriations Act, 2022." **https://www.congress.gov/bill/117th-congress/house-bill/2471/text**
18. CISA. (2024). "Cyber incident reporting for Critical Infrastructure Act of 2022 (CIRCIA)." **https://www.cisa.gov/topics/cyber-threats-and-advisories/information-sharing/cyber-incident-reporting-critical-infrastructure-act-2022-circia**
19. Ibid.
20. Gevertz, J., & Rissmiller, C. (2024). "US CIRCIA update: CISA proposed regulations released." *DLA Piper.* **https://privacymatters.dlapiper.com/2024/04/us-circia-update-cisa-proposed-regulations-released**
21. Finch, B. E., Zimmer, E., Chegini, A., & Reno, S. (2024). "New CISA rule would require widespread cyber incident reporting, updated timelines and penalties for critical infrastructure sector." *Pillsbury Winthrop Shaw Pittman LLP.* **https://www.pillsburylaw.com/en/news-and-insights/cyber-incident-reporting-critical-infrastructure.html**

22. SEC. (2023). "Cybersecurity risk management, strategy, governance, and incident disclosure." *Federal Register*, 88(149), 51896–52084. **https://www.federalregister.gov/documents/2023/08/04/2023-16194/cybersecurity-risk-management-strategy-governance-and-incident-disclosure**
23. Ibid.
24. California Department of Justice, Office of the Attorney General. (2024). "California Consumer Privacy Act (CCPA)." California Attorney General's Office. **https://oag.ca.gov/privacy/ccpa**
25. New York State Office of the Attorney General. "SHIELD Act." **https://ag.ny.gov/resources/organizations/data-breach-reporting/shield-act**
26. European Parliament and Council of the European Union. (2022). "Directive (EU) 2022/2555 of the European Parliament and of the Council of 14 December 2022 on measures for a high common level of cybersecurity across the Union, amending Regulation (EU) 910/2014 and Directive (EU) 2018/1972, and repealing Directive (EU) 2016/1148 (NIS 2 Directive)." *EUR-Lex*. **https://eur-lex.europa.eu/eli/dir/2022/2555/oj/eng**
27. European Parliament and Council of the European Union. (2022). "Directive (EU) 2022/2555."
28. European Parliament and Council of the European Union. (2022). "Directive (EU) 2022/2555, Article 23: reporting obligations." **https://www.nis-2-directive.com/NIS_2_Directive_Article_23.html**
29. European Parliament and Council of the European Union. (2022). "Directive (EU) 2022/2555, Article 34: general conditions for imposing administrative fines on essential and important entities." **https://www.nis-2-directive.com/NIS_2_Directive_Article_34.html**
30. Greenberg Traurig, LLP. (2025). "EU NIS 2 Directive: expanded cybersecurity obligations for key sectors." **https://www.gtlaw.com/en/insights/2025/8/eu-nis-2-directive-expanded-cybersecurity-obligations-for-key-sectors**
31. Freshfields Bruckhaus Deringer. (2025). "Germany implements NIS2 – what you need to know now." **https://technologyquotient.freshfields.com/post/102lwz4/germany-implements-nis2-what-you-need-to-know-now**
32. European Parliament and Council of the European Union. (2024). "Regulation (EU) 2024/2847 of the European Parliament."
33. European Commission. "The Cyber Resilience Act - summary of the legislative text." **https://digital-strategy.ec.europa.eu/en/policies/cra-summary**
34. Gevertz, J., & Rissmiller, C. (2024). "US CIRCIA update: CISA proposed regulations released."
35. SEC. (2023). "SEC adopts rules on cybersecurity risk management, strategy, governance, and incident disclosure by public companies." **https://www.sec.gov/newsroom/press-releases/2023-139**
36. European Parliament and Council of the European Union. (2022). "Directive (EU) 2022/2555."
37. Cyber and Infrastructure Security Centre (CISC). (2018). "Security of Critical Infrastructure Act 2018 (SOCI)." **https://www.cisc.gov.au/legislation-regulation-and-compliance/soci-act-2018**
38. Cyber and Infrastructure Security Centre (CISC). (2018). "SOCI Act 2018 for food and grocery." **https://www.cisc.gov.au/information-for-your-industry/food-and-grocery/legislation-regulation-and-compliance/soci-act-2018**
39. Cyber and Infrastructure Security Centre (CISC). (n.d.). "Government assistance." **https://www.cisc.gov.au/how-we-support-industry/government-assistance**
40. Codex Alimentarius. (n.d.). "Codex Alimentarius Commission (CAC)." **https://www.fao.org/fao-who-codexalimentarius/committees/cac/about/en**
41. BRC Global Standards. (2022). *Global Standard Food Safety Issue 9*. **https://www.brcgs.com/product/global-standard-food-safety-issue-9/p-13279**

42. International Organization for Standardization. (2022). *ISO/IEC 27001:2022. Information Security, Cybersecurity and Privacy Protection—Information Security Management Systems—Requirements*. **https://www.iso.org/standard/27001**
43. U.S. National Institute of Standards and Technology. (n.d.). “Cybersecurity framework.” **https://www.nist.gov/cyberframework**
44. International Society of Automation. (2026). “ISA/IEC 62443 series of standards.” **https://www.isa.org/standards-and-publications/isa-standards/isa-iec-62443-series-of-standards**

CHAPTER 29

Digital Equity for Small and Mid-Size Farms

Security Can't Just Be for the Giant Corporations

Vital Bircher was in his 70s when ransomware locked the milking robot on his Swiss dairy farm. We met him at the beginning of the book. The attackers wanted $10,000. Bircher refused to pay, and a cow died along with her unborn calf because the equipment wouldn't function without the decryption key. He spent thousands of francs on veterinary bills and system restoration, and then he spent months wondering what he could have done differently.

What I keep coming back to, after examining cybersecurity across this sector, is everything Bircher didn't have when the attack hit: an IT department, an incident response plan, cyber insurance, cybersecurity awareness training, or a vendor support contract with someone who'd answer at 2:00 AM. He was running a small operation with modern equipment, and when that equipment failed him, he was on his own.

The tools that might have helped weren't missing because Bircher was careless. Multifactor authentication (MFA), network segmentation, regular offline backups, access to incident response services, these things weren't available to him in any practical way. The cybersecurity industry is building for enterprises with dedicated staff and six-figure budgets. Products get designed for organizations that can deploy them, not for a 70-year-old dairy farmer trying to figure out why his milking robot stopped working. Small farms weren't part of the design.

According to the USDA's Economic Research Service, small family farms—those with gross cash farm income below $350,000—make up about 86 percent of all American farms.[1] Nearly 9 out of 10 operations look more like Bircher's dairy than like the corporate agriculture where security vendors focus their sales efforts. These farms hold 41 percent of agricultural land but produce only 17 percent of output by value. The economics mean they matter less to companies selling security products, even though they matter enormously to the communities that depend on them for food.

Who's Coming After Small Farms, and Why

Understanding the threat helps calibrate the response. The FBI has warned that agricultural operations of all sizes face cyber risks, although for different reasons: large operations because attackers assume they can pay substantial ransoms, small operations because they're easier to breach.[2]

Attacks are already happening. Among small-scale Midwest farmers surveyed in 2024, nearly one in four reported having experienced a cyberattack, and another 18 percent weren't sure whether they had been targeted.[3] Ransomware groups run automated scans looking for vulnerable systems: unpatched software, default passwords, exposed remote access ports. They purchase stolen credentials from brokers who've already breached organizations. Often the attackers don't know who they've compromised until they're already inside.[4] Your dairy probably isn't on anyone's target list, but if your network has a vulnerability that automated scanning can find, you become a target of opportunity regardless. A 2025 survey of small-scale Midwest farmers found that 45 percent disagreed that their farms were likely cyberattack targets, and fewer than one in six believed they were.[5]

Nation-state actors represent a different category, one more relevant to critical infrastructure than to individual farms. A family dairy isn't necessarily going to attract foreign intelligence services, but cooperatives are another matter. Cooperatives exist to solve economic barriers by pooling resources across member farms, and that model works. It also concentrates risk in ways individual farmers may not recognize until something goes wrong. When NEW Cooperative was hit by ransomware in September 2021, the BlackMatter group demanded $5.9 million. It shut down feed scheduling, soil data, and grain logistics during harvest across more than 60 Iowa communities.[6] Weeks later, Crystal Valley Cooperative in Minnesota suffered a similar attack.[7] Farmers who'd never been valuable enough to target individually got caught up in attacks aimed at the shared infrastructure they depended on, which ultimately cascaded across hundreds of member farms. A regional cooperative may operate on margins just as thin as its member farms, with IT staff consisting of one or two people responsible for everything from email to elevator operations. These aren't organizations that can afford round-the-clock security monitoring or incident response retainers. They've become high-value targets precisely because of the leverage that harvest-time disruption provides.

The practical takeaway is that you're mostly defending against automated, opportunistic scans, and the defenses that stop opportunistic attacks are basic measures that don't require enterprise budgets. Attackers scanning for easy victims tend to move on when they encounter friction. The goal isn't to become impenetrable; it's to be a harder target than the next operation down the road that's running an unpatched system with a default password.

What Actually Works at Small Scale

Enterprise security frameworks assume things that don't exist on small farms: IT departments, budget lines for security tools, and staff who can implement zero-trust architecture. Small farms have operators already working dawn to dusk, margins that don't accommodate consulting engagements, and connectivity that may not support cloud-based anything. The Federal Communications Commission (FCC) reported in 2024 that nearly 28 percent of rural Americans lack broadband service that meets current standards.[8] Running cloud backups gets complicated when your Internet cuts out every time it storms.

The good news is that most farmers already understand the basics. A survey found that 84 percent of respondents were at least somewhat familiar with cybersecurity concepts, and majorities reported installing antimalware software and backing up files.[9] The problem isn't ignorance; it's the gap between knowing something that matters in the abstract and believing that it applies to your own operation.

Passwords are the obvious starting point because they cost nothing to fix. A 12-character passphrase made of random words resists cracking better than an 8-character jumble of symbols you'll forget and write on a sticky note. The phrase "correct horse battery staple" became famous in security circles because it's both memorable and, at 28 characters, far more resistant to brute-force attacks than "P@ssw0rd!" ever was. Different accounts should have different passwords; when one gets compromised, you don't want attackers walking into everything else. Password managers track credentials across accounts, and most have free tiers adequate for personal use. If you're resistant to using a password manager, at a minimum, use different passwords for anything that involves money or access to equipment. On top of a strong password, MFA makes a disproportionate difference. Microsoft's security research found that MFA blocks more than 99 percent of automated account attacks. Yet only 45 percent of Midwest farmers surveyed reported using it, and fewer than 1 in 10 encrypted sensitive files. If a service offers MFA and you're not using it, that's protection sitting unused.[10]

Backups to disconnected storage provide a recovery path that doesn't require paying ransom. An external hard drive that stays unplugged except during weekly backups won't prevent an attack, but it means you can restore operations afterward. Keep the drive physically separate from your main systems so that a fire or flood that destroys one doesn't destroy both.

Phishing recognition matters because most attacks start with someone clicking a link in a deceptive message. Check sender addresses: attackers register domains that look almost right. Hover over links to see where they point before clicking. Messages demanding immediate action ("Your account will be closed!" or "Approve this payment now!") are designed to bypass deliberation. Legitimate organizations generally don't threaten customers over email. The same skepticism applies to text messages; phishing by SMS, called smishing, often arrives disguised as a delivery notification or bank alert. It's okay to delete; if it's a real person, they will find a way to contact you.

Your phone deserves the same attention as your computer, especially if you use it to manage equipment, check sensor data, or access farm accounts. Use a PIN or biometric lock. Enable automatic updates. Know how to remotely wipe the device if it's lost. Phones connect to public Wi-Fi, download apps from various sources, and receive messages from unknown numbers, all of which create exposure.

Recognizing Trouble Early

Attackers who've gained access often spend time exploring systems before deploying ransomware, which means the strange behavior you notice midweek might be reconnaissance for the attack that locks you out over the weekend.

Unexplained slowness can indicate malware running in the background, consuming resources to encrypt files or communicate with external servers. Hardware ages and software accumulates cruft, so not every slowdown signals a compromise, but sudden performance changes warrant investigation rather than dismissal. If your farm management software that's been responsive for years suddenly takes 30 seconds to load, that's worth looking into before assuming it's just getting old.

Watch for account activity that doesn't match your behavior: emails in your Sent folder that you didn't write, password reset notifications you didn't request, login alerts from unfamiliar devices or locations. If your equipment vendor's portal shows that someone accessed your account at 3:00 AM while you were sleeping, that access wasn't yours. Many services let you review recent login activity; checking periodically takes moments and can reveal problems before they escalate.

Take note of equipment behaving unexpectedly, settings changing without input, programs launching on their own, and files moving or disappearing, all of which suggest unauthorized changes. The instinct to ignore small oddities and get back to work is understandable when there's always something demanding attention, but catching a problem during reconnaissance beats discovering it when the ransom note appears on your screen.

Questions to Ask Before Buying

Most security advice assumes you already own the technology. But the better moment to think about security is before signing the purchase agreement, while you still have choices.

Find out what data the technology collects, where it goes, and who has access to it. Some equipment keeps operational information on the device; other systems stream everything to cloud platforms where it may be aggregated, analyzed, shared with business partners, or used to train AI models. The American Farm Bureau Federation has pushed for transparency through its Privacy and Security Principles for Farm Data, although the voluntary principles don't require vendors to follow them.[11] If a dealer can't clearly explain what happens to your data, that's information about the vendor's priorities.

Understand what happens if the service goes offline. A soil sensor that optimizes irrigation is helpful, but fields can be watered without it. A milking robot requiring cloud authentication to function puts animals at risk when connectivity fails or the vendor's servers go down. Some systems have local fallbacks; others require constant connectivity. If a platform failure means animals can't be fed or crops can't be harvested, you need either a backup plan or a different vendor.

Learn who else can access your systems. Remote service and diagnostics have become standard, which is convenient when something breaks. It also means people outside your operation can connect to equipment on your property. Understand who has this capability, under what circumstances, and whether you can revoke it. The same connectivity enabling remote troubleshooting could, if credentials are compromised, let an attacker disable equipment or extract operational data.

Finding Trustworthy Technical Help

Small farms often rely on informal IT support: a nephew or niece who's good with computers, a local repair shop, a neighbor, or the dealer's service technician. These people may handle consumer devices competently while knowing nothing about agricultural OT security. Those family members might understand home networking but never have touched an industrial control system. The dealer's tech knows the equipment mechanically but may not be trained in digital security features.

Farming communities have always shared knowledge about equipment and suppliers. Ask around: if another farmer dealt with a cyber incident and found competent help, that's worth knowing. Cooperatives sometimes work with providers who understand agricultural operations, and those relationships may extend to member farms. Equipment vendors occasionally offer support contracts covering security assistance; ask about what's included, response times for incidents, and whether coverage continues past warranty.

Trust your instincts when evaluating help. Someone who dismisses your concerns or offers generic solutions without asking about your specific situation probably isn't the right fit. Agricultural equipment runs continuously, controls processes with safety implications, and can't just be rebooted mid-operation. Support providers who don't understand these constraints sometimes create more problems than they solve.

The Systems Working Against Small Farms

Individual farmers can improve their practices, but they're operating within larger systems that aren't set up to help them. Precision agriculture platforms that optimize inputs and satisfy sustainability requirements are designed with security as an afterthought, if it's considered at all. Equipment manufacturers maintaining remote access to tractors and harvesters have the technical capability to embed security from the start. Chapter 7 documented how they can remotely disable equipment when motivated, but protecting customers from connectivity-created vulnerabilities hasn't been a priority.

Policy is slowly beginning to respond. The Farm and Food Cybersecurity Act, first introduced in 2024 and reintroduced in 2025 (H.R. 1604), would require the USDA to assess agricultural cyber threats biennially and conduct annual crisis simulations.[12] The bill has bipartisan backing from the American Farm Bureau Federation, National Cattlemen's Beef Association, and other agricultural organizations. But threat assessments don't put tools in farmers' hands, and simulations don't close the broadband gap that makes cloud-based solutions inaccessible to rural operations. The legislation addresses the problem at the policy level but lacks mechanisms to reach most farms. A more comprehensive approach might draw from existing agricultural support frameworks—crop insurance, conservation programs, disaster assistance—that treat digital risk mitigation as part of the agricultural support portfolio rather than a separate concern.

USDA data shows decision-support technologies like yield monitors and soil maps on 68 percent of large farms in 2023 but only 13 percent of small operations.[13] Small farms that do adopt precision agriculture often do so through cooperatives, which provides access while concentrating risk in the ways I described earlier. Meanwhile, sustainability mandates push farmers toward exactly the technologies creating cyber exposure. The EU's Common Agricultural Policy ties payments to precision farming practices.[14] The UK's Environmental Land Management schemes do the same.[15] Applying inputs precisely does reduce environmental harm, and the policy logic makes sense. But pushing farmers toward digital dependencies without investing in security infrastructure leaves them exposed. Climate-smart agriculture is being built on systems whose vulnerabilities haven't been seriously evaluated.

Generational differences shape how farm operators manage technology and cybersecurity. Older operators, comfortable with analog methods, value redundancy and can work without connectivity, whereas younger ones rely heavily on digital systems but may overlook their vulnerabilities. Yet digital security rarely features in generational handoffs; credentials, system access, and backups often pass along informally. As farms evolve, succession planning needs to include digital infrastructure, not just land and equipment, to ensure continuity and protection across generations.

The Cooperative Extension System, with roughly 2,900 local offices staffed by agents who've built trusted relationships in farming communities over decades, represents infrastructure that already reaches the farmers who need help most. Extension has always translated research into practical guidance, pest management techniques, soil health practices, and livestock care. The agents speak the language of agriculture. They understand local conditions. They know which concerns are real and which are overblown. Yet when researchers asked Midwest farmers where they get cybersecurity information, extension services were "seldom used." Instead, 84 percent learned about cyber risks through mass media, and more than half said they'd turn to friends and family for guidance on protective measures. The researchers noted that this reliance on informal networks is "a double-edged sword, while these sources are often considered trustworthy, they may not always provide accurate or up-to-date information.[16]The FCC and USDA Precision Agriculture Connectivity Task Force recommended in 2024 that land-grant universities and extension services develop digital literacy training for precision agriculture adoption.[17] That's a start, but digital literacy should explicitly include digital security. Teaching farmers to use connected equipment without teaching them to protect it sets them up for exactly what happened to Bircher. An extension agent explaining phishing using examples from equipment vendors and cooperative services, acknowledging the constraints of intermittent broadband and nonexistent IT budgets, will be more effective than generic corporate training designed for office workers. The pieces exist, but they haven't been assembled.

The Data Question

A question small farms rarely ask: who owns the information generated by their precision agriculture systems? Modern operations produce hundreds of thousands of data points: daily yield records, soil measurements, input applications, and equipment performance metrics. This data represents both operational intelligence and accumulated knowledge about specific land, knowledge that may have taken years to develop through trial and observation.

The legal status of agricultural data remains murky. Raw data generally can't be copyrighted under U.S. law because facts aren't creative expression. Ownership gets determined by contracts between farmers and technology providers, and those contracts often contain provisions that farmers sign without fully understanding. Some agreements allow data sharing with business partners or the use of data in training AI models. These clauses may appear in dense legal language rather than marketing materials. A farmer who thinks they own their yield data may discover that the platform provider can aggregate and sell insights derived from it.

The cybersecurity implications reach beyond privacy. Aggregated agricultural data reveals patterns about productivity, practices, and vulnerabilities across entire regions. In the wrong hands, that information could inform commodity speculation by revealing crop conditions before public reporting, enable attacks synced to vulnerable operational windows, or provide intelligence about food production capacity at national scales. As precision agriculture expands, control of agricultural data becomes control of strategic information about who can grow food, where, and how much.

Small farms that sign up for digital services rarely have the legal resources to negotiate contract terms. The imbalance between technology companies with substantial legal departments and individual farmers reviewing service agreements creates bargaining conditions that heavily favor one side. Read contracts before signing, even if it takes time. Ask vendors about data practices directly. If they won't answer clearly, weigh whether their platform is worth the uncertainty, and consider alternatives.

Looking Ahead

The path forward isn't about choosing between digital adoption and security. Sustainability mandates, economic pressures, and operational efficiency all push toward connected systems, and that pressure won't ease. The question is whether farmers adopt these technologies having thought through security from the start or have to learn its importance the way Bircher did.

Manufacturers could accept responsibility proportionate to the control they maintain. If they can remotely disable stolen equipment, they can build protectors for legitimate owners. If they collect operational data, they can secure it properly. Extension services could add cybersecurity to the topics they already cover effectively, given adequate resources. Policy could move beyond assessments and simulations toward direct support for operations that can't afford enterprise solutions.

Farmers themselves can ask questions about equipment before buying: what data gets collected? Who can access the systems? And what happens when services fail? These are the same kinds of questions farmers have always applied to weather risk and mechanical reliability. Digital systems aren't somehow exempt.

The next chapter looks at how sustainability and security are becoming entangled. Climate-smart farming depends on sensors, data analytics, and connected equipment. Carbon markets require verified digital records. The transition toward environmental sustainability is simultaneously a transition toward digital dependency, with the vulnerabilities that implies. How farms navigate between economic pressure, environmental mandates, and cybersecurity gaps will shape whether that transition works for everyone or leaves most of the sector behind.

Notes

1. Lim, K., & Lacy, K. (2025). "Small family farms accounted for 86 percent of U.S. farms and generated 17 percent of the value of production in 2023." Chart ID 110693. *USDA, Economic Research Service.* **https://www.ers.usda.gov/data-products/charts-of-note/chart-detail?chartId=110693**
2. FDA. (2021). "Cyber criminal actors targeting the food and agriculture sector with ransomware attacks." **https://www.ic3.gov/CSA/2021/210907.pdf**
3. Rodriguez, L., Micheletti, K., Qu, S., Masambuka-Kanchewa, F. (2025). "'Too small to be targeted:' Perceived cybersecurity vulnerabilities of small-scale U.S. farmers." *Journal of International Agricultural and Extension Education*, 33(1), 17–32.
4. Braley, J. (2025). Quoted in Marston, J. (2024). "More than half of all cyber attacks in agrifood are ransomware, 'any size' business at risk, says threat intelligence group." *AgFunderNews.* **https://agfundernews.com/more-than-half-of-all-cyber-attacks-in-agrifood-are-ransomware-any-size-business-at-risk-says-threat-intelligence-group**
5. Rodriguez, L. et al. (2025). "'Too small to be targeted.'"
6. Ribeiro, A. (2021). "BlackMatter ransomware group strikes NEW Cooperative, makes demand of $5.9 million." *Industrial Cyber.* **https://industrialcyber.co/news/blackmatter-ransomware-group-strikes-new-cooperative-makes-demand-of-5-9-million**
7. Jones, D. (2021). "Ransomware compromises customer data in farm co-op attack." *Cybersecurity Dive.* **https://www.cybersecuritydive.com/news/agriculture-food-ransomware-coop/607080**
8. FCC. (2024). "Inquiry concerning the deployment of advanced telecommunications capability to all Americans in a reasonable and timely fashion." **https://www.fcc.gov/document/fcc-increases-broadband-speed-benchmark-0**
9. Rodriguez, L., et al. (2025). "'Too small to be targeted.'"
10. Ibid.
11. American Farm Bureau Federation. (2024). "Ag data: core principles." **https://www.agdatatransparent.com/principles**
12. Ribeiro, A. (2025). "Farm and Food Cybersecurity Act reintroduced to protect food supply chain from cyber threats." *Industrial Cyber.* **https://industrialcyber.co/regulation-standards-and-compliance/farm-and-food-cybersecurity-act-reintroduced-to-protect-food-supply-chain-from-cyber-threats**

13. Lim, K., & McFadden, J. (2024). "Precision agriculture use increases with farm size and varies widely by technology." *USDA, Economic Research Service*. **https://www.ers.usda.gov/data-products/charts-of-note/chart-detail?chartId=110550**
14. European Commission. (n.d.). "Eco-schemes." **https://agriculture.ec.europa.eu/common-agricultural-policy/income-support/eco-schemes_en**
15. Department for Environment, Food & Rural Affairs. (2024). "Environmental land management in 2024: details of actions and payments." *Defra Farming Blog*. **https://defrafarming.blog.gov.uk/2024/01/04/environmental-land-management-in-2024-details-of-actions-and-payments**
16. Rodriguez, L., et al. (2025). "'Too small to be targeted.'"
17. FCC. (2024). "Task Force for Reviewing the Connectivity and Technology Needs of Precision Agriculture in the United States." **https://www.fcc.gov/task-force-reviewing-connectivity-and-technology-needs-precision-agriculture-united-states**

CHAPTER 30

Sustainability, Climate Tech, and Cybersecurity

When the Data Doesn't Match Reality

In October 2024, federal prosecutors in Manhattan unsealed an indictment against Kenneth Newcombe, the 76-year-old founder of C-Quest Capital. His company had raised a quarter-billion dollars distributing clean cookstoves across sub-Saharan Africa, Asia, and Central America. The charges: wire fraud, commodities fraud, securities fraud, and conspiracy. According to the Commodity Futures Trading Commission (CFTC), the Department of Justice, and the SEC, C-Quest personnel had systematically fabricated the environmental data underlying millions of carbon credits.

Internal communications told the story. Executives instructed staff to "revise" data through "back-end management if you know what I mean." Survey teams who discovered that households had stopped using stoves, or that structures had been demolished, were allegedly ordered to fabricate numbers anyway. Six-million fraudulent carbon credits entered the voluntary market, purchased by corporations claiming environmental benefits they never achieved.[1]

No hackers breached firewalls, no ransomware encrypted servers. But C-Quest exposed something that should trouble anyone watching agriculture's digital transformation: the systems we've built to verify environmental sustainability are only as trustworthy as the data fed into them. And nobody is really asking who secures that data.

What's Already Been Hit

C-Quest was a fraud, not a cyberattack. But dismissing cybersecurity concerns about sustainability infrastructure as theoretical ignores what's already happened to systems built on identical architectures.

In 2017, researchers discovered Triton malware in a Middle Eastern petrochemical facility. Triton targeted Schneider Electric's Triconex safety systems, the controllers designed to shut down industrial processes before they become dangerous.[2] The attackers weren't trying to steal data or demand ransom. They were attempting to disable safety systems while manipulating process controls, a combination that could cause explosions or chemical releases. Those same Triconex systems run in food processing plants, water treatment facilities, and agricultural chemical production.

Back in 2007, Idaho National Laboratory's Aurora Generator Test demonstrated that cyber commands alone could physically destroy industrial equipment. Researchers manipulated the connection timing between a diesel generator and the electrical grid until the generator shook itself apart, spewing smoke and metal.[3] The vulnerability they exploited—improper synchronization commands through legacy protocols—persists throughout industrial control systems today, including those that sustainability technology inherits from manufacturing and energy sectors.

Between 2022 and 2024, renewable energy saw a wave of attacks. Ransomware campaigns hit solar aggregators and distributed energy resource management platforms. Attackers exploited vulnerabilities in inverters. Supervisory control and data acquisition (SCADA) telemetry manipulation allowed unauthorized grid access.[4] Industry analyses in 2025 showed attacks on energy and utilities up 40 percent year-over-year. Vertical farms run on the same SCADA systems. Carbon monitoring sensors use the same Internet of Things (IoT) protocols. Precision agriculture equipment relies on the same GPS and telemetry infrastructure. The question isn't capability. It's timing and target selection.

Who and Why

Ransomware operators see sustainability technology the way they see any operational technology: leverage for extortion. Colonial Pipeline paid $4.4 million in 2021 after an attack halted fuel distribution across the Southeast.[5] A ransomware attack on a vertical farm during critical growth phases wouldn't just create inconvenience; it would destroy crops worth millions within the time it might take to negotiate. Attackers don't need to understand agricultural sustainability. They need to understand that organizations with time-sensitive operations and high-value digital assets make attractive targets.

Nation-state actors pursue strategic objectives through infrastructure targeting. Russian-linked groups have repeatedly attacked Ukrainian energy systems. Chinese-affiliated groups, including Volt Typhoon, have established persistent access to American critical infrastructure in what officials describe as prepositioning for potential conflict.[6] Agricultural sustainability systems may seem distant from geopolitical competition, but food security is national security. Disrupting a competitor's sustainable food production, or undermining confidence in their compliance systems, provides leverage.

Hacktivists present a more complicated picture. Some might target industrial agriculture, which they view as greenwashing through technology rather than genuine transformation. Others might attack sustainability verification to undermine climate policy. Both have precedent in physical activism; digital capabilities lower the barrier while increasing potential impact. A compromised carbon registry could discredit climate markets more effectively than any protest. And competitive actors, those interested in stealing proprietary growing algorithms from vertical farms or breeding data from insect operations, require less sophistication and may be harder to attribute.

Mandates Driving Adoption

Environmental regulation in agriculture isn't new. Farmers have navigated requirements around water quality, pesticide use, and land conservation for decades. What's changed is the shift from preventing harm to proving benefit—demonstrating that practices actively contribute to environmental goals rather than merely avoiding damage. And that shift demands data.

The European Union has been the most aggressive. The Common Agricultural Policy for 2023–2027 dedicates 25 percent of direct payments—roughly €270 billion over five years—to climate and environment-friendly practices, with precision farming explicitly supported. The EU Deforestation Regulation, which took effect in December 2025 for large enterprises, requires companies to prove that commodities like cattle, cocoa, coffee, palm oil, rubber, soy, and wood haven't contributed to deforestation since December 2020.[7] The mechanism is explicitly digital: geolocation data to six decimal digits, polygon coordinates for plots over four hectares, all uploaded to a centralized Information System for cross-reference against satellite imagery. Products that can't be traced don't enter the EU market. Penalties reach 4 percent of total EU turnover.

In the United States, the trajectory has been rockier. The Biden administration's Partnerships for Climate-Smart Commodities invested $3.1 billion across 141 projects, connecting 14,000 farms to climate-smart markets and enrolling 3.2 million acres.[8] In April 2025, the Trump administration canceled the program, calling it a "climate slush fund."[9] Remaining projects were folded into a renamed "Advancing Markets for Producers" initiative with new requirements that 65 percent of funds flow directly to farmers. Thousands who'd invested in conservation practices found their reimbursements frozen mid-project.

This policy whiplash matters for security because sustainability mandates require long-term investment in technology and verification infrastructure, whereas political cycles operate on much shorter timescales. Farmers face pressure to adopt complex digital systems without assurance that the regulatory framework requiring them will persist. None of these policy frameworks adequately address who secures the resulting data. The EU Deforestation Regulation specifies geolocation requirements in exquisite detail but says almost nothing about protecting the integrity of the information system that receives it. Carbon credit methodologies focus on calculation protocols without examining the cybersecurity of the monitoring infrastructure. We've mandated digital systems without mandating their protection.

Efficiency Without Resilience

Nowhere does sustainability ambition collide with digital vulnerability more visibly than in controlled-environment agriculture. Vertical farms' indoor facilities, stacking crops in climate-controlled warehouses, represent the theoretical pinnacle of sustainable food production. Ninety-five percent less water than conventional farming. No pesticides. Urban locations to help reduce transportation emissions. Year-round production regardless of weather. Complete dependence on digital control systems. If those systems fail, crops die within hours.[10]

The past three years have witnessed a remarkable implosion across the sector. In May 2023, a pathogen called phytophthora infiltrated Bowery Farming's New Jersey facility, spreading through integrated water systems and wiping out enormous quantities of produce.[11] The company, once valued at $2.3 billion, never recovered. It ceased operations in November 2024. AeroFarms filed for Chapter 11 bankruptcy in June 2023 after raising more than $300 million.[12] AppHarvest, which had raised $700 million and gone public at a billion-dollar valuation, filed that same year. Plenty, backed by nearly $1 billion from investors including Jeff Bezos and Eric Schmidt, filed in March 2025.[13] European operations fared no better: Germany's Infarm laid off hundreds, France's Agricool entered receivership, and the Netherlands' Future Crops Cooperative declared bankruptcy.

Proximate causes varied: energy costs spiking after Russia invaded Ukraine, construction expenses exceeding projections, difficulty achieving competitive price points. But underlying these financial problems was a structural vulnerability the industry had largely ignored. When Bowery's pathogen spread, it moved through water systems designed to be efficient rather than resilient. No manual fallback, no isolation mechanism, no way to contain contamination, because the integrated design that made the facility sustainable also made it fragile. A cyberattack would exploit identical vulnerabilities because a hacker is like a pathogen. Climate control, lighting schedules, nutrient delivery, and irrigation timing are all managed by connected systems that must function continuously. Ransomware encrypting control systems during critical growth could destroy everything.

Ÿnsect, the French insect farming startup that Robert Downey Jr. promoted on late-night television, was placed into judicial liquidation in December 2025 after raising more than €600 million.[14] The company's mealworm operation never achieved viable economics—insect meal remained 2 to 10 times more expensive than soy or fish meal. But Ÿnsect's failure also illuminates digital vulnerabilities in controlled-environment protein production. Temperature, humidity, feeding schedules, and lighting cycles must all remain within narrow parameters. Breeding data and growth protocols developed over years constitute core intellectual property. A breach stealing this data wouldn't merely violate privacy; it would transfer competitive advantage. A ransomware attack disrupting environmental controls would kill not just a single crop cycle but breeding stock representing generations of selective development.

Carbon Markets and Data Integrity

C-Quest exposed problems extending far beyond one company. The voluntary carbon credit market, estimated at $2 billion and projected toward $250 billion by 2050, depends entirely on verified data.[15] Studies have found that as few as 12 percent of existing offsets create real emission reductions. MSCI found that only 5 percent of 4,000 projects meet its criteria for high integrity. In June 2023, the CFTC issued a whistleblower alert seeking tips on carbon market misconduct, fraud in spot markets, misleading methodology statements, wash trading, double counting, and manipulation of tokenized markets.[16] The agency created an Environmental Fraud Task Force. Its first action, in July 2024, resulted in an $83 million judgment against an Oregon man running a carbon-offset Ponzi scheme. C-Quest followed three months later.

For agricultural carbon markets, the implications are stark. Farm-level credits depend on monitoring systems measuring soil carbon sequestration, emissions reductions from changed practices, and verification that claimed improvements occurred. These systems are nascent, often operated by startups without extensive security infrastructure, processing data with direct financial value. The incentive to manipulate is substantial: each credit represents a claim that can be sold. If monitoring systems are compromised, whether by external attackers or internal fraud, the resulting credits appear legitimate until forensic investigation eventually exposes them. C-Quest's manipulation continued from 2019 through 2023 before employees raised concerns. Four years of fraudulent credits in the market before anyone noticed.

Where the Data Lives

Earlier in the book, we examined the unsettled legal status of agricultural data. Sustainability data carries additional complications. Environmental compliance data has regulatory consequences; who controls it may determine who can sell in certain markets. Carbon credit data has direct financial value. Supply chain traceability data reveals competitive information about sourcing relationships.

Consider where this data resides. A farmer using precision agriculture equipment generates soil data, yield data, input application records, and carbon sequestration estimates. This flows to equipment manufacturer platforms, third-party agronomic services, carbon-credit verification systems, and cloud infrastructure operated by technology companies across multiple jurisdictions. A Kansas farmer may have soil data processed through servers in Ireland, carbon calculations run in Singapore, and compliance records stored according to different privacy laws in different countries. The farmer signed service agreements authorizing this, probably without fully understanding the implications.

When sustainability technology vendors go bankrupt, which has happened repeatedly, who holds historical compliance records? Can farmers move environmental data to new platforms, or are they locked into vendor relationships by migration difficulty? The EU Deforestation Regulation requires maintaining due diligence statements and supporting documentation but says nothing about what happens if systems holding that documentation disappear.

The Small-Farm Problem

The previous chapter examined digital equity challenges for small and mid-size farms. Sustainability mandates intensify these considerably. The EU Deforestation Regulation[17] requires geolocation data that smallholders in developing countries may lack the technology to provide. Carbon-credit programs require monitoring infrastructure that small farms can't afford. Precision agriculture equipment enabling variable-rate application costs hundreds of thousands of dollars.

Large operations can afford the technology required for compliance with infrastructure regulations. They hire staff to manage data systems, negotiate with carbon-credit aggregators, and implement traceability systems that supply chain mandates demand. Small farms face the same requirements with a fraction of the resources. They invest in technology they may not fully understand, join cooperatives that provide shared infrastructure, or find themselves excluded from markets that require documented sustainability credentials. International development organizations have recognized this. In May 2025, coffee producers in Honduras exported a fully traceable, deforestation-free shipment enabled by open-source digital infrastructure, but that success required substantial technical assistance and coordination unavailable to most smallholders operating independently.[18]

Small farms adopting technology under sustainability pressure often lack the expertise to secure it. A smartphone app collecting geolocation data, a sensor monitoring soil carbon, a platform aggregating production records: each is both a compliance tool and a potential vulnerability. Small farms can't hire security staff. They depend on vendors whose security practices they cannot evaluate.

What Consumers Can Do

If carbon credits underlying a "carbon-neutral" coffee are fraudulent, most consumers will never find out. Complete independence from regulatory oversight isn't realistic. But practical steps can reduce risk.

Look for certifications from rigorous independent bodies—Verified Carbon Standard, Gold Standard, Climate Community & Biodiversity Standards—that mandate additionality (projects must demonstrate that they wouldn't have happened without carbon funding) and require third-party audits. Cross-check certifiers' track records through public registries, which should provide project details and retirement proofs. Legitimate carbon-neutral products should identify specific offset projects, methodologies, and audit reports rather than making vague assertions. Pricing far below market rates suggests corners being cut. Claims of 100% offsets without acknowledging leakage risks or impermanence defy how actual carbon projects work. Some brands now use blockchain-enabled traceability platforms that provide public ledgers to track credits from origin to retirement. QR codes linking to verifiable chains, although not foolproof, prevent double-counting through immutable records.

For Security Practitioners

If you're doing security work in food and agriculture and your organization is adopting sustainability technology, you need to know where to point your assessments. Most of what follows won't surprise you technically; the vulnerabilities are familiar. What's different is context: where these systems sit, what data flows through them, and why manipulation might go undetected for years.

Start with the control systems. Vertical farms and controlled-environment facilities inherit their industrial control architecture from the manufacturing and energy sectors. You'll find PLCs from Siemens and Rockwell Automation managing climate control, lighting, irrigation, and nutrient dosing. The protocols are what you'd expect: Modbus TCP (Transmission Control Protocol) still shows up frequently, which means unencrypted traffic and no authentication. Open Platform Communications (OPC) UA (Unified Architecture) offers better security, but implementations vary wildly in quality. HMIs running Windows are common, and they're running the same vulnerable configurations you've seen in every other industrial environment.

What makes these facilities different is the consequence timeline. In a manufacturing plant, a ransomware attack might halt production until you recover. In a vertical farm, crops begin to die within hours of an environmental control failure.

The IoT layer deserves particular attention. Carbon monitoring sensors, soil sensors, and environmental monitors typically use Message Queuing Telemetry Transport (MQTT) over Wi-Fi or Long Range Wide Area Networks (LoRaWANs), feeding edge gateways that push to cloud platforms. I've seen deployments where the MQTT broker has no authentication at all. Sensors from commodity manufacturers often ship without Transport Layer Security (TLS) being enabled and with unsigned firmware that accepts any update. If you're assessing one of these environments, check publish-subscribe configurations first. A misconfigured topic structure can let an attacker inject false readings or subscribe to data streams they shouldn't see.

For geolocation traceability systems, the kind that companies are implementing for EUDR compliance, the architecture usually involves distributed ledgers. IBM Food Trust and TE-FOOD chain GPS-tagged farm data through APIs to Hyperledger Fabric or Ethereum smart contracts. The vulnerabilities cluster in predictable places: API gateways, the Oracle feeds that bridge real-world data to blockchain, and the consensus mechanisms themselves. But don't overlook the basics. I'd check for SQL injection in traceability dashboards before worrying about smart contract exploits. The unsexy vulnerabilities are still the ones that get exploited.[19]

If your organization sources commodities covered by the EUDR, understanding this data flow matters for your risk assessment. Geolocation data originates with smallholders using GPS-enabled smartphones or Geographic Information System (GIS) applications like QField that record coordinates in raw formats (comma-separated values (CSV), Keyhole Markup Language (KML), shapefiles). That data moves to tier-one suppliers or cooperatives, often via email or USB drive, where it gets validated against satellite imagery and assigned plot identifiers. Compliance platforms then standardize everything to GeoJSON format. (GeoJSON is an open-standard format for encoding geographic data structures—such as points, lines, and polygons—using JavaScript Object Notation [JSON].) Importers embed the GeoJSON in due diligence statements and upload through API to the EU Information System.

Count the intermediaries in that chain. Typically, three to six parties handle the data before it reaches the EU system. Each handoff is a potential manipulation point. Format conversion from raw coordinates to GeoJSON could be exploited to shift polygon boundaries.

Supplier uploads over unencrypted email could be intercepted. Platform feeds that function as oracles could be spoofed to replace legitimate plot data with fraudulent coordinates. The API submission itself is vulnerable if the TLS implementation is weak. No end-to-end integrity verification governs this chain; there's no blockchain standard to ensure that the coordinates uploaded in Brussels match those captured in Ghana.

If you're assessing EUDR compliance infrastructure, focus on GeoJSON parsers (how do they handle malformed input?), supplier communication channels (is anyone still using unencrypted email for plot data?), and GNSS firmware on capture devices (susceptibility to GPS jamming or spoofing). The upstream vulnerabilities matter most because manipulation there propagates through the entire chain.

Know what normal looks like. Detection in environmental monitoring systems requires understanding baseline patterns. CO_2 sensors follow predictable rhythms: higher readings at night due to plant respiration, seasonal variation, and background noise of plus or minus 5 to 10 percent from normal calibration drift. Weather events cause legitimate spikes that correlate with other metadata, such as humidity, temperature, and barometric pressure. Industrial control traffic has its own baseline: steady polling intervals, consistent packet sizes, less than 1 percent loss under normal conditions.

Compromise looks different. Watch for statistical outliers beyond three standard deviations, especially readings that defy physics: CO_2 concentrations dropping when they should rise, or gradients showing impossible spatial patterns like decreasing concentration moving uphill in a greenhouse. Data that's too smooth, lacking the expected noise profile, often indicates that someone replaced real readings with fabricated values. Replay attacks show up as repeated historical packet sequences. Latency anomalies suggest that traffic is being proxied through something it shouldn't be. When sensors that have no physical reason to correlate suddenly show synchronized patterns, that's injection, not environment.

The C-Quest fraud evaded detection for four years partly because verification happened in silos; nobody cross-referenced reported data against independent sources. If you're monitoring agricultural carbon systems, build in cross-validation against satellite measurements. Normalized difference vegetation index data from Landsat or Sentinel can be used to sanity-check ground sensor claims about crop health and land use. Mismatches exceeding 15 percent warrant investigation. Audit chain-of-custody logs, looking for edit timestamps that cluster suspiciously—legitimate corrections don't usually happen in batches at 2:00 AM. Deploy Security Information and Event Management (SIEM) monitoring on ICS traffic, watching for volumetric changes that might indicate bulk data manipulation.

Your organization's sustainability claims ultimately flow through third-party verification systems, and their security posture affects your risk whether you control it or not. Verra, which operates the Verified Carbon Standard registry, uses risk-based Know Your Customer (KYC) procedures, sanctions screening, and transaction monitoring. Verified Carbon Units get serial numbers, enabling traceability through issuance, transfer, and retirement. What I haven't found in publicly available documentation: evidence of regular penetration testing, ISO 27001 certification, or SOC 2 Type II attestation. Third-party validation and verification bodies audit individual projects but don't examine registry-wide security controls. If Verra's administrative systems were compromised, an attacker could potentially issue fraudulent credits at scale.

The EU's EUDR Information System operates through the TRACES platform with TLS-secured API uploads, role-based access controls, and NIS2 compliance requirements. Independent oversight comes through the EU Court of Auditors or ENISA reviews, although I haven't seen results specific to EUDR systems published. Neither Verra nor TRACES employs blockchain-based immutability that would make retroactive tampering evident.

What can you do about dependencies you don't control? Request System and Organization Controls (SOC) reports or International Standards on Assurance Engagements (ISAE) 3000 attestations through procurement channels; legitimate verification bodies should be able to provide them, even if they're not publicly available. Build monitoring on your downstream API connections that watches for anomalous issuance patterns: unexpected batch sizes, credits appearing from projects you don't recognize, timing patterns that don't match normal verification cycles. You can't audit Verra's security, but you can detect when something strange comes through your connection to it.

When Something Goes Wrong with the Compliance Data

Discovering that your carbon monitoring system or sustainability compliance data may have been compromised creates problems that standard IT incident response doesn't prepare you for. These systems aren't just operational. They're tied to regulatory obligations and third-party contracts with their own disclosure requirements, and what you do in the first hours matters.

Environmental compliance data often has legal and financial implications beyond your organization. If compromised data is fed into carbon credits that were sold, or into EUDR due-diligence statements submitted to regulators, you may be looking at regulatory violations and potential fraud liability, even though you're the victim.

Your instinct will be to shut everything down, wipe affected systems, and restore from backup. Please don't do that. If you're going to have to demonstrate what happened to regulators or auditors or in legal proceedings, you need evidence intact. Disconnect compromised systems from your network, but don't wipe them. If you have IT support, have them image affected systems before any recovery work begins. If you don't have that capability, this is when you call for outside help. Forensic evidence disappears fast.

Work backward from the data: figure out what systems were affected and what compliance data flowed through them. Determine the time period. Did any of that data get submitted to regulators or verification bodies? Were carbon credits issued based on readings that might have been tampered with? The answers tell you who needs to know and how quickly.

Disclosure obligations are messier than you'd hoped. No clear global standard governs when you must report a sustainability data breach. Under EU NIS2, critical infrastructure operators may need to notify national cybersecurity authorities within 24 hours. GDPR kicks in if personal data about farmers or suppliers was compromised. Public companies with material exposure face SEC climate disclosure rules. And contractual obligations with carbon credit buyers often include breach notification clauses you agreed to and may have forgotten about.

For carbon registries specifically, Verra doesn't explicitly require disclosure of breaches but does require accurate data submissions. If compromised data resulted in credits being issued, self-reporting the inaccuracy is likely required, and credits may need to be retired or invalidated. That's painful, but it's better than having the discrepancy surface during an audit two years later.

When you're uncertain about disclosure, assume it's required. Over-disclosure is awkward. Failing to disclose when required compounds the original problem with legal exposure you didn't need.

If you're mid-verification when the incident hits, with an auditor reviewing your carbon project or EUDR submissions in progress, halt your submissions and contact them directly. Explain that you've identified a potential data integrity issue and are investigating. This feels terrible, but verification bodies have seen problems before. What damages relationships is submitting data you knew or should have known was compromised. Most frameworks include provisions for timeline extensions under extraordinary circumstances. Document when you discover the issue, what you find, and what you're doing about it.

Depending on what was compromised, you may need to invalidate carbon credits, which means either retiring them at your cost or dealing with buyers who purchased them. EUDR noncompliance can trigger penalties up to 4 percent of EU turnover, plus potential exclusion from market access. Insurance claims will require documentation of what happened and what you did in response. And if your organization has marketed itself on environmental credentials backed by data that turned out to be compromised, reputational damage may exceed the direct costs. Getting ahead of the story is usually better than having it emerge later.

Once you've contained the incident and met disclosure obligations, cross-validate your environmental data against independent sources, satellite imagery, third-party measurements, and historical baselines where possible. Determine which time periods are suspected and what can be reconstructed. Consider engaging an independent auditor to verify data integrity before resuming normal compliance reporting. If carbon credits are in question, work with the registry on remediation. Proactively retiring suspect credits demonstrates good faith and may preserve relationships you'll need going forward.

Document what you've learned. What allowed this to happen? What would have detected it earlier? Regulators and auditors will ask. Having answers ready matters.

The Missing Frameworks

No dedicated cybersecurity frameworks exist for sustainability technologies. NIST Special Publication 800-82 addresses industrial control system security generally but contains nothing specific to carbon telemetry or geolocation verification. ENISA's NIS2 implementations treat green infrastructure as critical sectors requiring supply chain audits, but threat reports focus on energy grids rather than compliance platforms.[20] The Integrity Council for the Voluntary Carbon Market's (ICVCM) Core Carbon Principles require "robust data security" without defining what that means in technical terms.

This absence confirms that practitioners are working in genuinely uncharted territory. Teams must hybridize existing frameworks by applying NIST Cybersecurity Framework 2.0 Govern and Identify functions while integrating EUDR evidentiary requirements and carbon market verification standards. Custom anomaly baselines for environmental sensor data must be developed from first principles because no standard profiles exist.

Agricultural sustainability depends on trust that environmental claims are verified. Trust that data systems work. Cyberattacks, data fraud, and system failures all erode that trust, and once it's eroded, it's difficult to rebuild. The vertical-farming companies that collapsed invested billions without investing in resilience. Carbon markets enabled C-Quest's fraud for four years. Sustainability policies proliferating across jurisdictions mandate digital systems without mandating protection. Each mistake assumes that good intentions are sufficient and that systems designed for sustainability will automatically be secure. They won't be. And the current moment offers an opportunity to build security into sustainability infrastructure before major incidents force reactive responses. That window won't stay open forever.

Technology enabling sustainable agriculture doesn't respect national borders any more than climate does. Precision agriculture platforms developed in the Netherlands serve Australian farms. Genetic data from Iowa corn varieties flows through Singaporean servers. Carbon credits generated in Brazil trade on European exchanges. Chinese-manufactured soil sensors monitor American fields. When a vulnerability in one system can cascade across continents within hours, the question of who secures the global food system becomes urgent. That's where we're headed next: how threats move through these global systems, who's responsible for stopping them, and why the answers aren't as clear as anyone would like.

Notes

1. Commodity Futures Trading Commission. (2024). "CFTC charges former CEO of carbon credit project developer with fraud involving voluntary carbon credits." **https://www.cftc.gov/PressRoom/PressReleases/8994-24**
2. Johnson, B., Caban, D., Krotofil, M., Scali, D., Brubaker, N., & Glyer, C. (2017). "Attackers deploy new ICS attack framework 'TRITON' and cause operational disruption to critical infrastructure." *Mandiant Threat Intelligence.* **https://cloud.google.com/blog/topics/threat-intelligence/attackers-deploy-new-ics-attack-framework-triton**
3. Rodriguez, J., & Gasper, P. D. (2007). *Understanding the Value of a Computer Emergency Response Capability for Nuclear Security* (Idaho National Laboratory). **https://inldigitallibrary.inl.gov/sites/sti/sti/6618301.pdf**
4. Dragos, Inc. (2026). *2025 OT Cybersecurity Year in Review.* **https://hub.dragos.com/ot-cybersecurity-year-in-review-2025**
5. U.S. Department of Justice. (2021). "Department of Justice seizes $2.3 million in cryptocurrency paid to the ransomware extortionists Darkside." **https://www.justice.gov/archives/opa/pr/department-justice-seizes-23-million-cryptocurrency-paid-ransomware-extortionists-darkside**

6. CISA. (2024). "PRC state-sponsored actors compromise and maintain persistent access to U.S. critical infrastructure." Advisory AA24-038A . **https://www.cisa.gov/news-events/cybersecurity-advisories/aa24-038a**
7. European Commission. (n.d.). "Deforestation regulation implementation." **https://green-forum.ec.europa.eu/nature-and-biodiversity/deforestation-regulation-implementation_en**
8. USDA. "Partnerships for climate-smart commodities legacy site." **https://www.usda.gov/climate-solutions/climate-smart-commodities**
9. USDA. (2025). "USDA cancels Biden era climate slush fund, reprioritizes existing funding to farmers." **https://www.usda.gov/about-usda/news/press-releases/2025/04/14/usda-cancels-biden-era-climate-slush-fund-reprioritizes-existing-funding-farmers**
10. Taylor, G. "Analysis: five reasons why vertical farming is still the future, despite recent business failures." *UCL News.* **https://www.ucl.ac.uk/news/2025/jan/analysis-five-reasons-why-vertical-farming-still-future-despite-recent-business-failures**
11. Bradbury, R. (2024). "Wilted greens: how a leaf-eating pathogen, failed deals brought down Bowery Farming." *PitchBook.* **https://pitchbook.com/news/articles/bowery-farming-collapse-leaf-eating-pathogen-failed-acquisition**
12. Szhutak, R. (2024). "Bowery Farming is ceasing operations." *TechCrunch.* **https://techcrunch.com/2024/11/04/bowery-farming-is-ceasing-operations**
13. Azevedo, M. A. (2025). "Vertical farming company plenty files for bankruptcy after raising nearly $1B." *TechCrunch.* **https://techcrunch.com/2025/03/24/vertical-farming-company-plenty-files-for-bankruptcy-after-raising-nearly-1b**
14. Heim, A. (2025). "How reality crushed Ÿnsect, the French startup that had raised over $600 m for insect farming." *TechCrunch.* **https://techcrunch.com/2025/12/26/how-reality-crushed-ynsect-the-french-startup-that-had-raised-over-600m-for-insect-farming**
15. Weinkle, J. (2024). "Is an equitable carbon offset industry possible?" *The Breakthrough Institute.* **https://thebreakthrough.org/journal/no-20-spring-2024/is-an-equitable-carbon-offset-industry-possible**
16. Commodity Futures Trading Commission. (2023). "CFTC issues alert seeking tips regarding potential carbon market misconduct." Release No. 8723-23. **https://www.cftc.gov/PressRoom/PressReleases/8723-23**
17. European Commission. (n.d.). "Deforestation." **https://environment.ec.europa.eu/topics/forests/deforestation_en**
18. SASI. (2025). "A shared infrastructure for traceable coffee: Honduras takes a bold step toward EUDR compliance." **https://www.sustainable-supply-chains.org/news/a-shared-infrastructure-for-traceable-coffee-honduras-takes-a-bold-step-toward-eudr-compliance**
19. Morpheus.Network. (2025). "EUDR compliance with blockchain traceability." **https://morpheus.network/eudr-compliance-with-blockchain-traceability**
20. European Commission. (n.d.). "NIS2 Directive: securing network and information systems." **https://digital-strategy.ec.europa.eu/en/policies/nis2-directive**

CHAPTER 31

Global Food Systems, Global Threats

Supply Chains Cross Borders: So Do Cyberattacks

In April 2022, Dmitry Medvedev posted on Telegram: "Our food is our silent weapon. Silent but formidable."[1] The former Russian president wasn't being metaphorical. Within months, Russian drones would systematically destroy Ukrainian grain infrastructure—ports, silos, storage facilities—while a cyberattack on the Viasat satellite network disabled agricultural equipment across Europe. The opening salvo of the war combined kinetic strikes on food with digital attacks on the systems that modern farming depends on.

On February 24, 2022, hours after Russian troops crossed the border, the Viasat KA-SAT network went dark. The intended target was Ukrainian military communications. But satellites don't respect borders. In Germany, 5,800 Enercon wind turbines lost remote monitoring.[2] Tens of thousands of broadband customers across Europe experienced outages. Security researchers at later found developmental links between the AcidRain wiper malware used in the attack and Russian VPNFilter malware. The EU and Five Eyes governments attributed it to Russia's GRU.[3]

The attack demonstrated something agricultural operators need to understand: satellite-dependent systems create vulnerabilities that cross borders without warning. A farmer in Poland using GPS-guided equipment, a grain elevator in Romania coordinating shipments, a processor in Germany monitoring cold storage, all exposed by an attack aimed at Ukrainian military targets. The cyber and the kinetic weren't separate campaigns. They were coordinated, and it's our new reality of warfare.

By August 2023, after Russia withdrew from the Black Sea Grain Initiative, the kinetic campaign intensified. Drones struck the port of Izmail on the Danube, destroying 40,000 tons of grain destined for Africa, China, and Israel. Over the following weeks, Russian strikes hit 26 port facilities, 5 civilian vessels, and roughly 180,000 tons of stored grain. The Black Sea Grain Initiative had moved 33 million metric tons during its 12 months of operation. Its collapse wasn't a failure of negotiation. It was a deliberate weaponization of the global food supply.[4]

Chokepoints and Concentration

Before the invasion, Ukraine accounted for 9 percent of global wheat exports, 46 percent of sunflower oil exports, and about 10 percent of corn exports. Together with Russia, these two countries supplied more than a quarter of the wheat traded worldwide. Russia also dominates fertilizer exports—the synthetic nitrogen without which modern yields collapse. When these supplies were threatened, there was no backup warehouse large enough to matter.[5]

The food system we've built optimizes for efficiency, not resilience. Just-in-time logistics minimize inventory costs but eliminate buffers. The Panama Canal, hit by drought in 2023, reduced transits and slowed food commodity flows. The Suez Canal, blocked by the *Ever Given* in 2021, showed how a single vessel could paralyze global trade; Houthi attacks in 2024 disrupted routes again. Even food that feels local depends on globally sourced inputs: GPS satellites, cloud platforms, equipment parts, and proprietary seeds developed across multiple countries.

This interconnection means cyberattacks ripple in ways that aren't obvious. Port logistics systems, shipping coordination, and commodity trading are digital. Commodity markets react to news of disruptions in real time, creating economic leverage that amplifies any successful intrusion.

Take commodity trading platforms. The Chicago Mercantile Exchange (CME) handles grain futures, contracts that let farmers hedge price risk and processors lock in costs months ahead. A grain elevator in Iowa sells futures to guarantee a price for unharvested wheat. A flour mill in Egypt buys contracts to secure supply at a known cost. The system matches buyers and sellers electronically, clears trades through central counterparties, and produces the price signals that global food trade runs on.

In August 2013, a software fault halted CME grain futures trading for several hours.[6] When trading resumed, wheat prices gapped 4 percent. American farmers had to sell on spot markets at losses while European mills paid premiums. That was a glitch, not an attack, and it lasted hours. A sustained intrusion, ransomware encrypting order books, distributed denial-of-service (DDoS) flooding trading systems, would trigger circuit breakers that halt trading entirely. Prices freeze. Open positions can't adjust. Physical hedgers face cash-flow crunches as futures diverge from spot markets. For import-dependent nations, Chatham House modeling suggests a 50 percent halt in major exporter capacity could double prices.[7]

Who Gets Hurt

When Russian missiles destroyed grain silos along the Danube, European consumers barely noticed. They have diversified supply chains and the money to absorb price increases. The immediate victims were in Africa. Twenty-five African countries import more than a third of their wheat from Russia and Ukraine.[8] Some are entirely dependent. Somalia gets all its wheat from those two countries: 68 percent from Ukraine and 32 percent from Russia.[9] Benin gets 100 percent from Russia. Egypt, the world's largest wheat importer, sources 82 percent of its wheat from Russia and Ukraine combined. After the invasion, Sub-Saharan Africa's share of Ukrainian wheat exports fell from 10 percent to 3 percent.[10] In Nigeria and Kenya, roughly 57 percent of household spending goes to food.[11] When prices spike, people don't switch to alternatives; they eat less.

The connection between food prices and political instability isn't theoretical. When the FAO Food Price Index hit an all-time high in February 2011, the Arab Spring was underway.[12] Food wasn't the sole cause; decades of grievances fueled those uprisings, but it was the accelerant.[13] In Egypt, the word for bread, *aeesh*, also means "life." Seventy percent of Egyptians depend on bread subsidies. Lebanon imports 95 percent of its grain from Russia and Ukraine; the 2020 Beirut explosion destroyed its grain silos, leaving essentially no storage capacity.[14] Tunisia sourced 80 percent of wheat from Ukraine before the war. Algeria's domestic production dropped 38 percent in 2021–2022.

Disruption to global commodity trading systems hits these populations directly. Disinformation about food availability spreads faster than contamination. The Kremlin repeatedly claimed that grain shipped under the Black Sea Initiative went primarily to the EU rather than developing countries, a falsehood designed to fracture the coalition supporting Ukraine.[15] In reality, China received nearly 8 million tons. The populations most vulnerable to food insecurity are also those with the least capacity to verify information or absorb price shocks.

When Ransomware Hits

On September 29, 2025, Asahi Group Holdings detected a ransomware attack. Asahi is Japan's largest brewer and a multinational food and beverage company; the Qilin ransomware group had encrypted data across multiple servers. Digital ordering went down. Shipment systems went down. Customer service went down. Operations at roughly 30 breweries were affected.[16]

The scope kept expanding as investigators dug in: 1.52 million customer records potentially exposed, 107,000 employee records, another 168,000 records belonging to employees' family members. Nearly 2 million people were affected. Production at multiple factories was interrupted. Beer production resumed within a week, but distribution problems persisted. Asahi's domestic beverage and food divisions saw October sales decline 10 to 40 percent compared to the prior year.[17]

CEO Atsushi Katsuki's response was blunt: "We thought we had taken sufficient measures, which were easily broken. It made me realize there's no limit to the precautions that can be taken." Asahi's European brands, Peroni and Fuller's, were unaffected because those systems were managed separately. The attack on one geography didn't cascade into others. That's a lesson in segmentation, although it's cold comfort for Japanese consumers facing empty shelves.

The Qilin ransomware infrastructure spans Russia, Hong Kong, and sanctioned networks. Whether that makes it a criminal operation with nation-state tolerance or something more directly connected is hard to determine and ultimately doesn't matter much to victims. The BlackMatter group that hit NEW Cooperative in 2021 (discussed in Chapters

5 and 7) was Russian-speaking with disputed state connections. These groups operate independently while benefiting from jurisdictions that don't pursue them.

Asahi wasn't unique. In October 2023, a ransomware attack disrupted Brazil's Port of Santos, which handles about 60 percent of the country's soybean exports.[18] An estimated million metric tons of grain shipments were delayed for weeks. Freight rates spiked 15 percent. Brazil and Argentina together supply 40 percent of global soybeans and 20 percent of traded corn; they're the buffer when Northern Hemisphere supplies falter. Brazil shipped 6.2 million metric tons of soybeans to China in the first quarter of 2026 alone.[19] If Ukrainian infrastructure can be targeted strategically, so can Santos. So can Argentina's Matba-Rofex exchange, the hub for South American grain futures.

The China Situation

China complicates any clean narrative about victims and attackers. The country imports more than $200 billion in agricultural goods annually—American soybeans, Brazilian corn, Australian barley—to feed 1.4 billion people on limited arable land. It received grain under the Black Sea Initiative. When trade war tariffs hit, soy prices surged 20 to 30 percent, forcing stockpiling and domestic rationing. China is as vulnerable to food supply disruptions as any major importer.[20]

China is also an active threat actor, and the FBI has prosecuted multiple cases of seed genetics theft. Between 2013 and 2016, Mo Hailong and accomplices dug up patented corn seeds from DuPont Pioneer and Monsanto fields in Iowa and Illinois, shipping seeds valued at $30 million to $40 million to Dabeinong Technology Group via microwave boxes and storage lockers. In a separate case, Weiqiang Zhang, a breeder for Ventria Bioscience, stole genetically modified rice seeds that represented $75 million in R&D investment; he received an 11-year sentence. These seeds represented decades of development in drought and pest resistance. Stealing these shortcuts R&D costs and erodes American agricultural advantages over time.[21,22]

Meanwhile, China has become a major agricultural technology exporter. DJI drones cover more than 173 million hectares of farmland domestically.[23] Chinese agricultural machinery exports reached $9.3 billion by mid-2025. These tools create dependencies in recipient nations while generating data about planting schedules, crop health, and operational patterns, and information with both commercial and strategic value. China isn't the only country blending import vulnerability with export leverage and intellectual property acquisition. But it's the largest, and any discussion of global food system threats that ignores it is incomplete.

The Expanded Attack Surface

Global food trade depends on infrastructure that most people never think about until it fails. Container ships are floating OT environments with navigation systems, cargo management, refrigerated container monitoring, engine controls with satellite connectivity, and minimal security oversight. When NotPetya hit Maersk in 2017, the world's largest container shipping company saw its entire IT infrastructure collapse within hours.[24] Terminals worldwide couldn't process shipments. Trucks lined up at ports with no way to load or unload. Maersk estimated $300 million in damages. For food shippers, such disruptions carry additional stakes: refrigerated containers maintaining cold-chain integrity depend on continuous power and monitoring. A shipping disruption isn't just a delay; it's spoilage.

International food trade also runs on documentation: letters of credit, bills of lading, certificates of origin, and phytosanitary certificates. These have increasingly moved from paper to digital systems. Documentary fraud has always existed, but digitization enables manipulation at scale. If someone compromises the systems that verify whether grain came from where it claims to have come from, the trust model breaks down. Bills of lading could be altered to mask sanctioned origins. Certificates of analysis could be fabricated to pass contaminated products. The International Chamber of Commerce has flagged trade finance documentary fraud as a growing concern.[25]

Sanctions compliance compounds the problem. Food companies trying to avoid buying from sanctioned entities or transshipping through prohibited routes face a digital verification challenge. Commodity tracking relies on platforms, ledgers, and trade documents that can all be manipulated. According to Treasury Department enforcement data, 72 percent of sanctions penalties in 2023 were tied to due diligence failures in verification systems.[26] The same shadow-fleet tactics that mask Russian crude oil could also mask grain origins. Hack vessel tracking data, alter certificates, or compromise screening platforms, and sanctioned entities can access global markets through falsified documentation.

What's Missing

International mechanisms for food system cooperation exist. INFOSAN, the FAO/WHO International Food Safety Authorities Network, was established in 2004 and includes more than 600 members from 188 member states.[27] It facilitates rapid information exchange during food safety emergencies. But its focus is contamination, outbreaks, and recalls—traditional food safety rather than cyber-physical threats. The Codex Alimentarius Commission sets guidelines for food safety communication but doesn't address cybersecurity at all.

There's no global equivalent of financial sector cyber information sharing for food and agriculture. The financial industry has robust mechanisms; a ransomware variant hitting one bank gets shared across the sector within hours. A ransomware variant hitting a grain processor in Australia might hit a meat packer in Brazil three weeks later, and neither company would have any formal way to share what they learned. The EU's NIS2 Directive advances requirements in Europe, but no global standard exists. A multinational food company faces a patchwork robust requirement in Europe, aspirational guidance in the United States, minimal or absent frameworks elsewhere.

Attribution challenges also hamper response coordination. Even when attacks are clearly linked to nation-state actors, diplomatic consequences rarely match operational damage. Russia faces criticism for grain infrastructure attacks, yet the strikes continue. China denies agricultural IP theft while FBI prosecutions mount. Developing nations face challenges and lack cybersecurity capacity, even as they adopt precision agriculture tools that create new exposures. A small farming cooperative in East Africa using satellite-connected irrigation has vulnerabilities similar to those of large commercial operations but lacks the resources to address them.

What Can You Actually Do?

If you're a procurement manager at a regional grocery chain or a food company sourcing from 12 countries, the preceding analysis might feel abstract. Global supply chains, nation-state attackers, and commodity market manipulation are systemic challenges no single company can solve. But there are practical steps that matter:

- *Diversify suppliers.* Identify two or three backups for each key product category. The goal isn't eliminating concentration risk, which may be impossible for some products, but reducing dependence on any single source. Companies with preestablished alternative relationships adapted faster during tariff escalations and supply disruptions than those scrambling under pressure. Audit backup suppliers before you need them urgently.
- *Build buffer stocks for high-risk imports.* You should have four to eight weeks of inventory for products from chokepoint-dependent regions. Prioritize based on risk: which products come through vulnerable routes, which have limited alternatives, which spoil fastest. Pair strategic stockpiling with just-in-time for stable items.
- *Verify foreign supplier controls.* Require certificates of analysis, conduct quarterly sampling, and ensure hazard plans align with international standards. Documentation isn't just compliance; it's evidence of due diligence when something goes wrong.
- *Monitor continuously.* Subscribe to alerts on commodity volatility, weather disruptions in key growing regions, and geopolitical developments affecting trade routes. Run scenarios: what if your primary supplier goes offline for two weeks? Build flexible pricing clauses into contracts. Document response plans so they're actionable, not aspirational.

These steps won't solve systemic problems. But companies that build resilience through diversification, buffer inventory, verification, and adaptive monitoring weather disruptions better than those that don't.

Looking Ahead

The chapters in Part IV traced forces that shape food system security beyond the technical: regulatory frameworks struggling with evolving threats, digital equity gaps leaving small farms exposed, sustainability mandates creating new dependencies without corresponding security requirements. And now the geopolitical dimension: nations weaponizing grain trade, cyberattacks as tools of statecraft, intellectual property theft eroding agricultural innovation over decades.

When Medvedev called food a "silent weapon," he was stating what military strategists have understood since antiquity. Control over food is control over populations. What's changed is the digital layer interwoven through production, processing, and distribution. A cyberattack on satellite communications disables agricultural equipment across a continent. Ransomware on a brewer creates national product shortages. A strike on port logistics strands grain destined for food-insecure populations.

Food depends on water, energy, transportation, telecommunications, and healthcare. When one sector fails, others feel it. Increasingly, they fail together with shared infrastructure, shared vulnerabilities, and attackers who've learned that hitting one can cascade into many. That's where we turn next.

Notes

1. Medvedev, D. (2022). Telegram post cited in Sebastian, C. (2023). "Russia wields 'silent weapon' of grain exports to weaken Ukraine." *CNN Business*. **https://edition.cnn.com/2023/10/05/economy/russia-wheat-exports-ukraine-war**
2. Malpas, A. (2023). "Russia hits Ukraine grain export route near Romania." *The Moscow Times*. **https://www.themoscowtimes.com/2023/08/02/russia-strikes-kyiv-and-grain-export-route-on-danube-a82029**
3. Welsh, C., & Glauber, J. (2024). "Food as the 'silent weapon': Russia's gains and Ukraine's losses." Center for Strategic and International Studies. **https://www.csis.org/analysis/food-silent-weapon-russias-gains-and-ukraines-losses**
4. CNBC. (2023). "Russia strikes Ukraine's Danube port, sending global grain prices higher." **https://www.cnbc.com/2023/08/02/russia-strikes-ukraines-danube-port-sending-global-grain-prices-higher.html**
5. Welsh, C., & Glauber, J. (2024). "Food as the 'silent weapon'."
6. Willuhn, M. (2022). "Satellite cyber attack paralyzes 11GW of German wind turbines." *pv magazine*. **https://www.pv-magazine.com/2022/03/01/satellite-cyber-attack-paralyzes-11gw-of-german-wind-turbines**
7. Wellesley, L., Preston, F., Lehne, J., & Bailey, R. (2017). "Chokepoints in global food trade: assessing the risk." *Research in Transportation Business & Management*, 25, 15–28. **https://doi.org/10.1016/j.rtbm.2017.07.007**
8. Bilger, V. (2022). "Could disruptions to Eastern European wheat spur displacement in Africa?" International Centre for Migration Policy Development. **https://www.icmpd.org/blog/2022/could-disruptions-to-eastern-european-wheat-spur-displacement-in-africa**
9. Reidy, J. (2022). "UN urges increased Black Sea ag exports." *World Grain*. **https://www.world-grain.com/articles/18671-un-urges-increased-black-sea-ag-exports**
10. Welsh, C., & Glauber, J. (2024). "Food as the 'silent weapon'."
11. Lacina, B., Heidland, T., Jävervall, S., Mahlkow, H., Mukasa, A. N., & Woldemichael, A. (2024). "Long-run impacts of the conflict in Ukraine on grain imports and prices in Africa." *African Development Review*, 36(S1), S10–S24. **https://ideas.repec.org/a/bla/afrdev/v36y2024is1ps10-s24.html**

12. Keating, J. (2014). "A revolution marches on its stomach." *Slate*. **https://slate.com/technology/2014/04/food-riots-and-revolution-grain-prices-predict-political-instability.html**
13. Tree, T. (2014). "Food insecurity and political instability during the Arab Spring." *E-International Relations*. **https://www.e-ir.info/2014/09/07/food-insecurity-and-unrest-in-the-arab-spring**
14. Cincotta, R. (2014). "High food prices an unlikely cause for the start of the Arab Spring." *New Security Beat*. **https://www.newsecuritybeat.org/2014/04/high-food-prices-arab-spring**
15. U.S. Department of State. (2023). "Russia must stop its attacks on food security." **https://2021-2025.state.gov/russia-must-stop-its-attacks-on-food-security**
16. Speed, J. (2025). "Asahi CEO says ransomware attack might have caused 1.9 million data leaks." *The Japan Times*. **https://www.japantimes.co.jp/business/2025/11/27/companies/asahi-beer-leak-presser**
17. Khaitan, A. (2025). "Asahi cyberattack forces shift to zero-trust model, delays earnings by over 50 days." *Cyber Express*. **https://thecyberexpress.com/asahi-group-cyberattack-cyber security-unit**
18. Porto de São Francisco do Sul. (2024). "Porto retoma operações em menos de 24 horas após ataque cibernético." **https://portosaofrancisco.com.br/saiba-mais/id/293**
19. USDA, Foreign Agricultural Service. (2026). *Brazil: Oilseeds and Products Update*. Report BR2025-0012. **https://apps.fas.usda.gov/newgainapi/api/Report/DownloadReportByFileName?fileName=Oilseeds+and+Products+Update_Brasilia_Brazil_BR2025-0054.pdf**
20. U.S. International Trade Administration. "China country commercial guide: agriculture." **https://www.trade.gov/country-commercial-guides/china-agriculture**
21. United States Attorney's Office, Southern District of Iowa. (2016). "Chinese national sentenced to prison for conspiracy to steal trade secrets." **https://www.justice.gov/usao-sdia/pr/chinese-national-sentenced-prison-conspiracy-steal-trade-secrets**
22. U.S. Department of Justice. (2022). "Chinese national sentenced to over 10 years for stealing trade secrets."
23. DJI Agriculture. (2024). "DJI report shows agricultural drone industry is booming." **https://www.dji.com/media-center/announcements/agricultural-drone-industry-insight-report-2023-2024-en**
24. Greenberg, A. (2018). "The untold story of NotPetya, the most devastating cyberattack in history." *Wired*. **https://www.wired.com/story/notpetya-cyberattack-ukraine-russia-code-crashed-the-world**
25. International Chamber of Commerce Germany. (2023). *Shutting Fraudsters out of Trade*. **https://www.iccgermany.de/wp-content/uploads/2023/09/Shutting_Fraudsters_Out_of_Trade_2023_FINAL.pdf**
26. U.S. Department of the Treasury. (2024). "Office of Foreign Assets Control: 2023 civil penalties and enforcement information." **https://ofac.treasury.gov/civil-penalties-and-enforcement-information/2023-enforcement-information**
27. World Health Organization. (n.d.). "FAO/WHO International Food Safety Authorities Network (INFOSAN)."**https://www.who.int/groups/fao-who-international-food-safety-authorities-network-infosan/about**

CHAPTER 32

Interconnected Risks: Cross-Sector Dependencies and Resilience

When the Satellites Went Quiet

On May 10, 2024, farmers across three continents watched their tractors drift off course. Solar eruptions had scrambled GPS signals for six hours, throwing precision guidance off by as much as 70 meters. In Nebraska, one farmer told reporters that every tractor on his operation was sitting at the field's edge, shut down.[1] Equipment wandered in circles, and rows planted during those hours were so crooked that harvest machinery would struggle to follow them months later.

The Gannon Storm, named for the late space weather scientist Jennifer Gannon, was the worst geomagnetic disturbance in two decades. Kansas State University economist Terry Griffin estimated the damage to American corn and soybean operations alone at more than $500 million due to delayed planting and lost productivity.[2] Aircraft landing systems reported errors exceeding four meters.[3] Similar losses hit operations in Canada, Germany, France, and the Netherlands, although nobody compiled a global figure.

No hacker caused this; the sun did. But the incident revealed something that most people outside the agricultural world never think about: modern food production depends on systems far beyond the farm field. Satellites. Power grids. Water utilities. Trucking networks. Cloud platforms hosting logistics software. Chemical plants making fertilizer. When any of this fails, consequences reach the food supply within hours or days. And increasingly, the systems food depends on are targets for cyberattack.

This chapter traces those dependencies across six sectors and examines what happens when failures in one domain cascade into another. The goal is practical: understanding which vulnerabilities matter most, how fast they can hurt you, and what food and agriculture organizations can do about dependencies they don't control.

The Global Stakes

Every industrialized nation has designated certain sectors as critical infrastructure, although they categorize them differently. The United States lists 16, the European Union's NIS2 Directive identifies 11, and Australia's Security of Critical Infrastructure Act names 22 asset classes.[4] Food and agriculture appear in virtually every national framework, and governments understand that food disruptions become political crises fast.

The numbers give a sense of scale. The Food and Agriculture Organization of the United Nations (FAO) estimates that global food imports reached $2.1 trillion dollars in 2024.[5] Roughly 35 percent of fisheries and aquaculture production crosses international borders between harvest and consumption.[6] Low- and middle-income countries have grown more dependent on imported calories over the past two decades, from about 60 kilocalories per capita per day in the early 2000s to 90 by 2020.[7] When agricultural supply chains break anywhere, prices move everywhere.

Agriculture's dual role as both a resource consumer and a provider makes it a powerful amplifier of risk across other critical sectors. These cross-sector dependencies merit closer examination.

Water

Agriculture accounts for roughly 70 percent of global freshwater withdrawals, according to FAO data. In South Asia, the figure reaches 91 percent.[8] India alone extracts about 230 cubic kilometers of groundwater annually, with more than 60 percent going to irrigation.[9] Growing a single bushel of corn requires approximately 4,000 gallons of water over the plant's lifecycle. Without reliable water, crops die on timelines measured in days during hot weather.

The cybersecurity dimension of water infrastructure has moved from theoretical to operational. In December 2024, Denmark publicly attributed an attack on a waterworks in Køge to a pro-Russian group called Z-Pentest. The attackers manipulated pump pressure, burst pipes, and left homes without water.[10] Physical damage was achieved through digital access.

Similar incidents have multiplied. Norway's Police Security Service attributed an August 2024 dam attack—hackers remotely opening outflow valves—to pro-Russian actors.[11] In the UK, Southern Water paid millions following a February 2024 Black Basta ransomware attack that the company initially described as affecting only IT systems; cybersecurity researchers noted that such claims often understate reality.[12] In the United States, the

Municipal Water Authority of Aliquippa, Pennsylvania, shut down OT in November 2023 after an Iranian-linked group called Cyber Av3ngers exploited vulnerabilities in Unitronics controllers at a booster station. Federal investigators found at least 10 other American water facilities that were compromised through similar methods.[13]

A December 2024 joint advisory from CISA, the FBI, the NSA, and international partners warned that pro-Russia hacktivists were actively targeting water utilities across North America and Europe.[14] The Government Accountability Office found that nearly 170,000 American water systems face cyber risks.[15] For farms and food processors that depend on municipal water or shared irrigation systems, this represents inherited vulnerability. You can run a perfect security program on your own equipment and still get hurt when your water supply is compromised.

The timeline for water disruptions hitting agriculture varies by season and crop. During drought or high heat, field crops show visible wilting within 24–48 hours without irrigation. Most corn and soybeans can survive three to seven days before yields suffer permanent damage at peak growth, but vegetables with shallow roots, such as lettuce, fail in one to two days.[16] A 72-hour irrigation outage during drought could cut yields 20–50 percent, depending on timing. These biological clocks don't care about your security posture or lack thereof.

Energy

The obvious dependency runs from energy into agriculture: tractors burn diesel, processing plants consume electricity, and refrigeration runs around the clock. But the relationship also goes the other way. Agricultural products have become significant feedstocks for energy production, creating a reverse dependency in which the energy sector needs agriculture as much as agriculture needs power.

Global biofuel production totaled roughly 175 billion liters in 2023, supporting about 3.9 million jobs and generating more than 150 billion dollars in economic activity.[17] The United States leads with corn ethanol and soybean biodiesel. Brazil follows with sugarcane ethanol. Indonesia has emerged as the world's largest biodiesel producer, driven by palm oil mandates.[18] The EU relies heavily on rapeseed and used cooking oil. Asia is also driving growth: China's production increased 15.3 percent annually in recent years, India's by 11.8 percent.[19]

What this means practically is that a sustained disruption to oilseed production anywhere affects fuel supply somewhere else. The same applies to sugar crops, corn, and the animal fats that also go into biodiesel. Latin America has become a net biodiesel exporter as Asian producers consume their own output to meet domestic mandates. The EU's Renewable Energy Directive has established binding targets that depend on continued access to agricultural feedstock. Sustained interference with that access would force difficult choices between food and fuel.

The cascade timelines for energy failures affecting food are brutal. A commercial refrigerator maintains safe temperatures for about four hours with the doors closed during a power loss. A full freezer holds for roughly 48 hours, a half-full one 24 hours.[20] Commercial facilities with backup generators can extend these windows, but backup power eventually runs out or runs dry. Every hour of delay past these thresholds risks food safety violations and mandatory disposal.

Transportation

Food moves constantly, and unlike manufactured goods, it can't wait. Delay fresh produce for two days during summer, and it arrives as compost. The scale of global food movement is staggering. Major exporters including Argentina, Australia, Brazil, Canada, Russia, and the United States combine low population density with high productivity to feed populations far beyond their borders.

In February 2022, a cyberattack on Port of Antwerp operations disrupted Zespri's kiwifruit shipments from New Zealand. The orchards kept producing. Pickers kept picking. But grading, packing, and shipping halted at the European end, stranding perishable goods.[21] The JBS ransomware attack in May 2021, examined in earlier chapters, revealed transportation cascades within national borders: when processing plants closed across North America and Australia, cattle trucks had nowhere to deliver animals, holding them in transports, and creating both welfare concerns and economic losses.[22]

Refrigerated transport operates on tight timelines. A truck carrying dairy or poultry reaches the danger zone within two to four hours if refrigeration fails. Full spoilage or sufficient bacterial growth, which triggers safety concerns, occurs in 12–24 hours.[23] Under FSMA rules, delays exceeding eight hours often require diversion or disposal rather than continued delivery.[24] For nonperishables like grain, timelines are more forgiving, but storage facilities can buffer only so much overflow before harvest operations must halt.

Information Technology and Cloud Concentration

Modern agriculture runs on data. Sensors measure soil moisture. GPS guides tractors to centimeter accuracy. Yield monitors track variations in productivity across individual acres. Supply chain platforms coordinate the movement of goods from millions of farms to billions of consumers. According to USDA data, automated guidance systems now operate on more than half of planted acreage for major American crops; adoption among large operations exceeds 70 percent.[25] European and Australian precision agriculture follow similar patterns.

The May 2024 solar storm showed what happens when this infrastructure fails. Tractors wandered meters off course. One John Deere dealership chain warned customers that systems were "extremely compromised" and that crops planted during the disruption would create problems at harvest.[26] The Viasat KA-SAT attack on the day of Russia's invasion of Ukraine demonstrated that agricultural connectivity can become collateral damage in conflicts farmers never anticipated: German wind turbines lost remote monitoring, and farm operations across Europe using the satellite service experienced outages.[27]

But the solar storm and satellite attack point to a more fundamental vulnerability that gets less attention: much of the digital infrastructure supporting food, water, energy, and transportation runs on the same handful of cloud providers. Amazon Web Services, Microsoft Azure, and Google Cloud collectively control more than 65 percent of the global cloud computing market.[28] These providers operate massive data centers, physical facilities housing thousands of servers, often clustered in specific geographic regions for latency and cost reasons. A food processor's enterprise resource planning system, a water utility's billing platform, a trucking company's fleet management software, and an energy provider's customer portal may all run in data centers within the same region, sometimes the same facility.

To be clear, not everything lives in the cloud. OT, PLCs running processing lines, the SCADA systems monitoring cold storage, the sensors tracking fermentation tanks, mostly remain on-premises, and for good reason. These systems often can't tolerate the latency of cloud connections, and many organizations keep them air-gapped or segmented from Internet-facing networks. But the business systems that coordinate operations increasingly do run in the cloud: ERP platforms, supply chain management, customer relationship databases, and financial systems. When those fail, the OT may keep running, but nobody can ship product, pay suppliers, or know what inventory they have.

Major AWS outages have simultaneously disrupted logistics tracking for food transporters, monitoring dashboards for water utilities, and IoT platforms for dairy cold-chain management.[29] When a cloud region fails, organizations operating in completely different sectors discover they share the same vulnerability at the worst possible moment. A food processor asking whether its cloud provider is reliable rarely considers whether that same provider hosts the processor's water utility's monitoring systems and its trucking company's dispatch software. The interconnection becomes visible only during failures.

Addressing this requires thinking at the systems level. Organizations can distribute critical workloads across multiple providers or maintain on-premises backup for essential functions. Regional failover planning should map which availability zones host not just your systems but those of critical upstream dependencies. At the policy level, advocates can push for treating major cloud providers as meta-infrastructure requiring enhanced scrutiny.

Healthcare, Pharmaceuticals, and Chemical Inputs

The healthcare connection to agriculture extends beyond the obvious fact that hospitals serve meals. Agricultural products serve as raw materials for pharmaceutical manufacturing in ways most patients never realize. Soy lecithin, a phospholipid mixture extracted from soybeans, functions as an emulsifier in drug delivery systems, helping medications become bioavailable. The global market for organic soy lecithin is projected to exceed $712 million by 2027, with roughly 11 percent for medical applications.[30] The economics favor agricultural sources: soy-derived phospholipids cost about €196 per kilogram versus nearly €980 for egg-derived alternatives.[31]

Eggs themselves connect food and healthcare. Global production reached roughly 91 million tons in 2023, about 1.7 trillion eggs, with eggs serving not just as food but also as a substrate for vaccine production.[32] When highly pathogenic avian influenza swept through poultry operations globally, with the United States losing more than 173 million chickens since 2022, the implications extended beyond egg prices to pharmaceutical manufacturing capacity.[33]

The chemical sector enables agriculture at its current scale. Synthetic fertilizers provide the nitrogen, phosphorus, and potassium plants require, and the geopolitics of fertilizer have become impossible to ignore since Russia invaded Ukraine. Russia holds significant shares of global fertilizer export markets, and sanctions and supply disruptions drove prices from a peak of $815 per ton in April 2022 to roughly $437 per ton by May 2025.[34] These swings rippled through planting decisions worldwide.

Chemical manufacturing facilities run ICSs similar to those in food processing and water treatment. If fertilizer production halted during planting season anywhere in the global supply chain, crops in importing nations would go unfed. The dependency runs in both directions: agriculture provides raw materials that the chemical industry transforms into solvents, adhesives, and even firefighting foams—soy-based formulations have emerged as PFAS-free alternatives adopted by fire departments from North America to Australia.[35]

Overlooked Corners of the Food System

When we discuss cybersecurity in food systems, certain sectors are often overlooked because they don't fit into the usual categories. Pet food manufacturing might seem peripheral, but the FDA regulates it under the same framework as human food, and many pet food facilities use ingredients from the same supply chains that feed humans.[36] An attack compromising a pet food plant's sanitation control could affect human food operations at the same facility.

Food banks and charitable distribution operate on shoestring budgets, volunteer labor, and donated technology of uncertain provenance. They're essential during natural disasters and economic crises, exactly when communities are most vulnerable. A ransomware attack during a disaster would be catastrophic for populations with no alternative, yet food banks rarely appear in discussions of food system cybersecurity.

Institutional food service faces a challenge specific to its setting: captive populations. When the hospital kitchen goes down, patients can't walk to a restaurant. When a school nutrition program gets disrupted, children don't have other options for lunch. Prisons, military bases, nursing homes, same dynamic. Dietary restrictions and medical nutritional needs tracked through digital systems create genuine safety risks when those systems fail.

Ghost kitchens and delivery-only restaurants were built from the ground up on digital platforms. They've never operated without apps and software. There's no paper process to fall back on, no institutional memory of how things worked before computers. Platform dependency is existential for these operations, yet many are launched without considering what happens when technology fails.

Deliberate Cross-Sector Targeting

The cascade effects just described might seem like unintended consequences—collateral damage from attacks aimed elsewhere. Evidence increasingly suggests otherwise. Sophisticated threat actors have started treating cross-sector dynamics as features to exploit.

China's Salt Typhoon campaign in 2024 compromised telecommunications infrastructure across more than 20 countries.[37] The operation targeted systems providing surveillance and call data across government, energy, and transportation simultaneously. Leaked assessments revealed target lists that prioritized what analysts called "high-impact nodes with verified downstream effects." The attackers had modeled how compromising shared communications would cascade into intelligence opportunities across multiple sectors.

Russia's TRITON malware, first discovered in 2017 after it targeted petrochemical safety systems, selected entry points through Schneider Electric equipment used not just in petrochemicals but also in water treatment and food processing.[38] The attackers understood that compromising a common vendor provided pathways into multiple sectors. The Cl0p group's 2023 MOVEit exploit targeted a file transfer tool known to be ubiquitous across energy, healthcare, and logistics. Post-incident analysis suggested that attackers had conducted reconnaissance on MOVEit's client base before launching, selecting the target specifically for its cross-sector footprint.[39]

These patterns shift how defenders should model threats. Assessing each sector in isolation misses how adversaries think. Security professionals need to identify shared chokepoints—common cloud platforms, industrial control protocols, software dependencies—that attackers might exploit to achieve effects across sectors simultaneously.

Cross-Sector Coordination Mechanisms

Several information sharing and analysis centers (ISACs) touch the global food system, each focused on different domains: the Food and Agriculture ISAC, BIO-ISAC, Water ISAC, Electricity ISAC, IT-ISAC, and others covering transportation, healthcare, and downstream sectors. These organizations exist to share threat intelligence among members within their respective sectors. Some coordinate across sector boundaries more readily than others, and membership tiers determine what level of intelligence access organizations receive. Smaller food processors, family farms, and regional utilities often find the membership fees difficult to justify, although trade associations sometimes provide pooled access for their members.

International coordination typically occurs through multilateral forums, Organisation for Economic Co-operation and Development (OECD) frameworks, NATO task forces, and the European Union Agency for Cybersecurity (ENISA), rather than direct operational channels. These enable policy alignment and strategic coordination, but incident response during active events still depends heavily on relationships built before crises occur.

The practical implication: coordination mechanisms exist and provide value to members, but organizations shouldn't assume that these structures will automatically protect them. Knowing the security contact at your water utility or your primary logistics provider matters at least as much during an incident as having access to sector-wide alerts.

Practical Steps for Cross-Sector Resilience

The foregoing survey of vulnerabilities leads to an obvious question: what can a food or agriculture organization do about dependencies it doesn't control? Several approaches have practical values.

Dependency mapping sounds bureaucratic, but it matters. Most organizations have never documented what external systems they rely on. ENISA offers a free Dependency Mapping Tool aligned with ISO standards.[40] The OECD recommends a four-step process: inventory assets, survey suppliers, score criticality, visualize interdependencies.[41] Start with tier-one relationships: direct providers such as municipal water, electricity, cloud platforms, and logistics companies. For each, document the provider, service type, recovery time objective, and emergency contact. Extend to tier two (dependencies of dependencies) only where cascades would amplify risk: shared cloud regions, single-source suppliers, geographic concentrations. A reasonable threshold is the top 20 percent of dependencies by risk score.

Recognizing when an upstream failure might be a cyberattack rather than a routine disruption helps you respond appropriately. Routine issues typically follow predictable patterns: maintenance windows, weather events, equipment wear, and utilities communicating promptly about causes. Cyber incidents often look different: sudden service drops without prior notice, pressure or flow fluctuating erratically rather than degrading steadily, software errors that persist despite restarts and affect only certain functions. Watch for communication breakdowns where official word comes hours late, using vague "technical issues" language, while social media rumors surface first. Cross-check with peers in the same region; cyber cascades tend to hit multiple organizations simultaneously, whereas routine failures are usually isolated. If something feels wrong, ask directly regulated utilities, which may confirm a cyber event under public accountability requirements even if they can't share details.

Engaging upstream providers feels impossible for small operators, but the power asymmetry isn't absolute. Many utilities convene customer advisory panels. Regional emergency management offices host resilience forums. State-level infrastructure working groups often invite input from the food industry. These venues provide neutral ground to raise cybersecurity concerns. Direct questions work too, especially if framed as mutual risk reduction rather than accusations: "What standards do you follow for SCADA security?" Industry associations can amplify individual voices; the Food Industry Association and state farm bureaus negotiate on behalf of members, pooling leverage that no single dairy possesses.

Contracts represent an underutilized lever. Incident notification clauses can require providers to report breaches affecting your operations within 24–72 hours, including root cause and remediation timeline. Security attestation requirements can mandate SOC 2 reports, ISO 27001 certification, or alignment with CISA's Cyber Performance Goals. Audit rights can reserve the right to periodically assess provider systems. Trade associations like the International Dairy Foods Association provide template language.[42] Utilities under regulatory pressure often accept such clauses more readily than unregulated suppliers.

Build relationships with peers who do this work in your sector and adjacent ones. Go to conferences—not the big vendor showcases, but practitioner-focused gatherings where people discuss operational realities. Get to know the ICS and OT security professionals working in water, energy, and transportation; their challenges mirror yours more than you might expect. Community groups can be as valuable as formal structures. BEER-ISAC, for instance, isn't organized like a traditional information-sharing organization; it's a community of ICS and OT professionals who support each other through discussion, shared learning, and informal information exchange.[43] These relationships become invaluable when you're facing a 2:00 AM incident and need to know whether the problem is yours, your provider's, or something hitting the whole region.

The Web Beneath the Plate

The connections traced in this chapter span continents. Water flows through irrigation systems and becomes grain that becomes oil that becomes fuel that moves trucks that deliver food. Each transformation may occur in a different country. Each involves technology and infrastructure spanning multiple sectors.

The global food system's attack surface extends far beyond farms and processing plants. It includes water utilities, power grids, transportation networks, cloud platforms, and chemical facilities, regardless of which flag flies over each. Understanding this isn't cause for despair. Recognizing dependencies is the first step toward managing them. Systematic mapping reveals concentration risks before failures do. Relationships with upstream providers, built before crises occur, not during them, accelerate response when something breaks. Contract language extends influence beyond your own walls, and manual backup procedures preserve options when the screens go dark.

The 2024 solar storm knocked tractors offline across three continents. Within days, operations resumed. Some farmers went back to visual planting methods their grandparents would have recognized. The season proceeded, and the harvest arrived on schedule. That resilience came from somewhere: knowledge about how systems work, preparation for disruptions that might never come, and accumulated judgment about what matters when everything goes sideways. Cyber threats are newer than solar storms, but they reward the same habits. The organizations that come through incidents intact will be the ones that understand their dependencies before those dependencies fail.

But understanding dependencies is only part of the picture. The people who operate these systems, who make decisions under pressure, who notice anomalies before they become crises, who communicate across organizational boundaries when something goes wrong, matter at least as much as the technical infrastructure. The next chapter examines how to build teams and cultures capable of responding when cross-sector failures arrive at their door.

Notes

1. Griffin, T. (2024). Quoted in Koebler, J. (2024). "Solar storm crashes GPS systems used by some farmers, stalling planting." **https://www.nytimes.com/2024/05/13/us/solar-storm-tractor-break-nebraska.html**
2. Griffin, T. (2024). "How will the GPS outage on May 10 affect US farm profitability?" *Farmdoc Daily*. May **https://farmdocdaily.illinois.edu/2024/05/how-will-the-gps-outage-on-may-10-affect-us-farm-profitability.html**
3. Massimini, V. (2024). "Sunspots or age spots? *SatNav News*, 78(Summer). **https://www.faa.gov/sites/faa.gov/files/SatNavNews_vol78_Summer_2024_Final.pdf**
4. European Union. (2018). "The NIS 2 Directive, Final Text. **https://www.nis-2-directive.com/NIS_2_Directive_Preamble_1_to_10.html**
5. FAO. (2024). *Food Outlook: Biannual Report on Global Food Markets*. **https://openknowledge.fao.org/items/d7e220b8-b60d-433d-be1a-d5938b83fe9a**
6. FAO. (2024). *The State of World Fisheries and Aquaculture 2024*. **https://www.fao.org/publications/fao-flagship-publications/the-state-of-world-fisheries-and-aquaculture/en**
7. FAO. (2024). *Food Outlook*.
8. FAO. (n.d.). "AQUASTAT - FAO's global information system on water and agriculture." **https://www.fao.org/aquastat/en/overview/methodology/water-use**
9. International Groundwater Resources Assessment Centre. (n.d.). **https://www.un-igrac.org**
10. Paganini, P. (2025). "Russia was behind a destructive cyber attack on a water utility in 2024, Denmark says." *Security Affairs*. **https://securityaffairs.com/185885/hacking/russia-was-behind-a-destructive-cyber-attack-on-a-water-utility-in-2024-denmark-says.html**
11. Norwegian Police Security Service (PST). (2025). Public statement quoted in Adomaitis, N. (2025). "Norway spy chief blames Russian hackers for dam sabotage in April." *Reuters*. **https://www.reuters.com/technology/norway-spy-chief-blames-russian-hackers-dam-sabotage-april-2025-08-13**
12. Jones, C. (2024). "Southern Water takes the fifth over alleged $750K Black Basta ransom offer." *The Register*. **https://www.theregister.com/2025/02/25/southern_water_black_basta_leak**
13. CISA. (2023). "IRGC-affiliated cyber actors exploit PLCs in multiple sectors, including U.S. water and wastewater systems facilities." **https://www.cisa.gov/sites/default/files/2023-12/aa23-335a-irgc-affiliated-cyber-actors-exploit-plcs-in-multiple-sectors-1.pdf**
14. National Security Agency. (2025). "NSA, FBI, and others call out pro-Russia hacktivist groups targeting critical infrastructure." **https://www.nsa.gov/Press-Room/Press-Releases-Statements/Press-Release-View/Article/4355881/nsa-fbi-and-others-call-out-pro-russia-hacktivist-groups-targeting-critical-inf**

15. U.S. Government Accountability Office. (2024). *Critical Infrastructure Protection: EPA Urgently Needs a Strategy to Address Cybersecurity Risks to Water and Wastewater Systems.* Report GAO-24-106744. **https://www.gao.gov/products/gao-24-106744**
16. USDA. (2021). "Sensor-based water stress assessment for high-throughput plant phenotyping and optimized water management." **https://www.nal.usda.gov/research-tools/food-safety-research-projects/sensor-based-water-stress-assessment-high-throughput**
17. International Energy Agency. (2024). *Renewables 2024.* **https://www.iea.org/reports/renewables-2024**
18. Statista. (2024). "Leading biodiesel and renewable diesel producers worldwide in 2023, by country." **https://www.statista.com/statistics/271472/biodiesel-production-in-selected-countries**
19. International Energy Agency. (2024). "Tracking Biofuels Supply." **https://www.iea.org/energy-system/low-emission-fuels/biofuels**
20. Garcia, J. (2021). "Avoid foodborne illness during temporary power outages." *USDA.* **https://www.usda.gov/about-usda/news/blog/avoid-foodborne-illness-during-temporary-power-outages**
21. Fresh Plaza. (2022). "Cyber attack at port companies affecting Zespri's kiwifruit shipments." **https://www.freshplaza.com/north-america/article/9399298/cyber-attack-at-port-companies-affecting-zespri-s-kiwiifruit-shipments**
22. BBC. (2021). "Meat giant JBS pays $11m in ransom to resolve cyber-attack." **https://www.bbc.com/news/business-57423008**
23. FDA. (2025). "FSMA final rule for preventive controls for human food." **https://www.fda.gov/food/food-safety-modernization-act-fsma/fsma-final-rule-preventive-controls-human-food**
24. Code of Federal Regulations. (2015). "Current good manufacturing practice, hazard analysis, and risk-based preventive controls for human food." **https://www.ecfr.gov/current/title-21/chapter-I/subchapter-B/part-117**
25. Ebel, R., & Schimmelpfennig, D. (2011). "The information age and adoption of precision agriculture." *USDA, Economic Research Service.* **https://www.ers.usda.gov/amber-waves/2011/december/the-information-age**
26. LandMark Implement advisory. (2024). Quoted in Fuller, S. L., & Zimmerman, S. (2024). "Geomagnetic storm scrambles tractor navigation systems." *Agriculture Dive.* **https://www.agriculturedive.com/news/geomagnetic-storm-scrambles-tractor-navigation-systems/716175**
27. Viasat. (2022). "KA-SAT network cyber attack overview." **https://www.viasat.com/perspectives/corporate/2022/ka-sat-network-cyber-attack-overview**
28. Synergy Research Group. (2026). "Market intelligence for emerging IT & cloud." **https://www.srgresearch.com**
29. Keckarovska, V. (2025). "AWS outage disrupts logistics platforms." *Transport Intelligence.* **https://ti-insight.com/briefs/aws-outage-disrupts-logistics-platforms**
30. Allied Market Research. *Organic Soy Lecithin Market (2024 - 2033).* **https://www.alliedmarketresearch.com/organic-soy-lecithin-market-A62619**
31. Wu, Y. (2002). *Soybean Lecithin Composition, Fractionation, and Functionality.* PhD dissertation, Iowa State University. **https://dr.lib.iastate.edu/server/api/core/bitstreams/5c8ecf42-8fe6-4e34-b1b4-ea423bf11695/content**
32. FAO. (2023). "Agricultural production statistics 2010–2024." **https://www.fao.org/statistics/highlights-archive/highlights-detail/agricultural-production-statistics-2010-2024/en**

33. USDA, Animal and Plant Health Inspection Service. "Confirmations of highly pathogenic avian influenza in commercial and backyard flocks." **https://www.aphis.usda.gov/livestock-poultry-disease/avian/avian-influenza/hpai-detections/commercial-backyard-flocks**
34. World Bank. (2026). "Commodity markets." **https://www.worldbank.org/en/research/commodity-markets**
35. SoyBiobased.org. "New firefighting SoyFoam™ TF 1122 is First and only GreenScreen certified Gold™." **https://www.soybiobased.org/success-stories/new-firefighting-soyfoam-is-first-and-only-greenscreen-certified-gold**
36. FDA. (2022). "FDA's regulation of pet food." **https://www.fda.gov/animal-veterinary/animal-health-literacy/fdas-regulation-pet-food**
37. Industrial Cyber. (2024). "Cantwell seeks Mandiant records amid concerns Salt Typhoon still embedded in US telecom systems." **https://industrialcyber.co/critical-infrastructure/cantwell-seeks-mandiant-records-amid-concerns-salt-typhoon-still-embedded-in-us-telecom-systems**
38. FBI. (2022). *TRITON Malware Remains Threat to Global Critical Infrastructure Industrial Control Systems (ICS).* **https://www.ic3.gov/CSA/2022/220325.pdf**
39. National Cyber Security Centre. (2023). "MOVEit vulnerability and data extortion incident." **https://www.ncsc.gov.uk/information/moveit-vulnerability**
40. ENISA. (2020). "ENISA publishes a tool for the mapping of dependencies to international standards." **https://www.enisa.europa.eu/news/enisa-news/enisa-publishes-a-tool-for-the-mapping-of-dependencies-to-international-standards**
41. Organisation for Economic Co-operation and Development. (2019). *Good Governance for Critical Infrastructure Resilience.* **https://www.oecd.org/content/dam/oecd/en/publications/reports/2019/04/good-governance-for-critical-infrastructure-resilience_7d5a9993/02f0e5a0-en.pdf**
42. International Dairy Foods Association. (2020). "Cyber security - where do we start?" **https://www.idfa.org/resources/cyber-security-where-do-we-start**
43. BEER-ISAC. (n.d.). **https://beer-isac.org**

PART V

The Way Forward

CHAPTER 33

Culture, Communication, and Cyber-Aware Teams

It Starts with People, Not Tech

A few years ago, a mid-sized food processing facility made a significant investment in cybersecurity technology. Firewalls, intrusion detection, encrypted communications: the budget was substantial, and management felt confident that they had addressed the digital threat. Then reality set in. Workers propped open the server room door because it got too hot. Passwords were shared openly because the system felt too complicated to manage individually. Employees clicked on phishing emails because nobody had ever explained what phishing was. When an incident eventually occurred, all that expensive equipment couldn't compensate for the human gaps that had been there from the start.

Now consider a small dairy operation in the American Midwest, maybe 50 cows, run by a farmer who never studied information technology. He locks his phone. Uses different passwords for different accounts. Taught his teenage kids to recognize suspicious texts. Not because anyone required it or some auditor was checking, but because that's how we protect what's ours. The difference between these two operations has nothing to do with budget or sophistication. The difference is culture.

This chapter is about building that culture. After everything we've explored, ransomware attacks on grain cooperatives, vulnerabilities in precision agriculture, insider threats, fragile cold chains, one truth keeps surfacing: sophisticated technology cannot protect an organization whose people don't understand why security matters or how to practice it.

Why Culture Eats Technology for Breakfast

The food industry learned this lesson painfully with food safety. For decades, regulators and companies focused on checklists, inspections, and compliance requirements. Facilities passed audits. Paperwork was immaculate. Foodborne illness outbreaks kept happening anyway. What finally changed wasn't the introduction of more rules. It was recognizing that culture, the shared values and norms that shape how people behave, matters more than any checklist ever could.

Here's the insight cybersecurity professionals need to absorb: cybersecurity is not a new culture to impose on food operations, it's an extension of the safety culture that already exists. The same principles that keep pathogens out of the product can keep attackers out of the system. The same communication habits that ensure contamination gets reported immediately can ensure a suspicious email gets flagged. The same leadership commitment that makes food safety nonnegotiable can make cybersecurity nonnegotiable.

The problem is that many in the food and agriculture sector haven't made that connection yet. A 2025 study published in *ScienceDirect* examining cybersecurity attitudes in Canada's agri-food sector found that farmers generally view cybersecurity as a low priority, "in stark contrast to experts who perceive significant and growing vulnerabilities."[1] The cultural disconnect runs deep: farmers see technology as tools to get work done, not as attack surfaces needing defense.

I've heard this firsthand. While interviewing a wheat farmer for my podcast, I asked whether he ever thought about cybersecurity. His answer surprised me: "I do, and I think other farmers also have it in the back of their minds that we need to do something because we know we are at risk. But we don't have that expertise, we don't know where to start." That's not apathy. That's a gap between awareness and action. Culture is the bridge.

The good news is that the foundation already exists. Farmers and food workers have managed risk for generations. They understand that prevention beats reaction, that attention to detail matters, and that complacency kills, whether the threat is bacteria, pests, or market volatility. Cybersecurity fits naturally into that framework. It just needs to be recognized as belonging there.

The Psychology of Security: Why Shame Backfires

When something goes wrong, whether contamination events or cybersecurity incidents, organizations face a choice in how they respond. Many instinctively reach for blame. Someone clicked the wrong link. Someone left the door open. Someone didn't follow procedure. Find the person responsible. Make an example of them.

This approach feels intuitively satisfying, like natural justice. But research in psychology and organizational behavior tells a different story. A 2021 study published in the proceedings of the New Security Paradigms Workshop examined what happens when organizations respond to cyber incidents with blame and shame. The findings were stark: "Shame as a result of blaming can lead to self-protective actions and ultimately result in a downward spiral," the researchers found, "while a response of understanding, resulting in guilt, from the employer can lead to self-acceptance, action, and learning behaviors."[2]

Shaming employees who trigger security incidents doesn't make organizations more secure. It makes them less secure. Workers who fear punishment learn to hide mistakes rather than report them. They become defensive rather than collaborative. The organization loses the chance to learn from incidents because incidents go unreported. The researchers concluded that organizations responding supportively to those who cause incidents retain employees who gain "invaluable learning experience," whereas shame "reduces self-efficacy"—particularly damaging in cybersecurity, where confidence in recognizing and responding to threats matters enormously.[3]

This finding carries weight in agricultural communities. Farmers already shoulder enormous stress. Market uncertainty, weather events, labor shortages, regulatory burden, relentless physical demands: these take a cumulative toll. The American Farm Bureau Federation has documented that farmers have among the highest suicide rates of any profession in the United States, with 91 percent reporting that financial issues impact their mental health.[4] Adding the shame of a cyber incident to these existing pressures isn't just counterproductive for security; it can be genuinely dangerous.

Farming communities are tight-knit, and word travels fast. Public shaming doesn't just affect one farmer; it silences an entire region. If admitting you clicked a bad link means becoming the cautionary tale everyone whispers about at the co-op, people stay quiet. Attackers benefit enormously from that silence because it prevents the sharing of information that could protect neighbors from the same threat.

What works? The food safety world figured this out years ago with near-miss reporting systems borrowed from aviation. Report something that almost went wrong, and you're thanked for the information rather than punished for the close call. The same principle applies to cybersecurity. "Thank you for telling us—here's how we'll protect you while we fix this" creates a fundamentally different culture than "How could you let this happen?"

Who's Really Part of Your Team?

We talk about training "employees," but who operates and interacts with our systems? The reality of modern food and agriculture is that the workforce is far more complex than any org chart suggests.

According to the Federal Reserve Bank of Kansas City, about 70 percent of workers in the farm sector are foreign-born, with many holding visas or lacking documentation.[5] The U.S. Department of Labor's findings for 2019–2020 report that 63 percent of crop workers were born in Mexico, 30 percent in the mainland United States or Puerto Rico, 5 percent in Central America, and 2 percent elsewhere.[6] High turnover is the norm, especially in seasonal operations. Workers may not be present long enough for traditional training programs to stick. Language barriers are significant. Power dynamics create reluctance to report problems.

These are the people operating sorting equipment, packing lines, and harvest machinery. If ransomware hits during harvest and the seasonal crew can't use the equipment, what happens? If a worker notices something strange on a system but doesn't feel comfortable reporting it because they fear drawing attention to their immigration status, what then?

Practical responses exist. Training materials can be multilingual—not just translated word-for-word but culturally adapted to resonate with the audience. Visual and pictorial security guides work across language barriers. Quick-reference cards that stay with equipment provide just-in-time reminders. Training that accounts for high turnover needs to be short, repeatable, and focused on essentials rather than comprehensive but forgettable. Most importantly, reporting channels need to feel genuinely safe.

Seasonal workers aren't the only invisible members of the security team. Truck drivers log into logistics systems from rest stops over public Wi-Fi, use personal devices for work, and sometimes work for multiple companies simultaneously. They're rarely included in security training, yet they access systems tracking shipments worth millions. Food service staff in restaurants, cafeterias, and institutional kitchens deal with high turnover, young workforces, and password sharing as standard practice. Point-of-sale systems, inventory management, and scheduling platforms are all accessed by people who've never received a minute of security awareness training.

Consider hospital kitchens, where patient allergies are tracked digitally. A cyber incident can be a patient safety crisis. Those kitchen staff may not think of themselves as part of a cybersecurity perimeter, but they're handling systems where errors could harm vulnerable people. Cleaning and maintenance crews have after-hours access to facilities, know the door codes, follow security routines, and may work for contractors rather than being directly employed by the facility. They rarely see themselves as part of "security," yet they have physical access that could be exploited.

And consider the farmer's spouse and family. Especially in smaller operations, the spouse often manages finances, handles purchasing, and makes technology decisions. They use household technology that connects to farm networks. Yet they may be excluded from the "shop talk" where informal security tips are shared. The attack surface extends in every direction, including people who may not realize they're part of the system.

Speaking to All Decision-Makers

The 2022 Census of Agriculture, conducted by the U.S. Department of Agriculture's National Agricultural Statistics Service, tallied 1.2 million female producers, accounting for 36 percent of all producers in the country. Fifty-eight percent of all farms had at least one female decision-maker. Women were most involved in day-to-day record-keeping and financial management decisions, with 870,000 women reporting responsibility in that category.[7]

These numbers matter for security culture. Women increasingly make technology and purchasing decisions on agricultural operations. They often manage both farm and household technology, which, in a modern operation, may share a network infrastructure. Yet vendor communications frequently default to male-oriented messaging. Women may be talked down to or dismissed by technology salespeople. They may be excluded from informal networks where security tips circulate.

The implication for training and resources is straightforward: explicitly include all decision-makers in security communications. Ensure that vendor relationships respect everyone involved in purchasing and implementation. Recognize that "the farmer" may not look like the stereotype, and security awareness programs built on assumptions about who runs operations will miss significant portions of their intended audience.

The Generational Knowledge Gap

A generational transition is underway across agriculture. As we mentioned previously, the average age of farmers in the United States is 58. Many operations are passing from one generation to the next, and that handoff carries implications for cybersecurity running in both directions.

Older operators learned farming before GPS, smartphones, and cloud software. They know how to run things manually when systems fail. They can operate without the technology that younger farmers have never worked without. But they may not fully understand the systems they're now using, or the risks those systems create. Passwords on sticky notes, equipment that hasn't been updated in years, the assumption that rural isolation provides protection: these are artifacts of a generation that adopted technology without always adopting the security practices that should accompany it.

Younger operators face the opposite challenge: tech-fluent, but potentially blind to what happens when technology fails. They may connect everything to the Internet without understanding the implications. They never learned manual fallback procedures because, in their experience, why would they need them? Connectivity and reliable systems are their baseline, not advantages that can be taken away.

The danger lies in the transition itself. A retiring farmer hands over equipment with passwords on sticky notes and systems that haven't been updated in a decade. Or a tech-savvy young farmer takes over and "modernizes" everything without understanding what their parents knew about resilience and redundancy. Critical knowledge gets lost in both directions.

Resilient handoffs look like this: documenting manual fallback procedures while people who know them are still around, including "what to do when systems fail" in succession planning, ensuring that both generations understand digital and analog operations, and treating institutional knowledge as an asset worth preserving.

Starting the Conversation

So how do you bring up cybersecurity in settings where it's never been discussed? The key is framing it as a practical, shared concern, the same way farming communities already swap tips on equipment, weather risks, or theft prevention.

Text message scams provide a natural opening. Partners talk about them. A weird text someone received becomes a teachable moment over dinner or coffee. At the farm table, ease in with a relatable hook: "Hey, I was reading about how thieves are hitting farms for tractors and tools. Did you see that our neighbor down the road lost their GPS unit? We should chat about simple stuff like locking the shop better or adding motion lights around the barn. What about online, too? I heard hackers are going after farm software now; maybe we double-check passwords on the ag apps tonight?" This mirrors real farmyard security talk—lighting, locks, cameras—and invites input without jargon.

In a plant or facility setting, add it to the safety meeting agenda as "team resilience," linking physical and cyber risks to production uptime: "Quick add to safety: We've all swapped shop talk on fence fixes or livestock checks to keep operations smooth, same idea for cyber stuff. Recent farm thefts show we need eyes on equipment; online, it's phishing or weak logins hitting supply chain software. Let's brainstorm: Who rotates password checks on scales or IoT sensors? Any close calls lately we can learn from?" This fosters ownership and ties cybersecurity to what people already care about.

Spotting a Slipping Culture

Most operations fall somewhere between the facility with propped-open doors and the dairy farmer who locks his phone. How do you know where you stand? Warning signs often start small but compound quickly.

Watch for frequent physical shortcuts, propped doors, unlocked gates, despite known theft or contamination threats. Cyber hesitancy follows the same pattern: shared passwords on ag software, ignored phishing alerts. When team members avoid raising concerns about equipment glitches or suspicious emails because they fear blame, that's a red flag. High turnover or disengagement, workers quietly dropping extra tasks like overtime checks or tool inventories, signals burnout eroding vigilance. And when post-incident conversations focus on "who forgot the lock" rather than on group prevention, blame has replaced learning.

Healthy culture looks different day-to-day. Staff naturally lock phones and screens, secure shops, and verify logins during shift handoffs. Casual tips are shared over coffee: "Rotate that scale password?" Workers flag suspicious texts or loose gates immediately and get thanked rather than scrutinized. No after-hours scrambles, because teams cover for each other on audits and updates proactively. Simple recognition flows for spotting risks: "Good catch on that weak Wi-Fi."

Handling Resistance

When you're the one person who cares about security and everyone around you thinks you're overreacting, local stories make threats real. "Remember when old man Jones down the road lost his tractor GPS to thieves last harvest? Hackers pull the same thing on John Deere apps, shutting down planting for days. It happened to three farms in our county last year." Tie it to money: "One phishing click froze a neighbor's co-op payments; it cost them $5,000 in downtime. Quick password swap could've stopped it."

Position it as group smarts rather than lecturing: "What if we treated farm software like the grain bin locks, check them weekly? Who's had a weird text on the ag app lately?" Offer easy first steps: "Let's just add that phone code thing tonight; it takes 2 minutes, like oiling the combine. I'll show you on mine." Track small wins and mention them later: "No glitches since we started locking screens." Invite peer leaders to demonstrate: "Jim, you spotted that scam email, want to walk through it next time we're all together?" This spreads ownership without you pushing.

Fitting Security into Busy Days

Farmers and food workers are already overwhelmed. Where does security fit into a day that's overflowing with demands? The answer is piggybacking on existing routines, keeping efforts to 2–5 minutes per touchpoint.

Layer cyber conversations on top of what teams already do. During equipment checks: "While greasing the planter, quick phone lock demo, takes 10 seconds, like flipping the ignition key." At shift handoffs or coffee breaks, tag onto safety huddles with one tip: "Fence secure? Same for app logins today." During invoice reviews, spot-check for phishing: "This email look off to you?"

Swap formal training sessions for bite-sized, repeated nudges. While fueling up trucks, rotating one shared password takes a minute. During livestock rounds, scan for odd texts or calls, 30 seconds. At end-of-day cleanup, back up one key file, two minutes. Off-peak times like winter downtime or post-harvest work well for slightly longer group shares, maybe 15 minutes. Frame it as uptime insurance: "Saves a harvest delay like good tires do."

The Security Champion Model

One of the most effective culture-building tools available is the security champion, a frontline operator, not an IT person, who models security habits organically within their team. Think of trusted tractor drivers, packing line leads, barn hands. They champion cyber-physical vigilance through peer respect rather than top-down mandates.

Why does this work? A field hand saying "I lock my phone like the fuel shed" lands differently than an email blast from corporate. One champion per crew can influence 10–20 others through daily conversation, no formal training required. In agricultural settings, they naturally tie digital concerns (app logins) to physical ones (gate checks), building holistic resilience.

Select champions based on influence, not expertise; you can train them afterward. Look for people others already turn to for equipment fixes, people who notice problems proactively, people who join shop talk naturally and show enthusiasm for new tools. During safety huddles, ask "Who do you turn to for quick fixes?" and let teams self-select one or two per shift.

Equip them lightly: a 30-minute walkthrough on basics like password hygiene and recognizing suspicious texts, plus a reference card for their pocket. Give them visibility, maybe a lanyard tag, and a channel like a group text to share daily wins. Monthly peer huddles of 10 minutes where champions swap field stories ("Caught a fake supplier email today") sustain the momentum. Track impact through quick polls: "Heard a tip this week?" Expect significant uptake within months.

Integrating with Existing Safety Structures

Most food operations already have safety meetings, toolbox talks, preshift huddles. The question isn't whether to create new structures for cybersecurity, it's how to weave security into what already exists.

Adopt a rotating "security moment" slot of 1–2 minutes in every existing meeting. Rotate who leads to share one cyber-physical link, keeping it brisk. At preshift huddles, end with "Risk check, gates locked, apps logged out?" from a different team member daily. At weekly toolbox talks, tag 60 seconds onto ladder or chemical reviews: "Like spotting frayed ropes, flag odd supplier emails." During plant floor rounds, pause at equipment for "Quick lock: Screen off like the mixer guard."

Avoid adding "five extra minutes." Instead, repurpose existing airtime by blending topics. At daily huddles, close with "Weather clear, tools secured, same for herd app logins today?" At weekly safety meetings, the champion leads a one-minute tip. At monthly reviews, link last week's near-miss to security: "Weak password nearly froze inventory." Start with champions nominating tips from their own turf to ensure relevance. Track via sticky notes on the board: "Tip used this week?"

The Phishing Simulation Question

One common security awareness tool deserves direct discussion: simulated phishing. Done wrong, these create exactly the shame spiral the research warns against. Done right, they can build awareness. The question is whether and how to use them.

Used sparingly and with the right framing, phishing simulations can help; they mirror real threats like those hitting meat processors or ingredient suppliers. Some food companies report significant improvement in awareness over time. But skip them if your culture is fragile; start with education instead.

Implementation matters. Announce them up front in a toolbox talk: "Like practicing fire drills, we'll run fake phishing emails next month; zero blame, just to sharpen our eyes on harvest payment scams." Use ag-themed lures ("Urgent: Seed order delay, click to confirm") sent quarterly, never harvesting real credentials. When someone clicks, provide instant positive feedback: "Good try! Here's the red flag. Report button works great, nice work." Non-clickers get optional recognition shareable in shop chat.

Track group trends, not individuals. Aim for click rates dropping over time, plus rising reports of real phishing attempts. If shame creeps in—eyerolls, complaints—pause and pivot to champion-led story shares instead. The point is building awareness, not catching people out.

Measuring Culture Progress

How do you demonstrate to your leadership that culture investments are working? Without metrics, culture work gets cut when budgets tighten. The answer is simple: operational metrics that tie directly to uptime, costs, and risks.

A phishing reporting rate measures the percentage of the team flagging suspicious emails or texts each month: target 80 percent or higher, and watch it rise as the culture embeds. Time-to-report captures average minutes from spotting a risk to notification target: target under 30 minutes, and watch it shorten with trust. Near-miss submissions count voluntary reports of close calls, such as shared passwords or unsecured Wi-Fi near silos; target 2–5 per week per shift.

Roll these into a one-page monthly view for executives, linking to business impact. If the reporting rate rose from 20 percent to 65 percent, that means fewer harvest delays from scams. If time-to-report dropped from 2 hours to 15 minutes, that would halve ransomware recovery costs. If near-misses rose from 1 per month to 12, that means issues got spotted prebreach. Quarterly, frame wins concretely: "Culture shift avoided three phishing hits, equaling two days' production saved." When cuts loom, show the trend: "Metrics up 40 percent, incidents flat, invest $5k in champions, yield $50k risk reduction."

When Security and Production Conflict

A 2025 study observed that "patching a milking robot's OS may require shutting down production at peak times, an unacceptable trade-off" for many operators.[8] But what happens when you're the security person and operations says no? How do you navigate that conversation in a way that preserves relationships while still managing risk?

Enter as a partner, not an enforcer. Schedule a quick 10-minute sidebar with the ops lead during a natural break, armed with specifics. Start with empathy: "I get peak lactation is non-stop; shutting down isn't ideal. What if we map options together?" Quantify the stakes with real cases: "JBS meat had ransomware halt operations for weeks, versus a 15-minute patch window."

Use collaborative framing. For patch timing: "Off-peak like 2:00 AM Saturday?" and ask "When's your next low-output window or next maintenance window for production?" For downtime concerns: "10 minutes now versus 2-day hack recovery?" and ask "What's a production hour worth?" For critical equipment: "Test on one unit first?" and offer "Pilot on the backup line?"

Propose tiered options showing flexibility. Tier one: virtual patching or agent-based fixes requiring no reboot. Tier two: staggered shutdowns across shifts, covering output gaps with manual backups. Tier three: compensating controls like network segmentation until ops allows, monitored with alerts. End with: "We'll track output impact next time, sound fair?" Celebrate joint decisions in the next huddle: "Ops plus security kept us at 100 percent uptime." Over time, this shifts "no" to "how," embedding culture where production sees security as uptime insurance rather than a hurdle.

Building Common Language

One persistent challenge in food system cybersecurity is the translation problem between IT teams and operations. IT speaks in acronyms: VPN, MFA, EDR, zero-trust architecture, endpoints. Operations speaks differently: throughput, yield, uptime, changeovers, sanitation cycles. Neither side naturally understands the other's priorities, and the result is mutual frustration.

Before any security communication goes to operations staff, run it through a simple test. What does this mean for production? What does this mean for food safety? What do I need them to do? If you can't answer all three clearly, the communication isn't ready.

Translation examples help. "We need to patch the PLC" becomes "The computer running your mixer needs an update to stay safe. Can we schedule 30 minutes during the next sanitation break?" "Implement MFA" becomes "Add a second step when you log in, like a code sent to your phone, two locks on the door instead of one." "Network segmentation" becomes "Keep the computers running equipment separate from the office computers. If someone clicks a bad link in an email, it won't reach the production line."

The goal isn't to dumb things down, but to connect security concepts to outcomes that matter to the audience. Operations cares about uptime, product quality, and worker safety. When security is framed in those terms, it stops feeling like an IT imposition and starts feeling like a shared priority.

Creating Community, Not Compliance

The food and agriculture sector already has networks built on trust and mutual support. Cooperatives share market information and pool resources. Equipment dealers provide service and advice. Extension services deliver practical knowledge. Farm bureaus advocate for farmers' interests. These existing networks are natural vehicles for cybersecurity awareness, far more effective than building something from scratch.

Peer-to-peer approaches matter because farmers trust other farmers more than outside "experts." Mentorship programs that pair security-aware operations with those just starting out can spread good practices organically. Regional learning circles where incidents are discussed without blame create opportunities for collective learning. Success stories shared through trusted community channels prove particularly powerful. When a farmer says, "We had something weird happen with our system, I reported it right away, and it turned out to be the start of an attack we stopped before it did damage," that's more persuasive than any awareness campaign. In September 2023, the American Farm Bureau Federation announced a partnership focused on cybersecurity awareness and incident reporting for the agricultural sector, recognizing that building security culture requires working through established trust networks.[9]

The Dignity Dimension

There's something deeper at stake here. Farmers and food workers take pride in feeding people. When someone asks what you do for a living, and you can say "I grow the food that ends up on your table" or "I make sure the milk in your refrigerator is safe," that carries weight. It's work that connects to something fundamental about being human.

And then an attacker in another country holds your systems hostage. An equipment manufacturer won't let you fix your own tractor because they've locked down the software. A vendor won't patch a known vulnerability in a system you depend on. A data broker sells information about your operation to parties you've never met. The indignity isn't just financial. It's losing control over something that matters deeply on a personal level.

This is what I mean when I say that cybersecurity in food is about safeguarding lives, not just data. The data matters because it's connected to real consequences: animals that need to be managed, crops that need to be harvested, products that need to reach consumers safely. When systems fail, real people get hurt, from the farmer who loses animals to the consumer who can't find food on the shelf.

The work we've done in this book, understanding threats, examining vulnerabilities, exploring how different parts of the food system connect, is ultimately in service of reclaiming that dignity. Through knowledge, we understand what threatens us. Through preparation, we build resilience. Through community, we support each other. Through advocacy, we demand better from vendors and policymakers.

The Culture We're Building

What does success look like? A culture where reporting a suspicious email is praised rather than questioned. Where asking for security help is as normal as asking about equipment repair. Where vendors are held accountable for the security of what they sell. Where the farmer who spots a threat protects neighbors by sharing what they know. Where cybersecurity is just another part of being a good steward of the operation, alongside soil conservation, animal welfare, and worker safety.

Building this culture takes time. It takes leadership willing to model the right behaviors. Investment in training that reaches the people who need it. Patience with those still learning. Humility from security professionals who need to understand food operations before they can protect them. Recognition that culture change has no end date—it's an ongoing commitment. But it starts with people, not tech. It always has and most likely always will.

In the next chapter, we'll explore why resilience matters more than compliance, why checklists alone can't keep us safe, and what it means to build systems that absorb shocks and keep functioning. Culture is the foundation. Now we build it.

Notes

1. Russell, C., Botschner, J., Duncan, E., Dehghantanha, A., & Fraser, E. D. (2025). "'I grow food, IT people do cybersecurity': addressing cybersecurity risks in Canada's agri-food sector." *Smart Agricultural Technology*, 10, 100866. **https://doi.org/10.1016/j.atech.2025.100866**
2. Renaud, K., Searle, R., & Dupuis, M. (2021). "Shame in cyber security: effective behavior modification tool or counterproductive foil?" In *NSPW '21: Proceedings of the 2021 New Security Paradigms Workshop* (ACM), 70–87, **https://doi.org/10.1145/3498891.3498896**

3. Ibid.
4. American Farm Bureau Federation. (2021). *Farmer and Rural Perceptions of Mental Health.* **https://www.fb.org/files/Farmer_and_Rural_Mental_Health_AFBF.pdf**
5. Hill, A., & Scott, F. (2025). "Some segments of the agricultural economy are particularly sensitive to changes in the foreign-born farm labor supply." *Federal Reserve Bank of Kansas City.* **https://www.kansascityfed.org/research/economic-bulletin/some-segments-of-the-agricultural-economy-are-particularly-sensitive-to-changes-in-the-foreign-born-farm-labor-supply**
6. U.S. Department of Labor. (2021). *Findings from the National Agricultural Workers Survey (NAWS) 2019-2020: A Demographic and Employment Profile of United States Farmworkers.* **https://www.dol.gov/agencies/eta/research/publications/findings-national-agricultural-workers-survey-naws-2019-2020**
7. USDA, National Agricultural Statistics Service. (2022). "Female producers." **https://www.nass.usda.gov/Publications/Highlights/2024/Census22_HL_FemaleProducers.pdf**
8. Neethirajan, S. (2025). "Safeguarding digital livestock farming - a comprehensive cybersecurity roadmap for dairy and poultry industries." Frontiers in Big Data, 8, 1556157. **https://doi.org/10.3389/fdata.2025.1556157**
9. American Farm Bureau Federation. (2023). "American Farm Bureau Federation partners with Food and Ag-ISAC on cybersecurity." **https://www.fb.org/news-release/american-farm-bureau-federation-partners-with-food-and-ag-isac-on-cybersecurity**

CHAPTER 34

Resilience over Compliance: Strategies for Long-Term Security

The week before Thanksgiving 2024, shoppers across New England walked into Stop & Shop expecting the usual abundance and found produce shelves bare, dairy sections with gaps, and pharmacy systems refusing to process prescriptions. A cyberattack on Ahold Delhaize, the Dutch parent company behind Stop & Shop, Hannaford, Food Lion, and Giant, had crippled more than 2,000 stores.[1] Online ordering went dark. Distribution systems seized up. For nearly two weeks, some locations couldn't keep basic items stocked during one of the busiest shopping times of the year.[2] Ahold Delhaize had cybersecurity controls in place. Policies documented, procedures established, compliance obligations met. None of it prevented the attack or meaningfully accelerated recovery. A Boston cybersecurity expert watching the situation called it plainly: "Cybersecurity is a dinner table issue, quite literally."[3]

Three years earlier, in September 2021, ransomware hit NEW Cooperative in Iowa during harvest season. Systems went dark, attackers demanded nearly $6 million, and the situation looked dire. Within hours, staff pivoted to paper tickets for documenting grain hauls. Farmers kept delivering their harvest. The cooperative had practiced operating without digital systems and run tabletop exercises about what to do when things broke down. That practice, not a policy binder, kept them functioning when a less-prepared operation would have frozen.[4]

Passing an audit doesn't guarantee you'll survive an attack. Compliance creates a wall. Whether you're still standing after attackers come through it depends on something different.

The HACCP Lesson

Food safety has learned this lesson, although the learning came hard. When HACCP became the regulatory framework for food manufacturing in the 1990s, many companies treated it as paperwork, identifying the hazards, documenting monitoring procedures, passing the annual audit, and moving on to other priorities. Then the outbreaks kept happening. Plants with impeccable HACCP documentation shipped contaminated products anyway. Over time, the industry recognized that the binder wasn't the point. If workers weren't trained, if supervisors didn't enforce standards on the production floor, if management didn't fund the systems that made compliance work in actual practice, people got sick regardless of how thorough the documentation looked.[5]

Cybersecurity in food systems is tracing that same arc. Organizations implement controls because auditors require them, document policies because regulations demand it, and check boxes because boxes need checking. Then ransomware encrypts everything, and nobody knows what to do because the response plan exists only in a filing cabinet, not in muscle memory.

The track record is consistent. JBS, the world's largest meat processor, paid $11 million in ransom after a 2021 attack shut down facilities across three countries.[6] Dole disclosed costs of around $10 million following its 2023 incident.[7] Both had dedicated security resources. Meeting compliance requirements didn't prevent either attack, and it didn't make recovery faster. Requirements codify what mattered at some point in the past; attackers are always working with what matters right now.

When Recovery Isn't Possible

In 2023, ransomware locked Vital Bircher, a Swiss dairy farmer, out of his milking parlor's data systems. The robots kept running, the physical milking process continued, but Bircher lost visibility into the animal health indicators the system tracked. He didn't notice when a pregnant cow began showing signs of distress because the alerts that would have flagged the problem had been encrypted along with everything else. She died. So did her calf.[8]

Better backups would have helped, and backups matter. But the deeper issue is that some losses can't be recovered through any amount of data restoration. You can rebuild yesterday's herd health records. You can't bring back the dead cow. You can restore production schedules, but not the week of harvest that spoiled while systems were down. A food processor can restore financial records from backup; if temperature monitoring failed during an attack and product was shipped in an unsafe state, no data recovery addresses that liability.

This is something professionals coming into food and agriculture from other sectors often miss. In banking, ransomware freezes transactions, and the money is still there when you recover. In hospitals, you revert to paper charts and catch up on digital records when systems return; the work waits for you. In food systems, biology operates on its own timeline. Resilience here has to account for consequences that backups can't reverse: animal welfare failures, spoiled inventory, and biological windows that close and don't reopen. Recovery time isn't measured purely by how fast you restore data. It's measured by how fast you resume the processes that keep animals healthy, products safe, and supply chains moving.

Equipment You Own but Don't Control

Farmers have been trying to repair their own equipment for years. John Deere has been restricting access to diagnostic software and tools for just as long. In January 2025, the FTC and attorneys general from Illinois and Minnesota sued Deere, alleging the company illegally monopolized the repair market by withholding essential software.[9] FTC Chair Lina Khan framed it as economic harm: repair restrictions "can mean farmers face unnecessary delays during tight planting and harvest windows."[10]

The economic framing is accurate, and there's a cybersecurity dimension that deserves separate attention. When manufacturers lock down equipment software, farmers lose the ability to verify what's running on machines they own. They can't confirm whether security patches have been applied, inspect systems for vulnerabilities, or determine what data their equipment transmits, to whom, and how often. The restrictions that prevent unauthorized tampering also prevent the oversight that would let farmers understand their own exposure.

Colorado passed agricultural right-to-repair legislation effective January 2024.[11] More than a dozen states introduced similar bills in early 2025. The FTC lawsuit continues working through federal courts. In the meantime, farmers and operators should be asking equipment vendors direct questions: what software runs on this machine, and when was it last updated? Can it keep functioning if connectivity fails? What data is it transmitting, and where is it going? Vague answers or salespeople who look confused by the questions tell you something important about how much thought the vendor has given to your security.

Records That Double as Legal Evidence

Operational data in food systems carries regulatory weight that data in most other industries doesn't. HACCP logs, batch traceability records, temperature monitoring histories: these aren't just operational conveniences. They constitute proof that food was produced safely, and regulators expect that proof on demand. FSMA requires companies to demonstrate control through documentation. Ransomware that destroys those records doesn't just create an operational problem. It can become a legal crisis, potentially an existential one, if regulators conclude that missing records indicate missing controls.

The regulatory environment is tightening, not relaxing. The EU now layers three separate frameworks on top of food and agriculture cybersecurity. General Food Law mandates traceability documentation at every stage of production and distribution. The NIS2 Directive, effective 2024, explicitly designates food production and processing as essential sectors subject to mandatory cybersecurity requirements, 24-hour incident reporting, and supply chain security obligations.[12] The Cyber Resilience Act, entering force December 2024 with full application by 2027, extends requirements to all products containing digital elements sold in EU markets: agricultural sensors, milking robots, precision agriculture platforms, processing line controls.[13] Australia's Security of Critical Infrastructure Act lists food and grocery among 11 designated critical sectors, with late-2024 amendments expanding government authority to mandate remediation of deficient cybersecurity programs.[14]

A multinational food company that loses batch-traceability data faces simultaneous compliance failures across every market where it operates, with different regulators asking different questions under different legal frameworks. The jurisdictional complexity alone is reason to treat data protection seriously before an incident forces the issue.

What Insurance Actually Covers

Cyber insurance premiums have jumped 50 to 100 percent for some food and agriculture companies over the past few years.[15] The cost increase matters, but what's changed more fundamentally is what insurers require before they write policies at all.

Multifactor authentication has become essentially universal as a prerequisite. Backup testing, not just having backups, but periodically demonstrating that restoration works, appears on applications with increasing frequency. Endpoint detection tools, incident response plans that have been exercised rather than just written, vendor risk assessments: these have become table stakes for coverage. Regulations have been slow to mandate specific cybersecurity practices in this sector. Insurers haven't waited. A food processor might defer investment when no regulator is forcing the issue, but that same processor can't defer when the carrier refuses to renew without specific controls in place.

There's also a feedback mechanism that regulations lack. Insurers track which controls correlate with fewer incidents and faster recoveries. Over time, the controls that demonstrably work become requirements, whereas the ones that look good on paper but don't move the needle get deprioritized. The market converges, imperfectly but directionally, toward practices that reduce real-world risk rather than simply real-world exposure.

Traditional agricultural insurance is beginning to intersect with cyber coverage in ways that underwriters are still working out. Crop insurance has long covered weather events, pest damage, and adverse market conditions. As farming becomes more technology-dependent, new questions arise. If a cyberattack disrupts planting during the optimal window and yields suffer as a result, which policy responds? The lines between cyber exposure and conventional agricultural risk are blurring in ways that will take years to fully resolve—and producers caught in the middle shouldn't assume that their existing coverage answers the question.

What the first half of this chapter describes is the landscape: why compliance falls short, what irreversible losses look like, where the regulatory and market pressures are heading. None of that tells you what to do Monday morning. The sections that follow are about that.

Getting Started Without Getting Overwhelmed

The question I hear most often from food and agriculture operators is simple: "Where do I start?" Vendor vetting, backup procedures, dependency mapping, and business continuity planning pile up fast, especially for a farm or processing plant where one person handles everything from email to equipment troubleshooting.

Here's a workable sequence. In the first week, take inventory. Write down every device, software application, and account that touches your operation—IT systems like email and accounting, OT systems like milk sensors or vessel monitoring equipment, all of it. A spreadsheet works fine. While you're at it, enable multifactor authentication wherever the option exists, change default passwords to something non-obvious, and apply software updates that have been sitting in the queue. One person can do all of this in a week without disrupting operations.

Over the first month, bring your team in. Phishing training takes about 30 minutes and focuses on scenarios your employees are likely to see—fake vendor emails, payment change requests. Industry associations, cooperative extension programs, and some commercial platforms make these available at low or no cost. Implement "least privileged" access so that not everyone has administrator rights on production systems. Start daily backups of key data: production logs, customer records, financial information. Before the end of that first month, test restoring one backup. Can you get your data back? A surprising number of operations discover their backup process doesn't work only when they desperately need it to.

In month two, draft a one-page incident response plan. Nothing elaborate: who does what when something goes wrong, who calls authorities, who notifies the team. Write it down and put it somewhere everyone can find without Internet access. If your processing equipment shares a network with the break room computer, where someone might click a bad link, basic network segmentation using your existing router settings can eliminate that exposure without significant expense.

Beyond those first two months: set up basic monitoring for unusual activity, score each connected device for risk, and run a tabletop exercise against your incident response plan. Gather the relevant people for an hour, present a scenario—"Ransomware has locked your production records. What do you do?"—and walk through it. The gaps you identify will be more useful than any checklist you could purchase.

What does "good enough" look like at different scales? For a 50-acre vegetable farm or a single fishing vessel: phishing training, MFA on email and cloud accounts, automatic software updates, and weekly offline backups. For a regional distributor or a fleet with multiple vessels: add a written incident plan, network segmentation between office IT and operational controls, monthly backup tests, and documented conversations with key vendors about their security practices. The rough line between prudent and excessive is affordability and time; if it takes under an hour weekly and uses tools already available, it belongs in the "just do it" category.

How Fast Is Fast Enough?

"As fast as possible" is not a recovery plan. You need concrete targets tied to what your operation cannot survive without.

Dairy operations need milking-parlor controls and milk cooling back within 30–60 minutes. If the automated parlor can't start milking due to a control failure, you need a manual-mode override or a paper-based schedule that the team can execute without waiting for IT. Herd-health systems can tolerate up to four hours; you can record milking weights by hand for one shift and re-enter them later. Accounting and payroll can wait up to 24 hours before missed payments start causing real problems.

Fishing operations face different constraints. Vessel monitoring systems and electronic catch documentation need to be back within one to two hours while at sea. Many fisheries require you to cease operations if VMS is down, so compliance is embedded in the recovery calculation. Shore-side inventory and sales can run on clipboards and phone calls for four to eight hours.

Cold storage and processing follow similar logic. Temperature monitoring and SCADA for cold rooms need to be back within 30 minutes, with any outage triggering immediate physical rounds: staff walking the rooms and logging temperatures by hand until systems return. Compliance documentation exports can wait up to 24 hours.

The practical way to manage this is through tiers. Tier 1, measured in minutes, covers anything life-critical: animal welfare, temperature control, vessel safety. Tier 2, measured in hours, covers production scheduling, catch documentation, sanitation tracking. Tier 3, up to 24 hours, covers administration, billing, and historical reporting. Align your backup frequency to these tiers. Tier 1 systems need near-real-time backups and manual override procedures that your team can execute without IT involvement. Tier 2 needs daily backups tested through periodic drills. Tier 3 needs daily backups verified quarterly. A year from now, if your Tier 1 systems consistently restore within target, you have concrete evidence that resilience is improving.

Testing Backups So They Actually Work

The only way to know whether your backups work is to use them. "Backup completed" logged in the system is not the same as "backup works." A real test means restoring the data, running the application, confirming that records came through intact, and verifying that the whole operation fit inside your recovery time target. If it didn't, you've learned something more useful than a year of completion logs could tell you: either the target was too optimistic or the process has a gap.

For a 10-person operation, testing critical systems quarterly is manageable. For a 500-person plant, layer in additional testing after major changes: a new server, a software upgrade, a new vendor system. A full restoration, where you bring a system completely back up, run real users through actual workflows, and verify that records came through intact, should happen at least once a year per critical system. Spot checks are faster and sustainable at a quarterly cadence: restore one day's worth of data, confirm that it opens correctly and timestamps match, look for corruption errors. Four of these tests a year, with outcomes approaching your targets, puts you ahead of most operations of your size.

Documentation doesn't need to be elaborate. A one-to-two-page backup policy covering what's protected, where it's stored, how often, and your recovery target. A runbook for each critical system, noting which backup to pull and how to restore it. A test log with date, system, type, outcome, and anything that went wrong. The point of the log isn't the paperwork; it's that you'll find out about backup problems during a scheduled test rather than when something has just gone badly wrong.

Mapping What Matters and Vetting Vendors

Most operations don't have a complete picture of their own dependencies until something breaks and takes three other things down with it. The point of dependency mapping isn't documentation for its own sake; it's to answer one question before you're under pressure: if this system fails, what else fails with it, and how long can we operate before that becomes critical?

The questions themselves are straightforward: what must we absolutely keep running for 24–48 hours if something breaks? For a dairy, milking and cooling. For a fishing vessel, catch documentation and refrigeration. For a processor, temperature control and order fulfillment. For each of those processes, identify who runs it, when it happens, and what the manual fallback is if technology fails. Then identify the systems that support each process and the vendors behind those systems. The practical output is a spreadsheet with process name, owner, systems used, vendors, criticality rating, recovery time objective, and last test date. Review it every three to six months. You're done when you can look at it and say "If these three systems go down, here's how long we survive and what we do next."

Your security is only as strong as your weakest vendor, which shouldn't surprise anyone who's worked in food safety. We already ask ingredient suppliers about sanitation practices, inspection records, and traceability systems. If a processing plant discovered that its ingredient supplier had never heard of HACCP, they'd find a new supplier. But ask whether that same plant's equipment vendor patches security vulnerabilities in a timely manner, and you'll likely get a blank stare.

For equipment vendors, ask: what operating system runs on that machinery? How long will it receive security updates? Where does the data go? Can it keep functioning if internet connectivity fails? Who has remote access capability? For software platforms: where is data stored, and in what jurisdiction? What's the backup and recovery process? When and how will I be notified if they experience a breach? For smaller operations, five questions cover most of it: What data are you collecting from my operation? Who can see it? How long will you support this product? What happens if your systems get hacked? Can I take my data if I switch? Bring those questions to every conversation with equipment dealers or software salespeople. Contract language requiring notification within 72 hours of any security incident affecting your data won't prevent every problem, but it establishes expectations that tend to be remembered.

Concentration Risk and Cross-Sector Dependencies

Four companies—Tyson, JBS, Cargill, and National Beef—control approximately 85 percent of U.S. beef processing, up from 36 percent in 1980.[16] Three companies handle roughly 60 percent of pork. In global grain trading, four firms control around 70 percent of supply.[17]

Seafood shows similar patterns. When NotPetya malware hit Maersk in 2017, disrupting global shipping, seafood supply chains felt it for weeks.[18] When ransomware hit JBS in May 2021, forcing shutdowns across three countries, it removed roughly 20 percent of U.S. beef processing capacity overnight.[19] An attack on any of the other major packers would produce comparable results.

Individual operations can't solve industry-wide concentration, but they can reduce their own single-point dependencies. Diversifying vendors where practical. Identify backup suppliers before you need them. Maintain manual fallback procedures so that when digital systems fail, operations can continue, even in degraded form.

Food and agriculture don't operate in isolation from other critical infrastructure. The Colonial Pipeline attack in May 2021 shut down a 5,500-mile pipeline and disrupted nearly half the fuel supply for the U.S. East Coast.[20] What often gets overlooked is how that rippled into food systems: trucks couldn't make deliveries, distribution centers struggled to keep refrigeration running, fishing vessels couldn't refuel. An attack on a fuel pipeline became, indirectly, an attack on food security. Operations that only plan for cyber disruptions in isolation are underestimating their exposure.

Your farm data also has commercial value that may not be obvious. Banks and lenders use agricultural data to assess creditworthiness; your cyber posture may eventually affect your loan terms. Commodity traders aggregate data from thousands of operations to inform trading decisions, and manipulated data could move markets in ways that benefit whoever did the manipulation. The American Farm Bureau Federation recognized these concerns in 2014 and developed Privacy and Security Principles for Farm Data, with a 2024 update adding requirements that providers explain whether farm data will be used to train machine learning models.[21] Ag Data Transparent certification is worth looking for in vendor selections, and any contract involving your operational data is worth reading before you sign it.[22]

Old Equipment, Departing Expertise

The food and agriculture industry runs on old equipment: PLCs from the 1990s, SCADA systems on Windows XP, fishing electronics that haven't seen an update in a decade. The question isn't whether to modernize but what to modernize first and how to manage risk during transitions that can't happen all at once.

Priority is based on three factors: safety and compliance criticality, business impact of downtime, and technical risks—specifically, whether the system is unsupported, running an end-of-life operating system, or sitting on a flat network where any breach can reach it. Life-safety-adjacent OT—cooling systems, milking parlors, processing line controls, and anything with no vendor support and direct regulatory exposure—should be modernized or strongly isolated first. Legacy ERP that's painful but not safety-critical can wait.

Avoid big-bang cutover. Add IoT sensors or edge gateways to existing PLCs so you can monitor and log data without rewiring the plant floor. Pilot on a single line or facility, running old and new in parallel until the new system proves stable. Keep manual overrides and documented fallbacks in place throughout. If you have equipment the vendor no longer supports, isolate it from the rest of your network, document everything you know about it, and plan replacement during a scheduled maintenance window when operational risk is lowest.

Then there's the knowledge problem. What happens when your one IT-aware person retires, or the employee who set up your network moves on, or the deckhand who understands the electronics takes a job elsewhere? Focus on what would break if someone forgot it. For each critical system, document what it does, where it lives, how it works day-to-day, and how to recover it. A lot of IT knowledge is tacit: "I know the server gets flaky after a network ping, so I reboot it Tuesday mornings." That kind of knowledge is worth capturing; phone video works fine for short walkthroughs. Ask people to describe what normal looks like versus what something-is-wrong looks like, and what their workarounds are. Build a simple who-does-what matrix with at least one backup person for each critical task. Once a year, walk through the documentation with whoever currently handles IT and have the backup person execute a small task. Success looks like one central place where all IT notes live, a backup person for each critical task, and evidence that the backup person knows what they're doing.

When Companies Change Hands

The food industry is consolidating, and cyber risk consolidates with it. Buying a marginally profitable plant can mean inheriting a decade-old network, weak vendor contracts, and latent breach exposure that nobody disclosed because nobody looked. Cyber due diligence deserves the same rigor as financial due diligence.

Acquirers should examine whether written policies and incident response plans exist and have been tested; the state of OT and IT infrastructure, including when core systems were last patched and whether plant floor networks are separated from office networks; incident history over the last three to five years; critical vendor relationships and whether contracts include cyber notification clauses; and any unresolved compliance findings. These questions translate into a risk scorecard that can influence valuation, escrow arrangements, and integration timelines.

Sellers should document current security posture honestly, close obvious gaps before going to market, and organize records of past incidents. Buyers will find problems—the question is whether you've already identified the easy ones, or whether discovery reveals that no one was paying attention.

Making the Business Case

Cybersecurity professionals often struggle to translate risk into language that owners and boards engage with. Here's a shorthand that tends to work: *annualized loss expectancy*. Multiply the estimated likelihood of an event per year by the financial impact per event.

For a regional poultry processor, if historical data suggests that ransomware incidents mean an average of three days of downtime and lost production runs $100,000 dollars per day, the impact per incident is roughly $300,000. If such events happen roughly every four years, the annual likelihood is about 0.25. Multiply them: $75,000 in expected annual loss at current controls. That's a sentence a board can engage with: "We expect to lose about $75,000 per year from cyber-related outages if we don't invest in better controls." Then show what better backups, segmentation, and training would cost and how much they would reduce that exposure. The language shifts from "We need better security" to a risk-reduction calculation with an actual number attached.

The investments don't require large budgets. An external hard drive for offline backups costs under $100. Password managers are available for free. Enabling MFA on accounts that support it costs nothing. Writing down emergency contacts and keeping that list somewhere accessible when the network is down requires no expenditure. Mid-sized operations considering more structured programs—annual security assessments, managed detection services, documented and exercised response plans—might be looking at $5,000 to $25,000 annually. Compare that to the cost of even a few days offline.

Doing This Together

Small farms share equipment. They share labor during harvest. They share marketing through cooperatives. Cybersecurity is another resource that can be shared. Agricultural cooperatives can host workshops during planning meetings. Regional farm bureaus sometimes facilitate group purchasing security tools, and state-level grant programs occasionally cover initial setup for qualifying groups of operations.

A shared IT arrangement, one expert serving 10–20 farms quarterly, with costs split across cooperative members, can run $500 to $1,000 dollars per farm per year, far less than any single operation would pay going it alone. Joint tabletop drills and shared recovery checklists mean that when one operation gets hit, the others already know what to do. Getting started requires three to five peers willing to spend 30 minutes monthly comparing notes. The goal is to make each operation more resilient through coordination, not create another administrative layer.

Knowing Whether It's Working

You can tell if your resilience is improving, and you don't need a sophisticated monitoring platform to do it. A handful of indicators tells most of the story.

Track what percentage of your critical systems have been backed up and successfully tested in the last 30 days. For Tier 1 systems, that number should be 100 percent, consistently. Track restoration time against the RTOs from the earlier tiering exercise; consistently hitting your Tier 1 target of 30–60 minutes means the practice is embedded, not just documented. Track security incidents per month and their business impact in hours of downtime; the benchmark here isn't a fixed number but a direction—if the trend is flat over a year, something needs attention. For MFA coverage, 90 percent or higher across all user accounts is a reasonable target; anything under 80 percent on administrative accounts is a genuine gap. For software currency, 95 percent of devices running current software is the practical ceiling for most operations and a meaningful security indicator.

Once a year, step back and assess maturity overall. *Ad hoc* means no written policies and backups that may or may not happen. *Defined* means a one-page checklist and incident plan, regular backups, and at least one test per year. *Consistently implemented* means a monthly checklist review, quarterly backup tests, and staff who recognize phishing attempts. *Managed and measurable* means tracked metrics, reviewed in team meetings, with documented recovery times for key systems. Each year, ask: did recovery time shrink? Did the incident count go down? Did more boxes get checked? Two out of three is a real improvement. If none of those are true, you know where to push.

The Work That Matters

Ransomware attacks against the food and agriculture sector more than doubled in early 2025: 84 attacks in the first quarter, compared to 40 in the same period of 2024, with more than 200 documented incidents throughout 2024.[23] JBS paid $11 million in ransom. Dole's 2023 incident cost $10.5 million in direct costs alone.[24] Those figures don't include reputational damage, lost customer relationships, regulatory scrutiny, or what it feels like to watch something you built stop being under your control.

Every vessel captain who assembles a paper catch-logging procedure and keeps it somewhere waterproof has reduced their exposure by one small but real measure. Every farmer who asks one more question of their equipment vendor—What's running on this machine? Who can access it remotely? When it was last updated?—has done the same. Resilience doesn't accumulate through a single transformation. It accumulates through decisions that compound.

Vet your vendors more carefully. Know which of your systems would take everything else down with them if they failed, and have a plan for that. Keep your incident contacts somewhere you can reach when the network is down, which is exactly when you'll need them. Capture what the people who manage your systems know before they move on. Test your ability to recover before you're forced to prove it under pressure. You already manage weather risk, market risk, and pest risk. You lock the barn at night. This is the same instinct applied to a newer kind of exposure.

When something goes wrong, stop using affected systems, disconnect from the network while you still can, and get help. If you have cyber insurance, your carrier's incident response line is usually available around the clock and included in the policy. If not, keeping the contact information for an incident response firm on file before you need one is a reasonable precaution. Stop, disconnect, get help. The sequence works at any scale.

Food safety culture took decades to evolve from treating HACCP as paperwork into a genuine organizational discipline in which safe practices became embedded rather than bolted on. Cybersecurity in food systems is earlier in that evolution, but the trajectory is familiar. The Stop & Shop incident wasn't fundamentally a technology failure—technology did what it was designed to do until attackers found a way through, which eventually they do. What it exposed was the gap between having compliance documentation and having the organizational capacity to keep functioning when everything goes sideways. Checklists establish baselines and create accountability, but treating them as the destination rather than the starting point is how organizations end up with policies that look thorough on paper and fail on contact with an actual attack.

The next chapter takes up collective action: what becomes possible when operations across the food system work together rather than managing this alone. Protecting what feeds us ultimately requires all hands.

Notes

1. Ahold Delhaize USA. (2024). "Ahold Delhaize statement on Ahold Delhaize USA cybersecurity issue." **https://newsroom.aholddelhaize.com/ahold-delhaize-statement-on-ahold-delhaize-usa-cybersecurity-issue**
2. Enwemeka, Z. (2024). "Not just tech: Stop & Shop hack shows cybersecurity matters everywhere." *WBUR News*. **https://www.wbur.org/news/2024/11/22/stop-shop-food-shortages-cybersecurity-experts**
3. Ibid.
4. Bajak, F. (2021). "Iowa farm cooperative hit by ransomware, systems go offline." *KRDO*. **https://krdo.com/news/ap-national-business/2021/09/20/iowa-farm-cooperative-hit-by-ransomware-systems-go-offline**
5. Global Food Safety Initiative. (2018). *A Culture of Food Safety: A Position Paper*. **https://mygfsi.com/wp-content/uploads/2019/09/GFSI-Food-Safety-Culture-Full.pdf**
6. BBC News. (2021). "Meat giant JBS pays $11m in ransom to resolve cyber-attack." **https://www.bbc.com/news/business-57423008**
7. Dole Food Company. (2023). "Form 10-K annual report." *SEC*. **https:/oo/www.sec.gov/Archives/edgar/data/18169/000119312513103254/d442446d10k.htm**
8. Bayer, M. (2024). "Ransomware attack paralyzes milking robots—cow dead." *CSO*. **https://www.csoonline.com/article/3484349/ransomware-attack-paralyzes-milking-robots-cow-dead.html**
9. Federal Trade Commission. (2025). "FTC, states sue Deere & Company to protect farmers from unfair corporate tactics, high repair costs." **https://www.ftc.gov/news-events/news/press-releases/2025/01/ftc-states-sue-deere-company-protect-farmers-unfair-corporate-tactics-high-repair-costs**
10. Selyukh, A. (2025). "John Deere faces U.S. lawsuit over farmers' ability to repair tractors." *NPR*. **https://www.npr.org/2025/01/15/nx-s1-5260895/john-deere-ftc-lawsuit-right-to-repair-tractors**
11. Proctor, N., & Sherwood, W. (2025). "John Deere and right to repair over the years." *PIRG*. **https://pirg.org/resources/john-deere-and-right-to-repair-over-the-years**
12. European Commission. (2022). "Directive (EU) 2022/2555 of the European Parliament and of the Council of 14 December 2022 on measures for a high common level of cybersecurity across the Union, amending Regulation (EU) No 910/2014 and Directive (EU) 2018/1972, and repealing

Directive (EU) 2016/1148 (NIS 2 Directive)." *EUR-Lex.* **https://eur-lex.europa.eu/eli/dir/2022/2555/oj/eng**

13. European Commission. (2024). "Regulation (EU) 2024/2847 of the European Parliament and of the Council of 23 October 2024 on horizontal cybersecurity requirements for products with digital elements and amending Regulations (EU) No 168/2013 and (EU) 2019/1020 and Directive (EU) 2020/1828 (Cyber Resilience Act)." *EUR-Lex.* **http://data.europa.eu/eli/reg/2024/2847/oj**
14. Australian Government. (2018). "Security of Critical Infrastructure Act 2018." As amended November 2024. **https://www.cisc.gov.au/legislation-regulation-and-compliance/soci-act-2018**
15. Sellars, S., Nafchi, A. M., Mehan, S., Vandermark, L., Brennan, J., & Yang, X. (2024). "Cyber insurance for agriculture." *South Dakota State University Extension.*" **https://extension.sdstate.edu/sites/default/files/2024-08/P-00306.pdf**
16. McCracken, J. (2025). "Fact-checking Trump's call for an investigation into meatpacking companies." *Investigate Midwest.* **https://investigatemidwest.org/2025/11/18/fact-checking-trumps-call-for-an-investigation-into-meatpacking-companies**
17. Klemcke, C. (2023). "Part 3—Meat Mergers and the Rise of The Big 4." *1915 Farm.* **https://1915farm.com/blogs/education/part-3-meat-mergers-and-the-rise-of-the-big-4**
18. Maersk. (2017). "Cyber attack update." **https://investor.maersk.com/news-releases/news-release-details/cyber-attack-update**
19. Associated Press. (2021). "REvil, a notorious ransomware gang, was behind JBS cyberattack, the FBI says." *NPR.* **https://www.npr.org/2021/06/03/1002819883/revil-a-notorious-ransomware-gang-was-behind-jbs-cyberattack-the-fbi-says**
20. Easterly, J. (2023). "The attack on Colonial Pipeline: what we've learned & what we've done over the past two years." *CISA.* **https://www.cisa.gov/news-events/news/attack-colonial-pipeline-what-weve-learned-what-weve-done-over-past-two-years**
21. Farm and Dairy. (2014). "Privacy and security principles for farm data agreement." **https://www.farmanddairy.com/news/privacy-security-principles-farm-data-agreement/226798.html**
22. Ag Data Transparent. (2024). "Core principles." **https://www.agdatatransparent.com/principles**
23. Dragos, Inc. (2025). *OT Cybersecurity Year in Review 2024* (Dragos). **https://www.scribd.com/document/833874127/Dragos-2025-OT-Cybersecurity-Report-a-Year-in-Review**
24. Dole Food Company. (2023). "Form 10-K Annual Report."

CHAPTER 35

A Call to Action: Who Needs to Be Involved?

Back to Where We Started

Return one more time to Vital Bircher, the Swiss dairy farmer who lost a cow and her calf to ransomware. To JBS, paying $11 million to criminals while meat plants in three countries sat idle. To Dole, watching grocery store shelves empty of salads because the company couldn't prove its food was safe when systems went down. To the NEW Cooperative, scrambling with paper tickets while grain piled up during harvest season in Iowa. These stories represent failure, but they also represent learning. And learning, shared across a community, becomes resilience.

There are other stories too, ones that don't make headlines because they end well. The cooperative that practiced its backup procedures during the slow season and processed grain by hand when ransomware hit, keeping harvest moving for its member farmers. The plant manager who saw a suspicious email and remembered her training, reported it immediately, and prevented an attack that would have shut down production for days. The equipment dealer who started asking customers about cybersecurity during service calls and helped a dozen farms improve their basic security practices. The extension agent who added a 15-minute cybersecurity section to the annual farm management workshop and reached 200 farmers who had never thought about digital threats before. These quieter stories exist. The goal of this chapter is to make them the norm rather than the exception.

Throughout this book, we've traced the deep connections between food, culture, technology, and security. We've examined specific sectors and their vulnerabilities. We've explored challenges that cut across the entire food system. Now comes the question that matters most: what do we do about it?

The answer isn't simple, because the food system isn't simple. From the farm field to the restaurant table, from the trawler deck to the food bank distribution line, protecting what feeds us requires everyone who touches the system. Farmers and processors, truck drivers and warehouse workers, regulators and policymakers, researchers and educators, consumers and communities. I say, all hands on deck.

The Stakeholder Map on Which Everyone Has a Role

Picture concentric circles radiating outward from the center. At the center sit individual food operations: farms, processors, distributors, retailers, restaurants. The next ring holds vendors, equipment manufacturers, cooperatives, and trade associations. Beyond those come regulators, insurers, and standards organizations. The outer ring encompasses policymakers, researchers, educators, and consumers. Every ring affects every other ring. No single stakeholder can secure the food system alone.

Food Operations

Farms, processors, distributors, retailers, and food service establishments form the operational core where food moves, transforms, and reaches people. They need to own their risk rather than delegating it entirely to technology professionals. Basic cyber hygiene proportionate to their scale matters. Reporting incidents helps others learn. Demanding security accountability from vendors matters. Participating in information-sharing through cooperatives and associations creates collective protection.

What these operations need in return is substantial. Resources scaled to their actual circumstances rather than enterprise solutions designed for Fortune 500 companies. Training that respects their time constraints and speaks in language that makes sense on a production floor or in a barn. Support systems that respond constructively rather than shame when things go wrong. Vendors who partner on security rather than deflecting responsibility through contract language.

Equipment and Technology Vendors

The companies that sell tractors, milking robots, processing equipment, and farm management software hold enormous influence over the security posture of the operations they serve. Their consumers need them to build security in from the design phase rather than bolting it on afterward, provide long-term software support that recognizes that equipment lifecycles often exceed a decade, be transparent about data collection and transmission practices, respond quickly to reported vulnerabilities, and give operators meaningful control over their own equipment.

The current model, in which manufacturers retain extensive control over equipment software while farmers bear the operational consequences of security failures, needs rethinking. The same remote-access capabilities that disabled stolen tractors in Ukraine could theoretically disable equipment anywhere in the world. What protects in one context creates vulnerability in another. Farmers deserve transparency about what's running on their equipment and the ability to verify that security measures are in place.

Cooperatives and Trade Associations

These organizations occupy a unique position in agricultural communities, providing collective resources that individual operations cannot access alone. Pooling cybersecurity resources that members cannot afford individually. Facilitating information sharing about threats and lessons learned. Advocating for members' needs with policymakers and regulators. Developing sector-specific training and resources.

Cooperatives already share market information, coordinate logistics, and aggregate buying power. Adding cybersecurity to that portfolio makes practical sense. A regional grain cooperative that provides shared security monitoring or incident-response support offers protection that no individual farm could afford.

Regulators

Government agencies face the challenge of protecting critical infrastructure without creating compliance burdens that smaller operations cannot meet. Guidance needs to be appropriate to different scales of operation. Offer resources, not just requirements. Coordinate across agencies, because food safety and cybersecurity intersect constantly. Enforce consistently without punishing honest reporting. Learn from the industry before writing rules from a boardroom.

The food and agriculture sectors span an enormous range of operations, from family farms operating on tight margins to multinational processors with dedicated security teams. Effective regulation must account for this diversity. Guidance that helps a small dairy farmer understand basic password practices serves a different purpose than requirements for incident reporting at major processing facilities. Both are needed, but they require different approaches.

Policymakers

Elected officials and their staff shape the broader environment in which food system security operates. Listen to the sector before legislating. Provide funding rather than just penalties. Support rural connectivity infrastructure. Recognize food and agriculture as genuinely critical infrastructure deserving serious attention. Balance security imperatives with operational reality.

Dedicated funding for food and agriculture cybersecurity programs would advance the cause. Other considerations might include tax incentives for security investment, support for cooperative security initiatives, international coordination on food system security, and investment in agricultural extension cybersecurity programs. What doesn't help is unfunded mandates, requirements developed without industry input, or treating food security as a lower priority than other critical infrastructure sectors.

Researchers and Universities

Academic institutions, particularly the land-grant universities established specifically to serve agricultural communities, have essential roles to play. Translate research findings into practical guidance. Engage extension services to reach practitioners. Study food-specific threats and defenses. Train the next generation of professionals who understand both food systems and cybersecurity. Partner with industry on real-world problems.

The Cooperative Extension System, with offices in nearly all of the nation's approximately 3,000 counties, represents an existing infrastructure for delivering practical knowledge to agricultural communities.[1] These extension agents have spent more than a century building trusted relationships with farmers and rural communities. Adding cybersecurity awareness to their toolkit leverages that trust rather than asking farmers to engage with unfamiliar outside experts.

Consumers

The people who ultimately eat the food have roles to play as well, even if those roles look different from operational responsibilities. Understand, at a basic level, where food comes from and what keeps it safe. Support brands and producers that demonstrate commitment to security. Report suspicious activity when you observe it. Resist panic at every headline; instead, seek accurate information. Advocate for food system resilience through civic engagement.

Consumer pressure shapes markets. When consumers demonstrate that they value security, not just price, producers have greater incentive to invest. When consumers resist misinformation and demand accurate information, the information environment improves.

Insurance Companies

The insurance industry increasingly plays a role in shaping security practices as well. Cyber insurance premiums have risen sharply for food and agriculture operations, and insurers now commonly require specific security controls before writing policies. This creates a financial incentive for security investment that complements regulatory requirements and industry best practices. Operations that can demonstrate a good security posture get better coverage at better rates; those that can't find themselves unable to obtain coverage at all.

It's Monday Morning: What to Do Now?

These roles and responsibilities paint the landscape. But if you close this book tomorrow morning, ready to act, what's your first move?

The answer depends on where you sit in the food system. A grain farmer in Kansas faces different constraints than a seafood processor in Alaska, a beverage manufacturer in Wisconsin, a cold storage operator in Georgia, or a restaurant owner in Oregon. The specifics differ, but the pattern stays the same: connect with one hub that speaks your language, adopt one concrete practice you can implement this week, join one community where others are working on the same challenges.

For Primary Producers

Whether you're growing vegetables, raising livestock, managing an orchard, or running an aquaculture operation, your entry point is the agricultural support infrastructure that already exists. In the United States, that means your local USDA Service Center. Internationally, it's your national Ministry of Agriculture or rural development office. These aren't cybersecurity agencies, but they can connect you to programs, training, and resources scaled to agricultural operations. Call them tomorrow. Ask what support exists for small-scale or family producers related to business planning, food safety, or risk management. Cybersecurity increasingly shows up in those conversations, and if it doesn't, you just made them that aware it should.

Download one starter checklist. Cornell Small Farms, the Sustainable Agriculture Research and Education program, your land-grant university extension service, and FAO's Family Farming Knowledge Platform all offer free guides. Pick one short resource, and complete the self-assessment in one sitting. You'll learn what you're already doing right and identify one or two gaps worth addressing.

Join a producer network. State-level farm coalitions, regional food hubs, producer cooperatives, and farmer associations create communities where people share what works. These groups run workshops, circulate templates, and connect members with peer mentors. When someone in the network learns something useful about securing their systems, that knowledge spreads. You're not solving this alone.

For Processors

Food processors, small artisan operations, mid-size regional plants, and large multifacility companies: start with your trade association. The National Food Processors Association, specialty crop councils, regional processor groups, and international equivalents like GFSI member organizations provide sector-specific guidance on food safety management systems that increasingly incorporate digital security.

Your first Monday action might be implementing one baseline security practice: turn on MFA for cloud-based accounts, create a tested backup-and-recovery routine, or draft a one-page incident response card that lists who calls who when something goes wrong. These actions aren't complicated, but they're concrete, and they reduce your exposure immediately.

Check your compliance requirements. Your buyers likely have expectations around food safety certification, traceability, and, increasingly, cybersecurity. Review those requirements, and map them into your current practices. Where gaps exist, prioritize closing them before an audit finds them first.

For Distributors

Distribution, logistics, and cold storage operators, the people who keep food moving and prevent spoilage: your vulnerabilities center on OT and supply chain visibility. Temperature monitoring systems, warehouse management software, fleet tracking, and inventory platforms all create digital dependencies. Start by inventorying which systems control critical processes, identifying what happens if each system fails, and documenting manual fallback procedures that work when digital systems don't.

Connect with logistics and transportation associations that are starting to address cyber risk in cold-chain management and just-in-time delivery systems. A ransomware attack that takes down route optimization or refrigeration monitoring can spoil products worth hundreds of thousands of dollars in hours.

For Foodservice

Restaurants, institutional kitchens, and catering operations: your exposure runs through point-of-sale systems, reservation platforms, inventory management, and payment processing. Ensure that your POS system provider offers security support, not just sales support. Ask them about their patching schedule, their incident response plan, and what happens to your operation if their system goes down. If they can't answer those questions clearly, that tells you something important about your vendor.

Hospitality industry groups increasingly include cybersecurity in their risk management guidance. The National Restaurant Association, state-level hospitality councils, and culinary professional organizations can connect you to resources.

For Concerned Consumers and Community Advocates

Your leverage is different but real. Visit a local farm or farmers market, and ask one food safety or sustainability question to learn, not to challenge. "How do you manage water quality?" "Do you follow any food safety protocols?" "How do you keep records?" These conversations build understanding of what producers do to keep food safe, and they signal that consumers care about these systems. However, be careful how you ask these questions, because it could come off as social engineering rather than genuine curiosity.

Use consumer advocacy organizations that provide action toolkits. Groups like the Environmental Working Group, Slow Food chapters, farm-to-institution networks, and state-level healthy food coalitions offer sample letters to legislators, local library lectures, talking points for school board meetings, and guidance on how food policy decisions are made. Support local food systems by buying from producers who demonstrate commitment to good practices.

What Doesn't Exist Yet

Honesty about what's missing matters: several tools that should exist to make these first steps easier simply don't yet, and building them is part of this call to action. There's no universal, low-tech self-assessment that a small-scale operator anywhere in the world can use to score their basic cyber and food safety practices in 15 minutes and get a clear "do this next" recommendation.

There's no globally harmonized baseline practice set that buyers, insurers, and regulators across borders can reference when they ask, "What should a responsible food operation be doing?" Different frameworks exist, but they don't align, and that creates confusion and a compliance burden.

There's no model safe-harbor clause for incident reporting that jurisdictions can adopt to protect operators who report cyber incidents in good faith. Without that legal protection, many operations stay silent when they should be sharing lessons.

There's no clear contractual framework for vendor liability when insecure equipment becomes the attack vector. Operators bear the costs when vendors hide behind limitation-of-liability clauses.

If you're a regulator, a trade association staff member, a university researcher, or a policy advocate, these gaps are your Monday morning assignment. Pick one. Draft something. Circulate it. Get feedback. Iterate. These tools won't build themselves.

The path forward isn't perfect, and the resources aren't complete. But they're better than they were five years ago, and they'll be better still if everyone who closes this book takes one concrete step rather than waiting for someone else to go first.

Bridging the Digital Divide

Any serious conversation about cybersecurity in agriculture must confront a fundamental challenge: many of the operations most vulnerable to digital threats lack reliable digital infrastructure in the first place.

According to the Federal Communications Commission, approximately 28 percent of Americans in rural areas and more than 23 percent living on Tribal lands lack access to fixed broadband service meeting current standards.[2] The digital divide isn't merely about convenience; it shapes what kind of security is even possible. Operations without reliable connectivity may use insecure workarounds just to get things done; without nearby technical support, they may ignore security warnings they don't understand; and without a budget for security tools, they may rely on outdated, unpatched systems because that's what they can afford.

The paradox runs deep. Precision agriculture technologies require connectivity to function, and sustainability policies increasingly push farmers toward these technologies. But the connectivity infrastructure and security practices that should accompany technology adoption often lag behind. Farmers adopt GPS-guided equipment because it improves efficiency, only to discover that they've created dependencies they didn't anticipate and vulnerabilities they don't know how to address. Solutions must recognize these realities rather than pretending they don't exist.

Leveraging existing infrastructure makes more sense than building parallel systems. Extension services already reach rural communities. Equipment dealers already provide service and support. Cooperatives already serve as hubs for shared resources. Farm Bureau chapters already advocate and organize. Rather than creating new cybersecurity-specific outreach channels, integrate security awareness into existing relationships and institutions.

Designing for disconnected operations acknowledges that not everyone has reliable Internet access. Security tools that work offline, paper backup procedures that are genuinely practical, local knowledge and decision-making capability, resilience that doesn't depend on cloud connectivity: these matter for operations that can't count on being continuously connected.

Making resources accessible means thinking beyond online delivery: free and low-cost security options, resources available in formats that don't require Internet access, hotlines staffed by people who understand agricultural contexts, peer networks for mutual support.

Addressing cost barriers directly requires creative approaches: shared services through cooperatives, government subsidies for security investment, insurance incentives for good practices, and vendor responsibility for the security of products they sell.

The federal government has invested substantially in rural broadband infrastructure through programs like the ReConnect Program, which has deployed billions of dollars to expand connectivity in underserved areas.[3] These investments matter for cybersecurity because they create the foundation on which secure operations can be built. But infrastructure alone isn't sufficient. Connectivity must be accompanied by knowledge, tools, and support that enable secure use of that connectivity.

Digital equity in agriculture isn't just about fairness, although fairness matters. It's about recognizing that the food system depends on operations of all sizes, and vulnerabilities at any scale can create consequences that ripple far beyond the immediately affected operation. The small farm that becomes a vector for supply chain compromise affects everyone downstream. The regional cooperative that can't afford security monitoring creates risk for all its members. Addressing the digital divide is addressing systemic risk.

A Global Food System Requires Global Thinking

The food system that feeds us constantly crosses borders. Ukrainian wheat feeds populations in Africa and the Middle East. Brazilian soybeans become animal feed in Europe and Asia. New Zealand lamb appears on tables from London to Los Angeles. Chilean fruit fills grocery bins during northern hemisphere winters. When attackers target food systems, the consequences rarely stay local.

The JBS ransomware attack in May 2021 demonstrated this interconnection vividly. The attack originated in the company's Brazilian operations but immediately affected facilities in the United States, Canada, and Australia. Beef processing capacity disappeared from three continents simultaneously. A single company's vulnerability became a supply chain vulnerability affecting grocery stores in multiple countries.

Sub-Saharan Africa holds 60 percent of the world's uncultivated arable land and represents a frontier for agricultural technology development.[4] Mobile technology adoption has leapfrogged traditional infrastructure in many African countries, with agricultural apps connecting smallholders to markets in ways that weren't possible a decade ago. But this rapid technological adoption often outpaces security awareness and infrastructure. An attack on systems serving Kenyan flower exports or Ethiopian coffee cooperatives would have implications far beyond the region.

Latin America provides food to the Americas and beyond. Brazil ranks as the world's largest soybean exporter and second-largest beef exporter. Argentina produces major grain volumes. Chile exports fruit globally. Supply chain connections between Latin American agriculture and North American grocery stores are direct and immediate. What happens in Brazilian processing plants shows up on American dinner tables within days.

The EU has moved furthest in establishing regulatory frameworks for food system cybersecurity. The NIS2 Directive, which took effect in October 2024, recognizes food production, processing, and distribution as important sectors requiring mandatory cybersecurity risk management measures.[5] Companies must assess risks, implement security measures, and report significant incidents quickly. Supply chain security requirements push accountability beyond individual operations to their vendors and partners. European regulatory approaches may increasingly set de facto global standards as companies operating internationally align with the highest requirements they face.

Smallholder farmers, who operate 84 percent of the world's farms and produce approximately one-third of the world's food, represent both vulnerability and an opportunity.[6] Many smallholders lack access to the technology that creates cyber risk in the first place. But as technology adoption spreads, as mobile apps connect farmers to markets, as data platforms aggregate information across millions of small operations, the attack surface expands. Security approaches that work for large commercial operations won't translate directly to smallholder contexts, but the need for protection is no less real.

International coordination on food system security remains underdeveloped, but the bodies that should lead this work already exist. The Food and Agriculture Organization and the WHO coordinate food safety systems globally; cyber-related disruptions to food safety and supply chains represent a natural extension of their mandates. The International Telecommunication Union and the Organization for Economic Co-operation and Development run cyber policy dialogues that could incorporate food-system-specific annexes without creating parallel bureaucracies. NATO and regional security alliances increasingly treat food system resilience as a hybrid threat issue, recognizing that adversaries who want to destabilize societies can target food production and distribution as effectively as they target energy or communications infrastructure.

The pathway forward involves plugging food cybersecurity into existing frameworks in the short term while building sector-specific reference models that can inform national regulations and trade agreements over time. This means leveraging current diplomatic channels for food safety, critical infrastructure protection, and cyber cooperation rather than starting from scratch with yet another international body. Mutual assistance agreements between food-importing and food-exporting nations could include provisions for sharing anonymized threat pattern data when cyber incidents affect shared supply chains. Development financing for agricultural infrastructure through institutions like the World Bank or regional development banks could incorporate cybersecurity risk management conditions, mirroring the environmental and climate-resilience requirements that already exist.

The goal isn't a binding treaty that takes a decade to negotiate, but practical cooperation that helps defenders share information across borders as easily as attackers already do. The architecture for that cooperation is in place; it needs adaptation, not invention.

The concentration of global food trade through a handful of major companies and corridors creates systemic risk that transcends any single nation's ability to address alone. When four companies control the majority of global grain trade and a handful of processors dominate meat markets, attacks on any of these nodes have consequences that cross oceans within hours. International cooperation is essential to match the scale of the threat.

Building the Hybrid Workforce

Food system security suffers from a talent gap that runs in multiple directions. Cybersecurity professionals often lack understanding of food operations, their regulatory frameworks, and their unique constraints. Food professionals often lack exposure to cybersecurity concepts and practices. IT staff may not appreciate the differences between office networks and OT environments. Operations staff may not recognize digital threats as relevant to their work.

What the sector needs are people who can translate between worlds: who understand both the biology of food production and the logic of digital systems. Food safety professionals who can assess how cyber risks affect safety systems. Operations managers who understand how IT and OT converge. Security professionals who understand biological timelines don't wait for patch schedules. Extension agents who can deliver cybersecurity guidance in a language that farmers trust. Farmers and operators themselves are developing basic security fluency as part of their professional toolkit.

Getting there requires deliberate effort on multiple fronts. Cross-training programs can help professionals from different backgrounds develop complementary skills. Food safety certification programs could include cyber awareness modules. Cybersecurity training could include a food system context. IT curricula could incorporate operational technology requirements. Agricultural education could address technology security as a standard topic rather than a specialty.

Different recruitment approaches would value hybrid experience. Someone with a food industry background and demonstrated technical aptitude may contribute more to food system security than someone with extensive cybersecurity credentials but no domain knowledge. Creating pathways for food professionals to gain security knowledge, and vice versa, expands the talent pool in ways that conventional recruiting misses.

Interdisciplinary education programs can develop the next generation of professionals who enter their careers already bridging these domains. Land-grant universities, with their historic mission of serving agricultural communities, are well-positioned to lead on agricultural cybersecurity curriculum development. Continuing education opportunities can help mid-career professionals expand their capabilities. Apprenticeship and mentorship models can transfer practical knowledge that formal education doesn't always capture.

On-the-job development recognizes that much learning happens through experience. Rotation programs between IT and operations functions help professionals understand each other's priorities. Embedding security personnel who work alongside food operations builds understanding of operational constraints. Communities of practice for hybrid professionals provide peer support and knowledge sharing. Recognition and career paths for bridging roles signal organizational value for this kind of work.

The workforce challenge connects to everything else. Without people who understand both food systems and cybersecurity, regulations will be written by people who don't understand operations. Training materials will be developed by people who don't speak the language of food workers. Security tools will be designed by people who don't grasp agricultural constraints. Building the hybrid workforce is foundational to everything else working.

The Future We're Building

Technology in food systems will continue to advance. More AI will shape decisions about planting, feeding, processing, and distribution. More sensors will monitor more parameters in more locations. More equipment will operate autonomously. More data will flow through more systems connecting more participants across longer distances.

This trajectory creates both opportunity and risk. The opportunity is substantial. Efficiency gains from precision agriculture can help feed a growing population on constrained land resources. Sustainability improvements from precise input application reduce environmental impact. Climate responsiveness from better monitoring helps operations adapt to changing conditions. Automation addresses labor shortages that threaten agricultural viability in many regions. Continuous monitoring improves food safety by catching problems earlier.

The risk is equally real. Every connectivity gain has a security trade-off. Every efficiency improvement creates a dependency. Every automation decision removes a layer of human judgment. Every data stream creates something that could be stolen, manipulated, or held for ransom.

Finding the balance requires asking hard questions about new technologies before adopting them. What happens when this system fails or is compromised? Who controls it, and who can take that control away? What does an attacker gain by targeting it? Is there a manual fallback if the technology becomes unavailable? Does the benefit justify the risk introduced?

Rolling back technology adoption isn't realistic or desirable. The efficiencies that technology provides are real, and the pressures driving adoption, from labor shortages to environmental requirements to market expectations, aren't going away. But technology adoption without security consideration creates fragility. The choice isn't between digital agriculture and analog agriculture. The choice is between digital agriculture done thoughtlessly and digital agriculture done with awareness of what's at stake when systems fail.

The technologies emerging now will shape the food system for decades. Decisions made today about how to integrate security into agricultural AI, autonomous equipment, and data platforms will determine whether the food system of 2050 is resilient or fragile. Getting it right requires engaging with these questions now, while the technologies are still being developed and deployed, rather than trying to retrofit security after vulnerabilities have been built into millions of devices and systems.

How We'll Know We're Winning

Calls to action fail when no one measures whether anyone answers. Five years from now, how will we know if this worked? The metrics that matter aren't compliance statistics for their own sake. They're the operational realities that tell us whether fewer animals are dying as a result of ransomware, fewer harvests are rotting because systems are down, fewer families are going hungry because supply chains failed. The numbers matter because the lives behind them matter.

Incident rates and recovery times tell us whether we're getting safer. The number of cyber incidents and food safety disruptions per thousand operations tracked over five years. How long it takes to detect an incident, respond to it, and restore operations. Shorter detection windows, faster response, less downtime, even when attacks occur. These are standard metrics in cybersecurity and food safety management, and they translate directly into real-world impact.

Reporting and transparency rates tell us whether the culture is shifting. What percentage of facilities have documented incident reporting processes and use them? How many operations report incidents, including near misses, to their sector hubs or regulators? More voluntary reporting means operators feel safe sharing problems rather than hiding them because they know the system rewards honesty rather than punishing it. When reporting increases while incident severity decreases, you're building a learning system rather than a blame system.

Baseline practice adoption tells us whether operations are doing the work. What percentage of farms, processors, distributors, and food service operations have implemented fundamental security practices: MFA, backup and recovery testing, basic access controls, incident response planning, regular training? How many critical systems get patched within reasonable timeframes when vulnerabilities are discovered? High adoption rates even among small-scale operators who lack dedicated security staff means the practices have been made accessible and the support systems are working.

Investment and ownership tell us whether security is becoming institutional rather than optional. What percentage of operations have named security owners, even if it's an existing staff member taking on the role part-time? How many have actual budget lines for security, not just ad hoc spending when something breaks? Security showing up on financial statements and organizational charts indicates that it is integrated into normal operations rather than treated as an afterthought.

Market signals tell us whether the economic incentives are shifting. Are insurance premiums diverging between operators who meet baseline security standards and those who don't? Are insurers offering discounts or better coverage terms for documented security programs? Do technology contracts include security requirements, vendor accountability clauses, and incident response service-level agreements? When security investment visibly improves market access, insurance terms, and competitive position, the economic case becomes self-evident rather than theoretical.

Policy adoption tells us whether the regulatory and funding environment is evolving. How many jurisdictions require or incentivize basic cyber risk assessments, participation in information sharing, or security readiness as conditions for accessing grants, loans, or disaster assistance? How many food system operations receive funding specifically for security improvements rather than being expected to bootstrap everything from operating budgets? Does the policy framework actively support security investment instead of creating unfunded mandates?

Hold stakeholders accountable by making these metrics visible. Large operators and buyers should publish annual resilience reports that include incident statistics, response times, training completion rates, and security investment as a percentage of revenue. Governments and funding agencies should report what percentage of their programs include security readiness conditions and how many operations they've helped improve their posture. Trade associations and information-sharing organizations should track member adoption of baseline practices and the quality of threat information being shared.

A Vision of Hope

It would be easy to end this book on a note of anxiety. The threats are real, and the vulnerabilities are extensive. The gaps between current practices and needed practices are wide. Attackers are sophisticated, well-funded, and relentless. Resources for defense are limited, especially at smaller operations. But anxiety isn't the right note to end on, because anxiety paralyzes rather than motivates. And the food system has reasons for optimism that other sectors might envy.

The sector has fed humanity through ice ages and warm periods, plagues and pandemics, wars and revolutions, droughts and floods, economic collapse and prosperity. Agriculture is the oldest industry, and it has adapted to every challenge that history has thrown at it. In three generations, farming went from animal power to tractors to GPS-guided autonomy. Processing went from hand labor to assembly lines to computer-controlled precision. Distribution went from local markets to global supply chains. Each transformation brought disruption, and each time, the sector adapted.

The cyber challenge is significant, but it's not beyond human capability to address. Knowledge, tools, and frameworks exist. What's needed is the will to apply them, the resources to support that application, and the community to sustain the effort over time. Success looks different on different scales.

At the individual operation, success might look like a farmer reporting a suspicious email and receiving praise rather than skepticism. The alert goes out through the cooperative network. Three neighboring farms check their systems and find similar attempts. The threat is stopped before anyone is harmed. The farmer who spoke up is recognized as protecting the community.

At the processing plant, success might look like ransomware hitting on a Friday afternoon, but the incident response team convening within an hour. Practiced procedures kick in: "Stop, Block, Tell." Manual production continues through the weekend while systems are being evaluated. Backups prove reliable. By Monday, operations are restored. Total impact: one day of reduced capacity, zero food safety incidents, no ransom paid.

At the industry level, success might look like threat intelligence shared as naturally as market prices. New employees receive cybersecurity training as standard as food safety training. Regulators partner with industry on practical guidance. Vendors compete on security as much as on features. Consumers understand and appreciate what it takes to keep them fed safely.

At the societal level, success might look like food system resilience recognized as national security. Investment in agricultural infrastructure that includes cybersecurity as a matter of course. Rural communities that have reliable connectivity. The next generation sees food-system cybersecurity as a career worth pursuing. Attacks still happen, because they always will, but they don't succeed in causing mass disruption or eroding trust.

The Final Message

Back to where we started.

A Swiss dairy farmer lost a cow and her calf because ransomware locked him out of the data that would have told him something was wrong. He didn't pay the ransom. The hackers got nothing financial from the attack. But Vital Bircher paid anyway, in animal lives, in veterinary bills, in sleepless nights, in the weight of grief that comes from losing animals in your care.

A corporate meat processor paid $11 million to criminals while plants in three countries sat idle. Farmers couldn't sell their cattle because the buyer's systems were down. Consumers wondered whether there would be meat on the shelves. A single ransomware gang caused disruption measured in billions of dollars of economic impact.

A produce company watched grocery store shelves empty of salads because it couldn't verify that its food was safe when systems went down. The company made the right call, holding product until it could confirm safety. But the incident revealed how thoroughly digital systems had become embedded in food safety assurance.

These stories explain why this book was written. Not to create fear, but to create understanding. Not to paralyze, but to prepare. Not to assign blame for what's gone wrong, but to build capacity for what comes next.

Safeguarding lives, not just data.

That phrase has run through this book like a thread, because it captures what makes food system cybersecurity different from securing other sectors. Every encrypted file represents something real. A field that needs watering. Animals that need feeding. Families that need to eat. Workers whose livelihoods depend on running operations. Communities whose food security depends on functioning supply chains.

This isn't abstract; it's breakfast, lunch, and dinner. This is the birthday cake and the holiday meal. This is the midnight feeding and the school lunch box. This is the hospital meal that helps someone heal and the food bank distribution that gets a family through a difficult week. This is the restaurant where people celebrate and the cafeteria where workers grab lunch. This is the morning coffee and the evening glass of milk. This is your favorite pint at the pub with a packet of crisps.

Food has always been about more than calories. It's culture, trust, memory, religion, and community. It's the thing we celebrate with, grieve with, show gratitude for, and comfort each other with. Protecting what feeds us means protecting all of that.

The threat actors think they understand our sector better than we do. They research their targets, profile their victims, and time their attacks for maximum leverage. They exploit the fact that food operations have been too busy feeding people to think deeply about digital threats. Let's prove them wrong!

Let's build a food system that's as resilient to cyber threats as it has been to every other challenge in human history. Let's create communities where reporting problems is celebrated, where knowledge is shared freely, where everyone who touches the food system understands their role in protecting it. Let's train the next generation of professionals who bridge food expertise and security expertise. Let's develop technologies that build security in from the start rather than bolting it on as an afterthought. Let's create policies that support rather than burden the operations that feed us.

Let's secure what feeds us, together.

Because the cow that dies because of ransomware matters. The harvest that rots because systems are down matters. The family that goes hungry because supply chains fail matters. The trust that erodes when food safety can't be assured matters. And because the farmer who protects their neighbors by sharing a threat warning matters. The worker who reports suspicious activity matters. The vendor that builds security into their products matters. The researcher who translates findings into practical guidance matters. The policymaker who listens before legislating matters. The consumer who understands what it takes to keep food safe and not spread misinformation matters.

All hands on deck. All of us together. Protecting what feeds us.

That's the call to action. That's what this book has been building toward.

Now it's time to answer it.

Notes

1. USDA, National Institute of Food and Agriculture (n.d.). **https://www.nifa.usda.gov**
2. FCC. (2024). *Inquiry Concerning the Deployment of Advanced Telecommunications Capability to All Americans in a Reasonable and Timely Fashion.* **https://www.fcc.gov/document/fcc-increases-broadband-speed-benchmark-0**
3. Congressional Research Service. (2024). *USDA's ReConnect Program: Expanding Rural Broadband.* Report R47017. **https://www.congress.gov/crs-product/R47017**
4. World Bank. (2025). "Food security update." **https://www.worldbank.org/en/topic/agriculture/brief/food-security-update**
5. European Commission. (2026). "NIS2 Directive: securing network and information systems." **https://digital-strategy.ec.europa.eu/en/policies/nis2-directive**
6. FAO. (2021). "Small family farmers produce a third of the world's food." **https://www.fao.org/newsroom/detail/Small-family-farmers-produce-a-third-of-the-world-s-food/en**

Acknowledgments

This book came from frustration and from years of conversations, cold forgotten coffee, a breaking incident that added to the narrative, scribbles in notebooks, highlighter marks on my hands from annotations, and hours nobody else saw. It is for the people who grow, raise, process, and protect what feeds us: the people the rest of the world forgets, because our modern world allows it. I hope it helps.

To my publishing team at Wiley—Navin Vijayakumar, Annie Melnick, Pete Gaughan, Jim Minatel, Krysta Winsheimer, and everyone who touched this manuscript along the way—thank you for taking a bet on a book about cybersecurity in food and agriculture. There wasn't a book like this. Now there is.

The *Bites and Bytes* podcast has been a years-long conversation with some of the best minds in food, agriculture, food protection, data science, research, and OT/ICS cybersecurity. Much of what I've learned from those conversations lives in these pages. To every guest who gave me an hour of their time and their unfiltered thoughts: this book exists in part because of you. Thank you for being part of this journey.

To the operators, food safety professionals, producers, farmers, and the cybersecurity professionals working specifically in food and agriculture that I've had the privilege of working alongside: you know what is at stake when these systems go down. That knowledge cannot be learned from a textbook, and it is what I tried to bring to every chapter. The work you do goes unnoticed by most, but it is the backbone of the food on our plates. Thank you for your acceptance, friendship, candor, and continued lessons.

A special thank you to my BEER-ISAC community. You have been a light in some dark moments when this work felt isolating. Thank you for the enthusiasm, for the humor, and for holding space when I was overwhelmed. I am grateful for every one of you.

To my Farrar family: you may not always understand exactly what I do, but your encouragement has meant everything regardless. Thank you for understanding when I was somewhere else entirely, even sitting right in front of you, or staring through a video call, looking thoroughly unhinged from overwork. We have celebrated and said goodbye over many meals these past few years. I did this work to secure the food that brings us together, always. Much love to my parents, my sisters, their husbands, and my niece and nephew.

To my King family: I'm sorry for the missed concerts, wrestling matches, meals, and video calls while I was up to my eyeballs in writing. Andrew and Jacob, thank you for filling our home with laughter and board games, and for being obsessed with my crispy chicken recipe. Eleanor, you serve food to people every day; naturally, you were on my mind throughout these pages. Christopher, you inspire me to be my own rockstar. Des and Sharon, thank you for always asking after me, and for your support, although I stand firm on avoiding that abnormally large pigeon in your back garden. Love you all, and I'm looking forward to our next great meal together.

And to my husband, Stuart: I honestly don't think I could have written this book without you. No one prepares you for what writing takes, and you never waver in your support. Thank you for loving me through the chaos, for your patience, and for reminding me that persistence matters more than perfection. We met at a food production company, which means you were woven into every page of this in some way. I love you a f' ton.

About the Author

Kristin King is widely recognized as a leading voice at the intersection of cybersecurity and food systems, a conversation she has been building since before the industry knew it needed one. A cybersecurity and operational technology risk management expert specializing in food and agriculture infrastructure, she is the founder of AnzenSage and co-founder and CEO of AnzenOT, an operational technology cybersecurity risk intelligence platform, helping industries that grow, feed, and inspire the world protect what matters most. With experience spanning environmental management, gorilla research, and more than 25 years in technology, Kristin brings a grounded, systems-based approach to her work. She also hosts the award-winning *Bites and Bytes Podcast*, which explores the intersection of food, agriculture, and cybersecurity.

Kristin lives in wine country in Northern Virginia with her husband, Stuart King, and their cats, Percy and Athena. She is originally from New England, has strong feelings about fried scallops and shoestring onion rings, and is allergic to bananas.

This is Kristin's first book.

Index

G

H

J

K

L

M

Q

R

S